Epidemiology of **Electromagnetic Fields**

Biological Effects of Electromagnetics Series

Series Editors

Frank Barnes
University of Colorado
Boulder, Colorado, U.S.A.

Ben Greenebaum
University of Wisconsin–Parkside
Somers, Wisconsin, U.S.A.

Epidemiology of Electromagnetic Fields, *edited by Martin Röösli*

Electromagnetic Fields in Biological Systems, *edited by James C. Lin*

The Physiology of Bioelectricity in Development, Tissue Regeneration, and Cancer, *edited by Christine E. Pullar*

Advanced Electroporation Techniques in Biology and Medicine, *edited by Andrei G. Pakhomov, Damijan Miklavčič, and Marko S. Markov*

Epidemiology of **Electromagnetic Fields**

Edited by
Martin Röösli

CRC Press
Taylor & Francis Group
Boca Raton London New York

CRC Press is an imprint of the
Taylor & Francis Group, an **informa** business

CRC Press
Taylor & Francis Group
6000 Broken Sound Parkway NW, Suite 300
Boca Raton, FL 33487-2742

Printed on acid-free paper
Version Date: 20140424

International Standard Book Number-13: 978-1-4665-6816-7 (Hardback)

Library of Congress Cataloging-in-Publication Data

Epidemiology of electromagnetic fields / [edited by] Martin Roosli.
p. ; cm. -- (Biological effects of electromagnetics series)
Includes bibliographical references and index.
ISBN 978-1-4665-6816-7 (hardcover : alk. paper)
I. Roosli, Martin, editor. II. Series: Biological effects of electromagnetics series.
[DNLM: 1. Electromagnetic Fields--adverse effects. QT 34]

QP82.2.E43
612'.01442--dc23 2014015930

Visit the Taylor & Francis Web site at
http://www.taylorandfrancis.com

and the CRC Press Web site at
http://www.crcpress.com

Contents

Preface

The idea of a possible interaction of magnetism with biological systems is quite old, receiving broad attention for the first time in the eighteenth century when the practice of mesmerism became popular in Europe. Mesmer's treatment is even mentioned in Wolfgang Amadeus Mozart's opera *Così fan tutte*, where, in Scene 4 of Act 1, Despina uses a magnet to treat Ferrando and Guglielmo.

Besides therapeutic applications, modern research began to address possible health risks of magnetic fields in the 1970s. Wertheimer and Leeper (1979) published a study showing an association between electromagnetic field (EMF) exposure (expressed as wire codes; see Chapter 9) with childhood leukemia. Subsequently, most epidemiological research focused on this potential association. The rapid and worldwide increase in wireless communication in the 1990s, however, extended the focus of epidemiological studies to increasingly tackle this type of exposure as well.

Nowadays, the possible health effects of EMFs are a contentious issue among the general public and the scientific community. This debate motivates one aim of this book that is, to summarize a state-of-the-art overview on the scientific evidence. For some, this alone will make this book worthwhile to read. Nevertheless, the scope of this book is much wider, providing an introduction in the methodology of environmental epidemiology for all levels, from student to seasoned researcher.

The first part of the book offers an overview of the general principles and methodological concepts in environmental epidemiology, focusing on EMF examples. Important topics include epidemiological study designs (Chapters 2 and 3), exposure assessment options and implications for the study results (Chapter 4), selection bias (Chapter 5), and confounding and other biases, including reverse causality and ecological fallacy (Chapter 6).

For several reasons, EMF epidemiology is a particularly appealing field within which to explore environmental epidemiological methods in detail. The second part of the book focuses on this theme. First, due to the lack of an established biological mechanism for the interaction of EMFs with the human body in the low-dose range, epidemiological findings are often more critically discussed than in other fields of research. Rigorous sensitivity analysis and simulation studies have been conducted to evaluate the role of bias for observed associations. Some examples are outlined in Chapter 9 for childhood leukemia and exposure to extremely low-frequency magnetic fields and Chapters 12–15 provide examples for brain tumor and mobile phone use. The second reason is the ubiquitous nature of EMFs. Ubiquitously distributed throughout our everyday environment, EMFs originate from numerous sources, ranging from small electrical and communication devices to large infrastructures such as power lines and broadcast transmitters. Detailed information on the range of EMF sources and exposures is presented in Chapters 7 and 8. Also, as presented in Chapter 4, sophisticated exposure assessment methods are often needed to address these varied and often complex exposure situations. Exposure assessment is further complicated by the fact that (1) EMFs cannot be perceived at levels that typically occur in our everyday environment; (2) technical development is very quick, resulting in new and changing exposure scenarios within a short time (Chapter 19); and (3) EMFs interact with the body, making on-body measurements more difficult compared with other environmental exposures. Thus, although modeling of EMF is relatively simple from a physical point of view, researchers are faced with interesting and often impossible challenges in obtaining all the relevant input data for all EMF exposure sources (Chapter 8).

Third, novelty and steep increase of the exposure, in particular for radiofrequency (RF) EMFs from use of wireless communication techniques, are more likely to create inappropriate conception and anxiety in the population. It also results in inherent uncertainties about long-term health effects, particularly given that only a small proportion of the population have used a mobile phone for >20 years, a typical induction period for cancers (Chapters 12–15). This and other challenges have consequences for risk assessment because even relatively small individual risks would have major public health consequences for EMF sources that are used by almost the entire population. Particular examples include mobile and cordless phones and wireless local area network (LAN) as outlined in Chapter 18. In contrast, as illustrated in Chapter 16, false alarms can create economic burden and anxieties in the population that can seriously reduce health-related quality of life.

In the future (Chapter 20), when more is known about the interaction between EMFs and health, it is my hope that this book may also serve as an interesting historical perspective on how the risk assessment of an emerging exposure, with incomplete information and uncertainties, has been dealt with for the benefit of the population.

Reference

Wertheimer, N. and Leeper, E. 1979. Electrical wiring configurations and childhood cancer. *Am J Epidemiol* 109(3): 273–84.

Editor

Martin Röösli has a background in atmospheric physics and a PhD in environmental epidemiology. He is a professor at the Swiss Tropical and Public Health Institute in Basel and leads the Unit for Environmental Exposures and Health.

His research focuses on environmental health and includes exposure assessment studies, etiological research, and health risk assessment in the areas of electromagnetic fields, ionizing radiation, noise exposure, passive smoking, climate change, and ambient air pollution.

He has conducted several epidemiological studies on personal exposure and health effects of electromagnetic fields, including occupational studies in railway workers as well as population-based studies dealing with cancer, neurodegenerative diseases, and nonspecific symptoms of ill health. He is a member in various national and international commissions on environmental health risk and has published numerous scientific papers, reviews, and book chapters.

Contributors

Anssi Auvinen
School of Health Sciences
University of Tampere
Helsinki, Finland

STUK—Radiation and Nuclear Safety Authority
Research and Environmental Surveillance
Helsinki, Finland

Christos Baliatsas
Institute for Risk Assessment Sciences
Utrecht University
Utrecht, The Netherlands

Frank S. Barnes
Department of Electrical Engineering
University of Colorado
Boulder, Colorado

Maria Blettner
Institute of Medical Biometry, Epidemiology and Informatics
University of Mainz
Mainz, Germany

Joseph D. Bowman
National Institute for Occupational Safety and Health
Cincinnati, Ohio

Elisabeth Cardis
Centre for Research in Environmental Epidemiology (CREAL)
Universitat Pompeu Fabra (UPF)
CIBER Epidemiología y Salud Pública (CIBERESP)
Barcelona, Spain

Isabelle Deltour
Section of Environment and Radiation
International Agency for Research on Cancer
Lyon, France

Frank de Vocht
Centre for Occupational and Environmental Health
The University of Manchester
Manchester, United Kingdom

Maria Feychting
Unit of Epidemiology
Institute of Environmental Medicine
Karolinska Institutet
Stockholm, Sweden

Patrizia Frei
Swiss Cancer League
Bern, Switzerland

Ben Greenebaum
Department of Physics
University of Wisconsin-Parkside
Kenosha, Wisconsin

Anke Huss
Institute for Risk Assessment Sciences
Utrecht University
Utrecht, The Netherlands

Hans-Peter Hutter
Institute of Environmental Health
Center for Public Health
Medical University of Vienna
Vienna, Austria

Peter D. Inskip
National Cancer Institute
National Institutes of Health
Bethesda, Maryland

Leeka Kheifets
Department of Epidemiology
UCLA Fielding School of Public Health
Los Angeles, California

Michael Kundi
Institute of Environmental Health
Center for Public Health
Medical University of Vienna
Vienna, Austria

Susanna Lagorio
National Centre for Epidemiology
Surveillance and Health Promotion
National Institute of Health
Rome, Italy

Chelsea Eastman Langer
Centre for Research in Environmental Epidemiology (CREAL)
Universitat Pompeu Fabra (UPF)
CIBER Epidemiología y Salud Pública (CIBERESP)
Barcelona, Spain

Norbert Leitgeb
Institute of Health Care Engineering
Graz University of Technology
Graz, Austria

Mark P. Little
National Cancer Institute
National Institutes of Health
Bethesda, Maryland

Hiltrud Merzenich
Institute of Medical Biometry, Epidemiology and Informatics
University of Mainz
Mainz, Germany

Gabor Mezei
Exponent Health Sciences
Menlo Park, California

Aslak Harbo Poulsen
Danish Cancer Society Research Center
Danish Cancer Society
Copenhagen, Denmark

Martin Röösli
Swiss Tropical and Public Health Institute
University of Basel
Basel, Switzerland

G. James Rubin
Department of Psychological Medicine
King's College London
London, United Kingdom

Joachim Schüz
Section of Environment and Radiation
International Agency for Research on Cancer
Lyon, France

Rachel B. Smith
MRC-PHE Centre for Environment and Health
Department of Epidemiology and Biostatistics
School of Public Health
Imperial College London
London, United Kingdom

John Swanson
National Grid
London, United Kingdom

Mireille B. Toledano
MRC-PHE Centre for Environment and Health
Department of Epidemiology and Biostatistics
School of Public Health
Imperial College London
London, United Kingdom

Ximena P. Vergara
Electric Power Research Institute (EPRI)
Palo Alto, California

Roel Vermeulen
Institute for Risk Assessment Sciences
Utrecht University
Utrecht, The Netherlands

Danielle Vienneau
Swiss Tropical and Public Health Institute
University of Basel
Basel, Switzerland

1

Introduction

Maria Blettner and Hiltrud Merzenich

CONTENTS

After 30 years of intensive research, there is still an ongoing debate on the possible health hazards resulting from nonionizing radiation. Electromagnetic fields (EMFs) are ubiquitous, and exposures are inevitably a part of modern life. It is not surprising that there is public concern about the potential adverse health effects of EMFs.

EMF—A Definition

Nonionizing radiation is subdivided into four main categories based on frequency (hertz [Hz]): static fields (0 Hz), extremely low-frequency electromagnetic fields (ELF-EMFs; 0 to ~300 Hz), intermediate frequency electromagnetic fields (IF-EMFs; 300 Hz to ~100 kHz), and radiofrequency electromagnetic fields (RF-EMFs)/microwaves (up to 300 GHz) (Figure 1.1).

The increase in residential and industrial generation, transmission, and use of electricity for power, heating, and lighting over the past 130 years has inadvertently led to a concomitant increase in human exposure to ELF-EMFs (Ahlbom et al. 2001). Average exposure levels to ELF-EMFs for the general population are typically 5–50 V/m for electric fields and 0.01–0.2 μT for magnetic fields (IARC 2002). Detailed information on ELF-EMF exposure is presented in Chapter 7.

Exposures to RF-EMFs are related to wireless communication devices, a widespread technology with the number of mobile phone subscriptions currently estimated at 5 billion globally. During operation, the antenna of a cellular telephone emits RF-EMFs that can penetrate 4–6 cm into the human brain (Schüz et al. 2006). Environmental exposures are also associated with emissions from mobile phone base stations and TV and radio towers. Average exposure levels are typically 0.2–0.3 V/m, and field strengths of 1 V/m are usually detected in the proximity of mobile phone base stations. In the vicinity of amplitude modulation (AM) transmitters, field strengths of >10 V/m can occur; however, exposure decreases with increasing distance from the transmitter. Emissions from

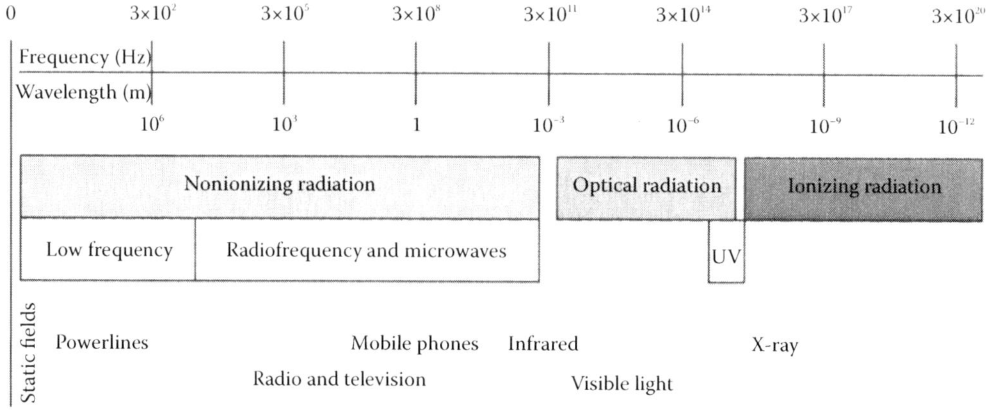

FIGURE 1.1
Electromagnetic spectrum. (Adapted from IARC Monographs on the Evaluation of Carcinogenic Risks to Humans, Volume 80, Non-Ionizing Radiation, Part 1: Static and Extremely Low-Frequency (ELF) Electric and Magnetic Fields, IARC, Lyon, 2002.)

frequency-modulated (FM) radio and TV transmitters are lower than those of AM transmitters, but they are usually located in urban areas with a high population density (Merzenich et al. 2008). Highest local exposures occur, however, when using a mobile or cordless phone. In addition to these environmental exposures, exposure to RF-EMFs from radar and microwaves can occur and must be considered in some occupational settings (IARC 2011). Detailed information on RF-EMF exposure is presented in Chapter 8.

EMF—The History of Research on Possible Health Hazards of EMFs

The discussion on possible health hazards of EMFs probably began in the 1960s. The first studies, as those presented in Chapter 3, looked at occupational exposures to EMFs in different occupational settings. In 1979, Nancy Wertheimer and Ed Leeper found that the risk of leukemia was doubled in children living near power transmission lines (see Chapter 9). Electrical current in the local water pipes or of AC magnetic fields was suggested as the possible source of the increased risk. In 1988, Savitz et al. published a case–control study on residential exposure to magnetic fields and the development of childhood cancer. Exposure was characterized through in-home electric and magnetic field measurements under low- and high-power use conditions and wire configuration codes, as a surrogate measure of long-term magnetic field level.

EMFs and Health: Biological Mechanisms

On initial consideration by early research, it was not obvious that EMFs would pose any hazard to human health. In contrast to ionizing radiation, for example, X-rays, nonionizing radiation has insufficient energy to damage DNA directly. The cellular, genotoxic, and nongenotoxic effects of such radiation on animal and human bodies have been extensively studied, collectively producing hundreds of publications and reports, and concluding that there is no clear evidence for cellular effects. Studies on genotoxic effects were mostly negative. However, it is now well established that there are biophysical mechanisms that might lead to adverse health effects.

The only known interaction between ELF-EMFs and the human body is the induction of an electric current that stimulates nerve and muscle cells. The electric current is proportional to the strength of the magnetic field, which, in turn, is not easily shielded by vegetation or buildings. For these reasons, the magnetic field rather than the electric field has been studied in most ELF-EMF epidemiological investigations (Ahlbom et al. 2001). Several epidemiological studies show a statistical association between exposure to ELF-EMFs and disease. This relationship is particularly so for childhood leukemia and magnetic fields >0.4 μT (Schüz and Michaelis 2001). Chiefly based on epidemiological studies of leukemia in children, the World Health Organization (WHO)/International Agency for research on Cancer (IARC) classified ELF-EMFs as "possibly carcinogenic to humans" (IARC 2002).* To date, however, there is no known mechanism to explain how magnetic field exposure may induce leukemia, and the effects have not been successfully replicated in animals (Schüz and Ahlbom 2008).

Radio-frequency electromagnetic fields (RF-EMFs) are absorbed by biological systems and can lead to warming of tissues. The physical basis of this thermal effect is well known and beyond dispute. Most relevant of biological effects of RF-EMFs is the absorbed energy in the body. The reference quantity is the specific absorption rate (SAR) measured in watts per kilogram and defines the power (energy per unit of time) absorbed per kilogram of body mass. A whole-body exposure to an RF-EMF corresponding to an SAR value of 4 W/kg results in a temperature rise in humans of approximately 1°C (Federal Office for Radiation Protection, Germany). Existing exposure limits, 0.08 W/kg for the general population as issued by the International Commission of Non-Ionizing Radiation (ICNIRP), are set such that they protect the population against possible adverse health effects related to EMFs. The threshold limit for (local) heating effects in the most sensitive tissues is limited to 4 W/kg (ICNIRP 1998). It is questionable, however, if there are additional nonthermal biological effects. Their existence has not been shown despite many further studies. As classified by IARC, RF-EMF is deemed "possibly carcinogenic to humans," based on an increased risk for glioma, a malignant type of brain cancer, associated with wireless phone use (IARC 2011). One epidemiological study of past cell phone use based on data collected before 2004 showed a 40% increased risk for gliomas in the highest exposure category (i.e., heavy users) (INTERPHONE Study Group 2010; see Chapters 12–15).

Detailed information on biological mechanisms is presented in Chapter 17.

EMF—Epidemiology

A possible adverse health effect can be indicated or detected by case reports or the observation of "clusters," for example, a suspected spatial aggregation of childhood leukemia cases near technical installations. Such observations and the investigation of temporal

* Static electric and magnetic fields and extremely low-frequency electric fields are not classifiable as to their carcinogenicity to humans (IARC 2002).

trends in disease development have to be evaluated very carefully, and they are typically the motivation for an epidemiological study (see Chapter 2).

From the viewpoint of an epidemiologist, the current issue in EMF research is the possibility of public health effects related to **long-term exposure at low levels** (background exposure). Due to the modern way of life with electrification, broadcasting, and mobile telecommunication, exposure to all sorts of EMFs is ubiquitous, and "zero exposure" does not exist in most parts of the world. With regard to the rapid development of new technical installations (see Chapter 19), even small risks may have public health consequences (see Chapter 18). Important disease **outcomes** to be investigated are not restricted to cancers but also include cardiovascular and neurodegenerative diseases (see Chapter 11), reproductive outcomes, and electromagnetic hypersensitivity (see Chapter 16).

Another important issue for epidemiologists is the question of **susceptibility** to EMFs. Most children, especially young children in the leukemia-relevant age groups, are exposed to a background level of EMFs. Risk to these young children increases with residences adjacent to a high-voltage power line or broadcasting station, which could increase EMF exposure distinctly above the background level. Studies have shown that in some houses indoor wiring may also result in field levels comparable with those from nearby power lines (Schüz and Ahlbom 2008).

Observational epidemiological studies on a possible relationship between EMFs and, for example, rare cancers need to be undertaken on a large enough population to ensure an **adequate study size** to find the "true" effect. Thus, rare events cannot be observed in small studies with a limited number of (highly) exposed persons.

To ensure a good epidemiological study, capable of detecting health effects against often low background rates of disease, there needs to be a **valid assessment of the exposure**. In general, EMF exposure is complex; hence, exposure assessment in an epidemiological study is difficult for several reasons. First, there is a lack of knowledge about a relevant metric (output power, frequency of the emitted field, proximity to the source of exposure). Second, very little is known about the possible induction period or a susceptible time window. Third, assessment of exposure in the past is difficult, including the problem of exposure sources probably not being characterized in a clear way, and it is not yet known how to combine exposures from different sources into one metric. Finally, exposure varies substantially over short- and longer time intervals (Ahlbom et al. 2001). Detailed information on epidemiological exposure assessment is presented in Chapter 4.

What is an **adequate study design**? Given that exposure to EMFs is ubiquitous, a randomized controlled trial (RCT) to investigate EMF effects on human health is not feasible nor ethical (see Chapter 2). Within an ecological study, it might be possible to investigate a disease outcome (e.g., incidence rate, cancer mortality) in a defined region, but with this study design, information on the individual's exposure and individual disease cannot be assessed. As a result, ecological studies are often difficult to interpret. The most appropriate study designs for an assessment of possible adverse effects related to EMFs are thus cohort studies or case–control studies with an individual assessment of an outcome and the exposure.

Case–control studies are prone to various sources of error and bias. We may, for example, see a low participation rate among the controls that might indicate a selection bias with regard to the exposure (see Chapter 5). Mobile phone users, for example, could be overrepresented among the nonparticipating subjects. A differential participation rate between users and nonusers of mobile phones and differences in usage level between cases and controls should be considered. Selection bias with regard to severity of illness may also exist among cases. Another form of bias is recall bias; this type of bias can be introduced due to self-reports of historical exposure (e.g., mobile phone use). If such errors occur randomly, they usually bias risk estimates toward the null. Cases may introduce a differential in the bias, because they tend to recall potential exposure situations with more accuracy than controls (Cardis et al. 2007).

Exposure misclassification is also an important methodological issue. In a Danish cohort study of private cellular telephone subscribers (Schüz et al. 2006), the exposure assessment was based on individual records (provided through subscriber information held by the provider). However, subscriber and user may not be the same person, and the risk of exposure misclassification was present in that study as well. Confounding by other potential risk factors of the disease may also be a problem in any cohort study. For example, in a study on brain cancer and mobile phone use, the socioeconomic status might be correlated with mobile phone use and also with the risk of brain cancer. Other potential confounders are family history of brain cancer, past medical radiation exposure, smoking history, and occupations with potential for ionizing and nonionizing radiation exposure (Cardis et al. 2007).

Chapters 2–5 provide a comprehensive overview of the advantages and disadvantages of epidemiological study designs, limitations of epidemiological studies, and strategies to cope with potential sources of bias. Using the example of mobile phone use and brain tumors, the advantages and disadvantages are discussed in detail for cohort studies (Chapter 12), case–control studies (Chapter 13), and ecological studies (Chapter 14), including an overall synthesis on the topic (Chapter 15).

Public Concern and Risk Communication

Very often, new technologies are met with enthusiasm followed by skepticism and, sometimes, even fear. Regarding EMFs, local pressure groups have been founded with the aim of removing mobile phone base stations from their neighborhoods even if they are in agreement with national guidelines. These initiatives sometimes question the scientific basis of health protection guidelines by claiming that radio waves are harmful below these hold levels that produce thermal effects. In these situations, the concern of the public has to be taken seriously. As a result, epidemiological studies should not only aim at "scientific results" but also toward supporting and establishing a qualified risk communication with the interested public. The main areas that epidemiologists have to tackle are as follows:

- How to make complicated issues accessible to the public in an adequate manner
- How best to communicate possible risks with laymen, politicians, and scientists alike

Difficulties in risk communication are usually grounded in a strong emotional bias in risk perception of the concerned people, accusations of the scientists' financial dependency on the investor, and unfortunately on a one-sided and often negative coverage in the mass media. Against this background, a factual and effective communication is rarely ever possible.

Recent history has shown that concerned people sometimes suspect "clusters of disease" in the surroundings of a technical installation. By definition, a cluster is an aggregation of cases of a particular disease in time and space, where the number of cases is statistically greater than what one would expect, when the natural history of the disease and chance fluctuations are taken into account. Here, the scientist has to assess the potential cluster with statistical methods (considering the number of cases, size of the population, incidence rate, standardized incidence ratio [SIR]). It is important to note that a statistically significant excess must have epidemiological relevance in terms of a cause-and-effect relationship between the cluster and a potential exposure (see Chapter 2). Hence, the management and investigation of potential disease clusters need very careful risk communication.

General recommendations for a successful risk communication are: to include the concerned people in the discussion, inform them of existing results, display the basic epidemiological principles in an easily accessible manner, explain the limitations of the research, and provide a communication of respectability and scientific acknowledgment (RKI 2009).

References

Ahlbom A, Cardis E, Green A, Linet M, Savitz D, Swerdlow A. Review of the epidemiologic literature on EMF and health. *Environ Health Perspect* 2001;109(Suppl 6):911–933.

Cardis E, Richardson L, Deltour I, Armstrong B, Feychting M, Johansen C, et al. The INTERPHONE study: design, epidemiological methods, and description of the study population. *Eur J Epidemiol* 2007;22(9):647–664.

Federal Office for Radiation Protection, Germany. Available from: http://www.bfs.de/en/bfs.

IARC. *Press Release No 208: IARC Classifies Radiofrequency Electromagnetic Fields as Possibly Carcinogenic for Humans*. Lyon: IARC; 2011.

IARC Monographs on the Evaluation of Carcinogenic Risks to Humans. *Volume 80, Non-Ionizing Radiation, Part 1: Static and Extremely Low-Frequency (ELF) Electric and Magnetic Fields*. Lyon: IARC; 2002.

ICNIRP. Guidelines—for limiting exposures to time-varying electric, magnetic and electromagnetic fields (Up to 300 GHz). *Health Phys* 1998;74(4):494–522.

INTERPHONE Study Group. Brain tumour risk in relation to mobile telephone use: results of the INTERPHONE international case-control study. *Int J Epidemiol* 2010;39(3):675–694.

Merzenich H, Schmiedel S, Bennack S, Brüggemeyer H, Philipp J, Blettner M, et al. Childhood leukemia in relation to radio frequency electromagnetic fields in the vicinity of television and radio broadcast transmitters. *Am J Epidemiol* 2008;168(10):1169–1178.

RKI (Robert-Koch-Institut). Empfehlungen für den Umgang mit Beobachtungen von räumlich-zeitlichen Krankheitsclustern. *Bundesgesundheitsblatt Gesundheitsforschung Gesundheitsschutz* 2009;52:239–255.

Savitz DA, Wachtel H, Barnes FA, John EM, Tvrdik JG. Case-control study of childhood cancer and exposure to 60-Hz magnetic fields. *Am J Epidemiol* 1988;128(1):21–38.

Schüz J, Ahlbom A. Exposure to electromagnetic fields and the risk of childhood leukaemia: a review. *Radiat Prot Dosimetry* 2008;132:202–211.

Schüz J, Jacobsen R, Olsen JH, Boice JD, McLaughlin JK, Johansen C. Cellular telephone use and cancer risk: update of a nationwide Danish cohort. *J Natl Cancer Inst* 2006;98(23):1707–1713.

Schüz J, Michaelis J. Epidemiologie nicht-ionisierender elektromagnetischer Felder—Eine Übersicht. *Umwelt Forsch Prax* 2001;6(2):67–76.

Wertheimer N, Leeper E. Electrical wiring configurations and childhood cancer. *Am J Epidemiol* 1979;109(3):273–284.

Part I

Epidemiological Concepts and Principles

2

Epidemiological Study Design: Architecture for Research

Anssi Auvinen

CONTENTS

Introduction

A study design defines a general outline of procedures for carrying out a study and involves several decisions about methods to be used. It precedes the empirical part and provides a broad outline guiding data collection and can hence be considered a blueprint for the study or for preparing a manual of operations for the research to be conducted.

A defining feature of epidemiology is its *probabilistic view of disease occurrence* as a stochastic process *at the population level*. For an epidemiologist, the book of diseases is written in the language of **risks**, that is, intensities for transitions between health states or probabilities of an event during a time period. Such transitions include shifts from a disease-free status to a diseased (occurrence of disease) or deceased status (death from a given cause), and in prognostic research from diseased to disease-free state (cure or recovery from illness) or demised state. Such transitions constitute events, and simple event counts can not only be expressed as frequencies but also as more refined quantities, when expressed as risks by relating the frequency to a denominator indicating the size of the underlying *population time at risk*, that is, **study base**, from which the events arise (Miettinen 1985b). In an epidemiological study, the causal parameter of interest is probability of outcome given exposure relative to such probability in the absence of exposure (or lower exposure level). Hence, causal statements in epidemiology can be expressed as conditional probabilities and relationships between such conditional probabilities.

Any estimate for a measure of occurrence obtained in an empirical study needs a definition of the study population, from which the observations are obtained. Membership in the population can be defined by residence, occupation, age, sex, or other characteristics, but it always needs a temporal frame. Hence, the study base is composed of a defined segment of population time. The terminology for the population concepts in epidemiology

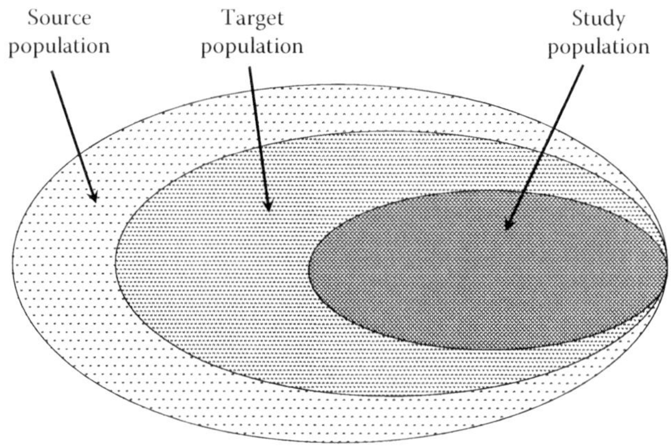

FIGURE 2.1
Population concepts. Source population refers to the population base; target population refers to those who were intended for inclusion; and study population refers to the subjects included in the analyses. Note that these can overlap to varying extent.

is far from consistent and well established, but the broad underlying population formed by particularistic bounds (typically space and time, such as geographical region and calendar period) can be called the **source population**. Within the source population, a researcher can define the *intended* group for the study, that is, the **target population**. The captured individuals, or subjects for whom information is available due to, for example, participation, can be called the **study population** (Figure 2.1). In general, the closer the target population is to the source population in terms of features relevant for the study (exposure distribution and disease risk), the better the generalizability of the results. Correspondingly, a close match in terms of the key features between the target and the study population means that there is less opportunity for selection bias. Yet, the *conclusions* are made at a more generalized, conceptual level with the results extended as universally applicable to all corresponding circumstances, for example, from observations derived from a sample of Finns to all of humankind as fundamentally similar biological beings.

Occurrence can be observed only at *population level*; hence, epidemiological studies need an outlook and a scale different from the clinical encounter of a physician and a patient. Observations are collected in a defined context, and the study design provides the architecture for collection of observations that are used for estimating the occurrence and deriving probabilities.

A key decision for designing an epidemiological study is choosing a format out of a handful of options for selection of study population (*sampling*) that provides a broad outline of the procedures for data collection. However, study design in the proper sense of the word is also a process involving other decisions about choice of methods for assessing exposure and outcome, sources of information, and data analysis (Figure 2.2). It involves *designing the purpose* (aims, hypotheses), *data collection*, and *data analysis*.

The process of study design should start with defining the aims of the study. The aims need to be carefully formulated and stated explicitly. Initially, they are expressed at a general conceptual level, and during the planning process interpretation in more concrete, applied terms is formulated (operationalization). The rest of the decisions in study design should be chosen to best suit the purpose of the study and be justified by maximizing validity within the bounds of feasibility (including cost and availability of information)

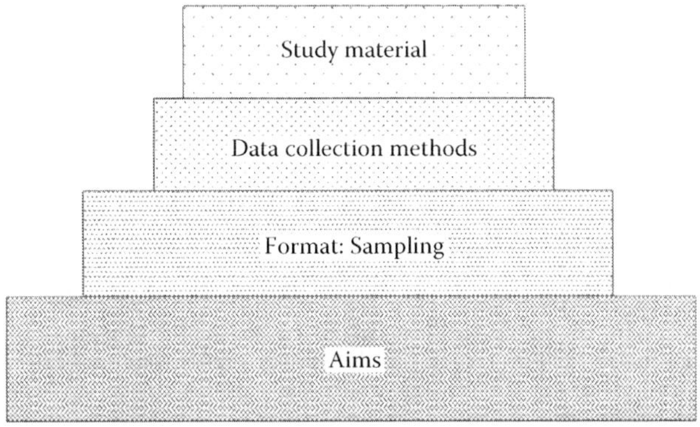

FIGURE 2.2
Choices in design of epidemiological studies and the hierarchy of issues addressed.

and efficiency (amount and quality of information available relative to sample size and resources required). The aims can be either **descriptive** or **analytical**. In a descriptive study, a single occurrence parameter is estimated with the goal to quantify disease occurrence within a particularistic context (bound by time and place); hence, the results are not even purported to be generalizable (but are nevertheless applicable, for example, planning and administrative purposes). Analytical studies use a comparative approach and estimate the relation of occurrence parameters between the exposure strata of the study, aimed at either *qualitative* inference (absence or presence of an effect, i.e., hypothesis testing) or a *quantitative* statement (estimation of effect size).

A general framework for analytical studies has been offered by the counterfactual model. It is based on Hume's definition of X as a cause of Y as a counterfactual conditional "if X had not occurred, Y would not have resulted." It defines causation as a contrast between the observed and a potential or hypothetical alternative course of events. According to the counterfactual model, the goal for study design is to provide as good approximation as possible of the *potential* experience among the exposed study population in the *hypothetical alternative situation* where the exposure had not taken place (i.e., the reference group should represent what *would have happened* if the exposed group had not been exposed). The validity of the comparison can be appraised as the extent to which this intention is fulfilled, that is, how closely the reference experience matches the ideal or index experience without exposure.

Cohort Studies

Cohort studies use *exposure-based sampling*, that is, groups to be compared are defined on the basis of exposure status. Then, rates of disease occurrence are compared between the groups with contrasting exposure—either exposed versus unexposed or groups with quantitatively different exposures. The time period, during which the events occur and from which observations are obtained, is called the follow-up time.

A cohort refers to a *closed (or fixed) population* that, in epidemiology, pertains to a permanent membership in the study population, that is, a subject becomes uninformative only through death or outcome event, unintentional attrition due to emigration or nonparticipation notwithstanding. Thus, a cohort study ideally involves "harvesting" of all endpoint events, exit only through death.

A cohort study should only include subjects at risk of the outcome; hence, those with the condition before the baseline should not be included in the study (provided that the event is irreversible or only its first occurrence is used as endpoint).

Largely, but not entirely, synonymous terms include follow-up studies and longitudinal studies. These terms differ from a cohort study in the sense that unlike a cohort study, the two definitions do not imply a closed population. Follow-up or longitudinal studies can use a dynamic (open) population with entries and exits from the study base (turnover of the study population through, for example, migration). They, as well as cohort studies, may be used for descriptive purposes without a reference group. Studies comparing a cohort of subjects to the general population, using standardized incidence ratio (SIR) or standardized mortality ratio (SMR) as the effect measure, are an example of studies using an open population as a reference group.

Cohort studies can be carried out in a **prospective** or **retrospective** manner. Prospective studies commence with exposure assessment, and the events are recorded as they are accrued during the follow-up period (in real time); thus, the study is carried out during the follow-up period. They also offer an opportunity for contemporaneous exposure assessment.

In retrospective studies, existing exposure data and events that have already been recorded are used, that is, the data are obtained by the researcher after the follow-up period. The flexibility for the researcher to choose the methods and perform assessment of concurrent exposure is the chief strength of prospective cohort studies, because retrospective reconstruction of exposure is often more uncertain and prone to errors.

Prospective studies are regarded as superior to retrospective studies, but it should be noted that even if they do offer *potential* advantage by allowing an opportunity to use methods chosen by the researchers to best suit the purposes of the study (for assessment of both exposure and outcome), this is not always used. Obviously, if similar sources of already collected information are used (e.g., registries or records), the timing of retrieval of the information has no bearing on its validity.

In a cohort study, the effect of an exposure is assessed by contrasting the occurrence of events (incidence of new cases) across exposure categories. Occurrence ratios (or incidence rate ratio or incidence density ratio, obtained by relating estimates between two strata of the study base, incidence ratio [IR] = I_1/I_0) are commonly used as indicators of relative effect (effect measures). Also, absolute measures of effect expressed as occurrence difference (incidence difference, ID = $I_1 - I_0$) can be used in cohort studies to give a more concrete indication of the magnitude of effect.

One way of looking at a cohort study is through a cross-tabulation of exposure and outcome (Table 2.1). A cohort study initially focuses on identifying the row totals (marginal frequencies by exposure), that is, the number of exposed subjects ($a+b$) and the size of the unexposed group ($c+d$). Subsequently, follow-up for disease status will reveal the cell frequencies, that is, the number of disease events (a and c) in the two groups.

Key aspects of validity for cohort studies include (1) accurate and valid exposure assessment covering the etiologically relevant time period; (2) comprehensive and nonselective follow-up, with a low attrition; and (3) valid assessment of confounders with appropriate analysis to control for their effects. Sufficiently long follow-up is a requirement related to both induction period (latency from exposure to manifestation of potential excess risk)

TABLE 2.1

Schematic Representation of the Tabular Data Obtained in an Epidemiological Study (with Dichotomous Exposure Classification)

	Disease		
	Present	**Absent**	**Row Total**
Exposure			
Present	*a*	*b*	*a*+*b*
Absent	*c*	*d*	*c*+*d*
Column total	*a*+*c*	*b*+*c*	*a*+*b*+*c*+*d*

and statistical power (accrual of sufficient number of outcome events for assessing the relation between exposure and outcome, avoiding type 2 error in statistical terms).

Comparability of populations and information are important considerations in cohort studies. A reference group should ideally be as close as possible to the index group in terms of all other determinants of the outcome except the exposure of interest (in accordance with the counterfactual approach). *Matching* is increasingly used in cohort studies to improve comparability of populations. Also, equally comprehensive, precise, and accurate exposure information and outcome data for the groups to be compared should be obtained. The minimum requirement is acquiring valid exposure information at baseline (entry to the cohort), but if exposure is not constant over time, repeated assessment should be considered. Exposure data are used for classifying follow-up time and events into strata and hence to obtain occurrence estimates for subgroups with contrasting exposure experience. It should be noted that subjects can switch exposure category during the follow-up. Commonly, an unexposed person can be exposed, or exposure level may increase with time. In principle, exposure may also decrease in case maximum latency, that is, time after which exposure no longer has an effect, is defined.

Comparability of endpoint data can also be phrased as having similar probability of being classified as having the outcome, conditional on its occurrence (i.e., similar completeness and validity of endpoint data across exposure strata).

Studies using the standardized mortality or incidence ratio SMR/SIR, observed relative to expected number of cases, with expected numbers calculated from the population rates in the appropriate age and sex strata and accrued population time, without an internal control group are rarely an optimal choice, because comparability of populations is often questionable, making them prone to bias. The healthy worker effect (see Chapter 3) is a classic example of such distortion due to selection of healthier population into any group with active employment. A reference group selected to resemble the study group in terms of underlying risk is preferable.

In electromagnetic field (EMF) research, cohort studies have been used for various research questions (Table 2.2), including extremely low-frequency electromagnetic field (ELF-EMF) exposure of railway workers and mortality (Röösli et al. 2007) (see Chapter 10), ELF exposure from power lines and childhood leukemia (Feychting et al. 2003) (see Chapter 9), or neurodegenerative diseases (Huss et al. 2009) (see Chapter 11). In the radiofrequency electromagnetic fields (RF-EMFs) range, cohort studies addressed the association of mobile phone use with brain tumors (Morgan et al. 2000; Frei et al. 2011) (see Chapter 12) and neurodegenerative diseases (Schüz et al. 2011), as well as mobile phone base station exposure and symptoms (Mohler et al. 2012) (see Chapter 16).

TABLE 2.2

Examples of Epidemiological Studies on Health Effects of EMFs Using Different Approaches (Study Formats)

Study Format	ELF-EMF	RF-EMF
Cohort	Baris et al. 1993; Verkasalo et al. 1993; Savitz et al. 2000; Feychting et al. 2003; Ray et al. 2007; Röösli et al. 2007; Huss et al. 2009	Morgan et al. 2000; Schüz et al. 2011; Mohler et al. 2012
Case–control	Savitz et al. 1988; Kleinerman et al. 2005; Kroll et al. 2010	Stang et al. 2009; Elliott et al. 2010; INTERPHONE Study Group 2010
Case-case	–	Hartikka et al. 2009; Sato et al. 2011
Case-specular	Ebi et al. 1999	Larjavaara et al. 2011
Ecological	Kraut et al. 1994; Ha et al. 2003	de Vocht et al. 2011; Deltour et al. 2012; Little et al. 2012
Cross-sectional	Poole et al. 1993; Yamazaki et al. 2006	Mohler et al. 2010; Heinrich et al. 2011
Randomized trial/ provocation	Lyskov et al. 2001; Nevelsteen et al. 2007	Rubin et al. 2006; Wallace et al. 2010; Sauter et al. 2011

Case–Control Studies

Case–control studies use *disease-based sampling*, that is, groups to be compared are defined on the basis of disease status. Effect estimates are obtained by relating exposure distributions among subjects with and without the target condition, or probability of exposure conditional on disease status. The principal difference between the cohort and case–control studies is that although a complete roster ("census") of disease events is used in both approaches, a case–control study involves only a sample of the subjects free of the outcome in the study base, whereas a cohort study extracts full information from the subjects regardless of disease status. In Figure 2.3, the shaded figures represent cases, subjects free of the target condition are shown as empty human outlines, and the exposed group is indicated. For a cohort study, the exposed group would ideally be recruited entirely, and an unexposed reference group would be selected as a sample of the unexposed population. For a case–control study, optimally all the cases would be enrolled, together with a sample of the noncases (independent of exposure, to represent exposure distribution in the population free of the disease). Case–control studies nested within a cohort are sometimes presented as a subtype of case–control study, but a case–control study should always be nested in the sense that it should be carried out within a well-defined study base (Miettinen 1986a).

Even if the conduct of a case–control study begins with assessment of disease status and only then exposure, the temporal sequence of events and direction of inference from exposure to outcome are similar to those in cohort studies. This logic should not be blurred by the fact that a case–control study does not directly provide outcome occurrence estimates, although they can be derived by relating the numbers of cases to the size of the study base.

Subject selection in a case–control study starts with defining the criteria of the endpoint (*case definition*) and applying them for identification of the case series (Wacholder et al. 1992). The controls are then sampled out of all subjects at risk in the study base. They should all be eligible as cases if they were to become diseased (and indeed a subject initially recruited as a control should be also included as a case if he or she develops the disease).

Incidence density sampling, with controls selected from the subjects alive (and at risk of the disease) at the time of the case's diagnosis (called the **risk set**) is regarded as the state of

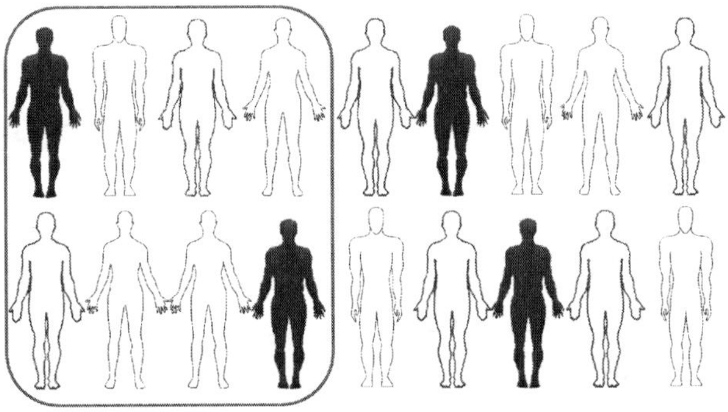

FIGURE 2.3
Study base for an epidemiological study: Population divided into an exposed and non-exposed segment, with cases (subjects with target disorder) shown as dark figures, those free of the outcome as empty figures. Cohort or case-control sampling can be applied to assess the association of exposure with outcome.

the art approach, because it produces unbiased estimates of the rate ratio (owing to the fact that drawing an independent sample at each time of case occurrence results in sampling probability proportional to person-time contribution to the study base) (Greenland and Thomas 1982). Other options include selection of controls out of those who remained free of the disease until the end of follow-up (cumulative incidence sampling) or at the start of follow-up (case-base study or case-cohort sampling).

Commonly, stratified selection is used to obtain similar distribution of most important determinants of risk as among cases, and is called **matching** (Breslow and Day 1980). Matching is often justified by control of confounding, but this notion has been challenged, because it has been shown that matching can also induce bias. For instance, *overmatching* occurs if features are used in matching that are not risk determinants but that correlate with exposure. Also, unnecessary matching complicates the selection procedure and decreases the number of eligible controls. Yet, matching on demographic factors is frequently used just for conformity, because it is commonplace and few studies omit it. Perhaps a better justification is efficiency, because obtaining similar distribution of potential confounders increases the power of the study.

Population-based case–control studies are regarded as the gold standard, because they use a comprehensive set of cases as well as selection of controls from the entire population base (with potentially most exhaustive and representative material, with good comparability of populations) (Wacholder et al. 1992). Comprehensive population registers are regarded as the best option for choosing controls, but other population rosters that have practically complete coverage can also be used, such as voter or tax lists, general practitioner patient lists, and, probably less ideally, driver's license databases or telephone catalogs.

In **hospital-based** case–control studies, the case series is derived from a hospital that may not have a clearly defined catchment population, which it serves (depending on the organization of health care provision). In such cases, selection of controls from patients of the same hospital with different diseases is commonly used, because they are assumed to represent the source population (this requires that the catchment populations for the diseases are similar). Furthermore, the impact of recall bias is assumed to be smaller when people with different diseases are compared. However, this may pose challenges to the validity of the results, because, for example, a factor associated with the control condition (higher frequency among controls than cases) will appear as a protective factor from the disease under study.

Because cases represent the manifestation of the (potential) effect of exposure, exposure assessment in case–control studies is always necessarily retrospective (although it may use contemporaneously recorded data such as monitoring records), unlike in prospective cohort studies (where concurrent recording is possible). This is a key challenge in case–control studies. Besides the time elapsed since exposure, the fact that cases are affected by the disease can threaten the comparability of exposure information (**recall bias** resulting from more comprehensive reporting of exposure by the cases than the controls, when exposure information is acquired from the study subjects).

Ascertainment and recruitment of incident cases continuously during the study period (similarly to prospective cohort studies) can also be used in case–control studies. This minimizes the lag between diagnosis and collection of exposure information, which can maximize completeness of case accrual in serious disease or when tracing is difficult, as well as improve the quality of the exposure information obtained from the subjects (minimizing the effect of disease on subsequent exposure, which can be reflected also on reporting of past exposure). Yet, it does not remove recall bias.

From the point of view of an exposure-disease cross-tabulation, a case–control study initially focuses on the column totals, that is, the number of cases ($a+c$) and controls ($b+d$) (Table 2.1). Subsequently, exposure assessment allows the researcher to break these totals down by exposure status to obtain the cell frequencies and estimate the odds ratio. The odds ratio is used as the effect measure in case–control studies, calculated as the ratios of exposure odds among case and controls. The odds ratio reflects the relation between exposure and outcome, but it is unaffected by the ratio of case versus control frequency (Cornfield 1951).

The key challenges in case–control studies include comprehensive identification and enrolment of cases, construction of a control group with minimal selection bias, as well as obtaining comparable exposure histories from both groups without recall bias (i.e., distortion of the results due to more comprehensive reporting of past exposures by the cases than the controls; Table 2.1).

Other Study Formats

Case-cohort studies use a cohort-based sampling and resemble nested case–control studies because they are conducted within large cohorts, but the reference cohort (a sample of the denominator series corresponding to controls) is formed at baseline, whereas the cases are accrued during the follow-up (Prentice 1986). Because the subcohort is formed before endpoint events, no matching is used. An advantage of the approach is that the same subcohort can be used as reference for several outcomes (various case series) within the base cohort. Typically, the endpoint is rare, so even if exposure and covariate data are collected for all cohort members, they are fully analyzed only for the cases and the subcohort. This approach is mainly used in studies using sample repositories such as biological sample banks and in some studies of occupational EMFs (Table 2.2).

A **case by case** or **case only study** involves contrasting distinct subgroups of diseased cases, with the aim to assess only if they differ in terms of etiology. The approach has been adopted more widely in infectious disease epidemiology (comparing, e.g., outbreak cases vs. sporadic cases), but it has also been applied in genetic epidemiology in assessment of gene–environment interaction. When conducting a case-case study, cases are divided into subgroups and comparisons are conducted between cases with different characteristics. Obviously, an *a priori* rationale is needed to justify for one subtype being related to exposure of interest but not the other (Prentice et al. 1984). An advantage for the approach is that selection bias and recall bias may be smaller when comparing subject with disease than for disease-free controls. This approach has been used in some studies of mobile phone use and intracranial tumors (Table 2.2; see Chapter 6).

Case-specular study is a modification of case–control study and involves also some features of case-case study. It uses a hypothetical control invoked from the case data. The rationale is to generate hypothetical, contrasting exposure experience based on the observed spatial exposure pattern among the cases. This can be achieved by creating a mirror image of the actual exposure setting by using the location of the source of exposure. It requires that exposure is characterized in spatial coordinates or as distance to an exposure source. Such approach has been used seldom in research on both ELF-EMFs and RF-EMFs (Table 2.2).

Case-crossover study is one of the few new formats intended for evaluating triggering, short-term effects. In a case-crossover study, the event and the reference exposure history are derived from the same subject. Exposure preceding the event during a specified time period is compared with frequency of exposure during a previous period for the same subject, that is, one exposure period is contrasted with another. For example, if exercise during 24 hr before onset of a condition is studied, the singular period preceding the event can be compared with several other corresponding periods of similar duration for the same subject. This approach has not been applied in EMF research to my knowledge.

Intervention studies such as **randomized trials** are the gold standard for assessment of effectiveness of interventions, but they are only exceptionally used for etiological research. In EMF research, short-term provocation studies are performed in the laboratory, but they are not generally considered epidemiological research. The validity of randomized studies is enhanced by the fact that the investigators are able to assign the intervention to a randomly chosen group to ensure comparability of populations and also administer a well-defined intervention (exposure). For ethical reasons, no exposure known to be harmful can be willfully administered to humans; therefore, only exposure reduction can be used for etiological agents. For practical reasons, intervention studies are in general not applicable to long-term exposure effects. Short-term provocation studies have been conducted on both ELFs and RF-EMFs (Table 2.2).

Ecological studies (also called correlational studies) use groups as unit of observation, that is, lack individual-level data, but they use only data on summary measures of exposure and disease occurrence at aggregate level comparing changes in their frequencies. Therefore, they are unable to relate exposure to outcome at the individual level and can only assess whether changes in the two are related in time, space, or both. They also differ from cohort studies in the sense that they are based on dynamic populations. Application of the results from ecological studies to individual level are prone to *cross-level bias* or *ecological fallacy* due to the fact that they cannot provide estimates of disease occurrence separately for the exposed and unexposed segments of the population (within-group misclassification) (Sheppard 2003). This inherent limitation also restricts the ability to control confounding (Greenland 2001). However, an ecological study may provide valid information on features predicting diseased occurrence at population level, even if such relation does not apply at individual level. The main advantage of ecological studies is their convenience—they can be characterized as the quick, cheap and easy approach to epidemiology, but their validity is generally low (unless exposures studies truly affect entire populations and the disease in question has few other risk determinants). Their applicability is best for population-level exposures affecting uniformly the entire population, such as recession, health care system, and legislation, but even for such phenomena individual-level data with, for example, multilevel analysis are preferable. Time series analysis and international comparisons are common examples of ecological studies and have increasingly been used to evaluate whether brain tumor incidence has paralleled the strong increase in mobile phone usage over the past two decades (Table 2.2; see Chapter 14).

Cross-sectional studies are based on exposure and outcome measures relating to the same time period, that is, prevalence of exposure is related to disease occurrence (prevalence or incidence) during the same period of observation. They can provide useful information for exposures that are constant traits, but for exposures that are not fixed, the fundamental shortcoming of cross-sectional studies is the inability to establish the sequence of events leading to temporal ambiguity and failure to assess the elementary requirement of causality. In the RF-EMF research area, cross-sectional studies have been used to investigate the association between environmental ELFs and RF-EMFs and symptoms (Table 2.2; see Chapter 16).

Several versions of hierarchies of research designs have been published (e.g., Guyatt et al. 2000; Petticrew and Roberts 2003). Very commonly, cohort studies are treated as inherently superior to case–control studies, always providing higher level of evidence with better validity, which is oversimplification of the matter. For instance, purely registry-based studies can be conducted using the same sources of information with either case–control or cohort format. It is obvious that there is no difference in validity for a study on, say, a given cancer type and job title recorded in census, whether it follows a case–control or cohort approach (although statistical efficiency, i.e., amount of information per subject for estimating the study parameter of interest may differ). This also applies to collection of the above-mentioned information in a prospective versus retrospective manner. Yet, a prospective cohort does offer an opportunity for a wider variety of choices for the investigator to make, hence potentially enabling application of more valid assessment methods. The point is that this is not inherent to a prospective cohort study but depends primarily on selection of methods other than the format in terms of cohort versus case–control study.

For EMF research, exposure assessment is one of the key concerns. In that respect, prospective studies have advantages. Exposure with behavioral determinants, such as EMFs from personal devices such as mobile phones and electric appliances has substantial variability that can be best captured when lag from occurrence of exposure to assessment is kept minimal. This probably applies also to workplace exposures, when tasks and work site including usage or distance to machinery or equipment is time-dependent and only inadequately captured by job title. EMF exposure from environmental and residential sources is generally more stable over time, so challenges for retrospective assessment are not as insurmountable.

Acknowledgments

I thank Riitta Luoto, MD, and Prof. Matti Hakama for constructive criticisms of an earlier draft of the manuscript. I am indebted to teachings of Prof. Olli S. Miettinen, who will undoubtedly recognize little of his dogma, save some elementary misunderstandings.

References

Baris D, Armstrong BG, Deadman J, et al. A case-cohort study of suicide in relation to exposure to electric and magnetic field among electricalty workers. *Occup Environ Med* 1996;53:17–24.

Breslow NE, Day NE. *Statistical Methods in Cancer Research.* Vol 1. The Analysis of Case-Control Studies. Lyon: International Agency for Research on Cancer; 1980.

Cornfield J. A method for estimating comparative rates from clinical data. *J Natl Cancer Inst* 1951;11:1269–1275.

Deltour I, Auvinen A, Feychting M, et al. Mobile phone use and incidence of glioma in the Nordic countries 1979–2008. *Epidemiology* 2012;23:301–307.

de Vocht F, Burstyn I, Cherrie JW. Time trends in brain cancer incidence in relation to mobile phone use in England. *Bioelectromagnetics* 2011;32:334–339.

Ebi KL, Zaffanella LE, Greenland S. Application of the case-specular method to two studies of wire-codes and childhood cancers. *Epidemiology* 1999;10:398–404.

Elliott P, Toledano MB, Bennett J, et al. Mobile phone base stations and early childhood cancer: case-control study. *Br Med J* 2010;340:c3477.

Feychting M, Jonsson N, Pedersen N, et al. Occupational magnetic field exposure and neurodegenerative disease. *Epidemiology* 2003;14:413–419.

Frei P, Poulsen AH, Johansen C, et al. Use of mobile phones and risk of brain tumours: update of Danish cohort study. *Br Med J* 2011;343:d6387.

Greenland S. Ecologic versus individual level bias in ecologic estimates of contextual health effects. *Int J Epidemiol* 2001;30:1343–1350.

Greenland S, Thomas DC. On the need for rare disease assumption in case-control studies. *Am J Epidemiol* 1982;116:547–553.

Guyatt GH, Haynes B, Jaeschke RW, et al. User's guides to the medical literature XXV. Evidence-base medicine: principles for applying the user's guides to patients care. *JAMA* 2000;284:1290–1296.

Ha M, Lim HJ, Cho SH, et al. Incidence of cancer in the vicinity of Korean AM transmitters. *Arch Environ Health* 2003;58:756–762.

Hartikka H, Heinävaara S, Mäntylä R, et al. Mobile phone use and location of glioma. A case-case study. *Bioelectromagnetics* 2009;30:176–182.

Heinrich S, Thomas S, Heumann C, et al. The impact of exposure to radio frequency electromagnetic fields on chronic well-being in young people—a cross-sectional study based on personal dosimetry. *Environ Int* 2011;37:26–30.

Huss A, Spoerri A, Egger M, et al. Residence near power lines and mortality from neurodegenerative diseases: longitudinal study of the Swiss population. *Am J Epidemiol* 2009;169:167–175.

INTERPHONE Study Group. Brain tumor risk in relation to mobile telephone use: results of the INTERPHONE international case-control study. *Int J Epidemiol* 2010;39:675–694.

Kleinerman RA, Linet MS, Hatch EE, et al. Self-reported electrical appliance use and risk of adult brain tumors. *Am J Epidemiol* 2005;161:136–146.

Kraut A, Tate R, Tran N. Residential electric consumption and incidence of childhood cancer in Canada. *Arch Environ Health* 1994;49:156–159.

Kroll ME, Swanson J, Vincent TJ, et al. Childhood cancer and magnetic fields from high-voltage power lines in England and Wales: a case-control study. *Br J Cancer* 2010;103:1122–1127.

Larjavaara S, Schüz J, Swerdlow AJ, et al. Location of gliomas in relation mobile telephone use: a case-case and case-specular analysis. *Am J Epidemiol* 2011;174:2–11.

Little M, Rajaraman P, Curtis RE, et al. Mobile phone use and glioma risk. *Br Med J* 2012;344:e1147.

Lyskov E, Sandström M, Mild KH. Provocation study of persons with perceived electrical hypersensitivity and controls using magnetic field exposure and recording of electrophysiological characteristics. *Bioelectromagnetics* 2001;22:457–462.

Miettinen OS. The "case-control" study: valid selection of subjects. *J Chronic Dis* 1985a;38:543–548.

Miettinen OS. *Theoretical Epidemiology—Principles of Occurrence Research in Medicine*. New York: Wiley; 1985b.

Mohler E, Frei P, Braun-Fahrländer C, et al. Effects of everyday radiofrequency electromagnetic-field exposure on sleep quality: a cross-sectional study. *Radiat Res* 2010;174:347–356.

Mohler E, Frei P, Fröhlich J, et al. Exposure to radiofrequency electromagnetic fields and sleep quality: a prospective cohort study. *PLoS One* 2012;7:e37455.

Morgan RW, Kelsh MA, Zhao K, et al. Radiofrequency exposure and mortality from cancer of the brain and lymphatic/hematopoietic systems. *Epidemiology* 2000;11:118–127.

Nevelsteen S, Legros JJ, Crasson M. Effects of information and 50 Hz magnetic fields on cognitive performance and reported symptoms. *Bioelectromagnetics* 2007;28:53–63.

Petticrew M, Roberts H. Evidence, hierarchies and typologies. *J Epidemiol Community Health* 2003;57:527–529.

Poole C, Kavet R, Funch DP, et al. Depressive symptoms and headaches in relation to proximity of residence to an alternating current transmission line. *Am J Epidemiol* 1993;137:318–330.

Prentice RL. A case-cohort design for epidemiological studies and disease prevention trials. *Biometrika* 1986;73:1–11.

Prentice RL, Vollmer WM, Kalbfleisch JD. On the use of case series to identify disease risk factors. *Biometrics* 1984;40:445–458.

Ray RM, Gao DL, Li W, et al. Occupational exposures and breast cancer among women textile workers in Shanghai. *Epidemiology* 2007;18:383–392.

Röösli M, Lörtscher M, Egger M, et al. Leukaemia, brain tumours and exposure to extremely low frequency magnetic fields: cohort study of Swiss railway employees. *Occup Environ Med* 2007;64:553–559.

Rubin GJ, Hahn G, Everitt BS, et al. Are some people sensitive to mobile phone signals? *Br Med J* 2006;332:886.

Sato Y, Akiba S, Kubo O, et al. A case-case study of mobile phone use and acoustic neuroma risk in Japan. *Bioelectromagnetics* 2011;32:85–93.

Sauter C, Dorn H, Bahr A, et al. Effects of exposure to electromagnetic fields emitted by GSM 900 and WCDMA mobile phones on cognitive function in young male subjects. *Bioelectromagnetics* 2011;32:179–190.

Savitz DA, Cai J, van Weijngaarden E, et al. Case-cohort analysis of brain cancer leukemia in electric utility workers using a refined magnetic field job-exposure matrix. *Am J Ind Med* 2000;38:417–425.

Savitz DA, Wachtel H, Barnes FA, et al. Case-control study of childhood cancer and exposure to 60 Hz magnetic fields. *Am J Epidemiol* 1988;128:21–38.

Schüz J, Steding-Jessen M, Hansen S, et al. Long-term mobile phone use and the risk of vestibular schwannoma: a Danish nationwide cohort study. *Am J Epidemiol* 2011;174:416–422.

Sheppard L. Insights of bias and information in group-level studies. *Biostatistics* 2003;4:265–278.

Stang A, Smith-Pokrzywniak A, Lash T, et al. Mobile phone use and risk of uveal melanoma. *J Natl Cancer Inst* 2009;101:120–123.

Verkasalo PK, Pukkala E, Hongisto MY, et al. Risk of cancer in Finnish children living close to power lines. *Br Med J* 1993;307:895–899.

Wacholder S, Silverman DT, McLaughlin JK, et al. Selection of controls in case-control studies. *Am J Epidemiol* 1992;135:1019–50.

Wallace D, Eltiti S, Ridgewell A, et al. Do TETRA (Airwave) base station signals have a short-term impact on health and well-being? *Environ Health Perspect* 2010;118:735–741.

Yamazaki S, Sokejima S, Misoue T, et al. Association between high-voltage overhead transmission lines and mental health: a cross-sectional study. *Bioelectromagnetics* 2006;27:473–478.

3

Occupational EMF Studies

Frank de Vocht

CONTENTS

Role of Occupational Epidemiology in EMF Research

To be able to detect effects of electromagnetic fields (EMFs), and more importantly establish exposure–response associations using epidemiological study methods, it is important to include a wide range of different exposure levels, and thus aim to include people with no or very low exposures as well people with very high received exposures. Typically, however, in the general population, 50- to 60-Hz EMFs [1] and radiofrequency (RF)-EMF [2] exposure levels are relatively low, and they are negligible for static (0 Hz) (electro) magnetic fields. Employees in certain industries or occupations receive greater exposures than the general population and can therefore make an important contribution to the study of exposures to, and epidemiological evidence on potential adverse effects on health from, electric and magnetic fields. Furthermore, exposure to EMFs has been increasing in the previous decades in many occupational settings and has expanded to new industries and occupations; thus, potential adverse health effects will have a greater population impact because they will be relevant to a larger number of workers [3]. More importantly, the information on exposure–response associations that can be obtained from occupational studies can be extrapolated to the general population to quantify the excess health risk from the (generally lower) exposure levels of the general population.

Depending on the industry, workers may be exposed to static magnetic fields, extremely low-frequency (ELF)-EMF (50–60 Hz) and RF-EMF (3 kHz–300 GHz) fields. Exposure to static magnetic fields occurs in industries such as aluminum and chloralkali production and also affects workers during certain welding processes, on train systems using direct current (DC) power supplies, and especially those in jobs with magnetic resonance (nuclear magnetic resonance [NMR] and magnetic resonance imaging [MRI]) systems [4]. High ELF-EMF exposure levels may occur in the occupations involved in the generation and transmission of electric power [5–7], but generally occupational exposure to ELF-EMFs from electricity is ubiquitous, comparable to the general environment, and occurs most notably through the use of computers. As such, "electric occupations" now also include many occupations in which electric power is used; for example, computer programmers, bookkeepers, clerks, and accountants [8], and also from other electric equipment, for example, those used by seamstresses, dressmakers, and tailors. Although the main source of exposure to RF-EMFs is the use of mobile telephones, occupations that may expose workers to RF-EMFs include radar technicians, radio and telegraph operators, telecommunication manufacturing occupations, dielectric heat sealing workers, workers using industrial heating equipment, workers manufacturing semiconductor chips or microelectronic devices, and workers involved in maintaining electromagnetic devices used to treat or diagnose diseases [4,9].

The first published study that aimed to investigate potential adverse health effects from occupational exposure to EMFs was published as early as 1966 and described a link between increased cancer risk and employment as high-voltage substation workers [10]. However, the start of "contemporary" occupational epidemiology on the effects of EMFs is arguable a study published by Milham in 1982 that also investigated cancer risks [11]. Since then, occupational studies have primarily focused on leukemia and leukemia subtypes, brain cancer, and breast cancer [1], but the results across studies so far remain inconsistent. Overall, the most recent comprehensive meta-analysis [12] including all relevant publications between 1993 and 2007 indicated a relatively small (14%) increased risk for brain cancer and leukemia (16%). It was, however, pointed out in this review that more recent studies show decreased risks compared with the previous studies, whereas also clear exposure–response associations seem absent, suggesting that EMF exposures may not be the causal factor for the observed increased cancer risks in the earlier studies [12].

In addition to cancer, other health outcomes for which an association with EMF exposure has been reported include neurodegenerative diseases [4], including Alzheimer's disease [13], Parkinson's disease [14], the development of dementia [15], and amyotrophic lateral sclerosis (ALS) [16]. The link with increased risk of ALS, however, is not uniformly accepted [17], but has been suggested to be associated with exposure to electric shocks rather than with ELF-EMF or RF-EMF exposure [16,18].

Short-term effects have also been reported among engineers [19] and nurses [20] working with high static field MRI systems, and they have been reported at exposure levels as little as 3 mT [21]. Occupational EMF exposure may further be associated with increased suicide rates [22,23], although this observation was based on job titles and death certificates only. Other biological effects, most notably immune system disturbances, have also been reported [24] and are discussed in more detail elsewhere. However, overall evidence of increased health risk with occupational exposure to static magnetic fields, RF-EMFs, and ELF-EMFs is inconsistent [4,16].

Epidemiological research to investigate whether exposure to EMFs in the occupational environment may be associated to increased risk of adverse health effects has involved the majority of available epidemiological study designs (see Chapter 2). The first studies aimed to investigate an increased cancer risk and relied on *cancer registry data* combined with information on occupation from death registries. The first to use this method to investigate increased cancer risk from exposure to ELF-EMFs and, therefore, described as the first study of the "modern era" of epidemiological studies of EMF, was a study conducted Milham in 1982 [11]. This study was then quickly followed by similar corroborative studies [25–27]. These studies all compared cancer mortality because this information was the only information routinely collected and readily available at the time. There are several limitations to the use of these data that are discussed later in this chapter, but an important limitation is that mortality data (and in more recent time also incidence data) that are routinely collected for cancer are not available for other diseases. After these first cancer studies, therefore, researchers quickly started to use other study designs to further investigate links between occupational exposure to EMFs and cancer as well as other health risks.

Occupational studies using *cross-sectional designs* alternatively enabled relatively fast and cheap answers. In these types of studies, the prevalence of a health outcome in a defined occupational group is compared with that of a nonexposed (occupational) group at a single point in time (see Chapter 2). In the context of occupational epidemiology, groups are typically groups of employees with comparable exposures, in its narrowest or broadest sense of the word, and have, for example, been defined as all employees from the same industry (i.e., the aluminum industry), employees in any occupation with "high" exposure (i.e., "electric" occupations), employees with the same occupation (i.e., electricians), the same task (i.e., electricians engaged in new house wiring), or any other definitions that sets a certain group apart in terms of exposure compared with the control group. The advantage of this design is that the study can investigate health outcomes other than cancer mortality and also be conducted relatively quickly to generate or further confirm new hypotheses. Also, no *a priori* hypothesis is required because all observed health outcomes can simply be compared between groups. For example, in a study published in as early as 1979, a variety of different health and biological outcomes (neurasthenic symptoms, psychological tests, electroencephalography, cardiovascular symptoms, blood pressure, electrocardiography, hemoglobin, red blood cells, reticulocytes, white blood cells, thrombocytes, sedimentation rate, and fertility) were measured in workers in high-voltage substations [28].

Cohort studies follow groups of subjects over time and are often used to identify potential risk factors for a health outcome of interest, or they are used to estimate the impact of treatment in real-world populations. These groups of subjects (employees) are called "cohorts" because in a sense when a cohort is established they resemble the distinct units in the Roman legions (see Chapter 2). Large cohorts of electricity utility employees [5,7,29–31] have contributed a significant share of the occupational epidemiological data. However, an important limitation of cohort studies is that the number of participants in the study will have to be very large to gain enough statistical power to detect increased risk for rare diseases (such as cancers of the brain and nervous system for which the world 1-year prevalence is 2% [32]). The reason for this is that at the start of the study when the cohort is constructed, it is unknown who will develop the disease, so to include enough cases for statistical methods to work, large amounts of people need to be included. Furthermore, for some diseases—most notably cancers—it takes a long time between the exposure that caused the disease and the point at which the disease is clinically detectable. In a prospective cohort study, taking this period, the latency time, into account would imply that the study would have to last for decades producing useful results. For certain industries or jobs therefore, these issues may preclude studies of certain rare cancers with long latency using a cohort study design.

Case–control studies are an alternative study design specifically suited for studies of diseases that are very rare in the population. Case–control studies do not, compared with the cohort design studies, divide people into high- and low/nonexposed groups and compare disease rates, but instead take a group of people with the disease (the cases) and a group without the disease (the controls) and then compare the exposure they have had before developing the disease (see Chapter 2). This approach allows for a more detailed, individual assessment of exposure to EMFs compared with cancer registry data where no contact with individuals exists or cohort studies that are generally very large. For example, in a recent study on occupational exposure to EMFs and risk of uveal melanoma [33], face-to-face or telephone interviews were conducted as soon as possible after diagnosis with every subject who agreed to participate. This enabled the researchers to include specific sets of questions on occupational exposure to EMFs and to ask detailed questions about each subject's occupational history. Women were found to be at increased risk compared with men; and specifically for women exposed to electrical transmission installations, a statistically significant fivefold increased risk was found. Similarly, investigating brain cancer risk associated with RF exposure using a case–control design enabled the use of a detailed questionnaire on occupational activities related to RF-EMF and other ELF-EMF exposures for each subject in the study (see Chapter 13). However, exposure assessment is done retrospectively, and it cannot be ruled out that disease status affects reporting of the exposure in that ill persons may put more effort to recall past exposures than healthy persons. This would result in a systematic difference between both groups that generally leads to inflated risk estimates and is a well-known pitfall in case–control studies known as recall bias (see Chapter 4).

Many case–control studies have been conducted to investigate whether EMF exposure may be associated to increased disease risk, and, for example, for female breast cancer these studies have been summarized in a 2010 meta-analysis [34]. Exposure assessment (discussed below) in these studies varied widely and included very crude "exposed/not exposed" as well as more sophisticated estimates based on measurements. Overall, 15 case–control studies were included to provide >24,000 cases and 60,000 controls, but no overall significant association between ELF-EMFs and female breast cancer was observed.

Strengths and Limitations of Occupational Studies

Many of the strengths and limitations of occupational epidemiological studies on EMFs are similar to those of studies in the general population and include various biases such as selection bias, information bias, recall bias, participation rates, and issues of confounding [35].

Strengths

As mentioned above, the main strength of using studies in the occupational environment is that, as a generality, occupational exposures are higher than those experienced by the general public. Because the adverse health effects of EMFs, if any, will be relatively small (compared with, say, risk of pleural mesothelioma and exposure to asbestos [36]), the studies in the occupational populations with high exposures allow much smaller (and thereby cheaper) studies to enable detection of statistically significant increased risks. Moreover, because the tasks conducted by people with the same occupation are generally comparable and the sources of exposure are similar, variability of exposure within a job title may be limited. Based on this idea, studies have been done using the occupational title as a proxy for exposure level. Indeed, the first studies in the late 1970s–early 1980s used this assumption to investigate cancer risks in certain electric occupations. Depending on the specific occupation, this assumption of comparable exposure within a job may or may not be valid, but for those occupations where it is this will provide information on excess risk in "high-risk occupations" rather than providing information on what exposure levels may pose increased health risks. The latter however is important because, ultimately, studies in occupational populations are most important if the results can be extrapolated to the, generally lower, exposure levels encountered by the general population. Therefore, in more recent studies, exposure assessment has been greatly improved by the use of job-exposure matrices (JEMs) that enable the linkage of job titles to specific estimates of exposure. These JEMs have been developed with various levels of detail, with the simplest matrices classifying occupations as exposed/nonexposed based on expert judgment and more sophisticated JEMs with increasing complexity to cross-tabulate occupations and time periods (with exposure estimates in each cell of the matrix), and ultimately with the inclusion of individual measurements of tasks, workdays, and other factors in the workplace environment [6]. JEMs may be specific to industries or occupations [37,38], or more generic for occupations of the general population [39,40]. JEMs have been developed for intensity and probability of exposure to EMFs (RF-EMFs and ELF-EMFs) and also for electric shocks [41]. More detailed work on JEMs included a shift from focusing on jobs and tasks in which exposure may occur to focusing on relevant sources of exposure [3,42,43].

Regardless, the big advantage of using job title as the basis for assigning EMF exposure is that this information is, generally speaking, easily accessible or easily obtained from study participants [1] and thus enables (large) epidemiological studies in the occupational environment that cannot be conducted with such ease in the general population. Furthermore, occupational records may be available from the employer that allows reconstructing exposure history in an unbiased way compared with personal interviews that are more likely to be biased (see Chapter 4).

Limitations

An important potential pitfall when comparing occupational populations with the general population is a well-known self-selective mechanism in which working populations are generally healthier than the general population, even if they are compared with a subsample of the population with comparable sex, age, and socioeconomic distributions. This is known as the "healthy worker effect" or "health worker bias" and has been specifically shown to have a marked effect on mortality for electricians [44].

A big challenge in occupational epidemiological studies is the assessment of exposure [3,45]. As discussed above, one of the important strengths of occupational studies is that received exposures in people with the same occupations may be comparable. However, this assumption may not always be correct [46], and therefore actual quantitative measurements of workers' personal exposure in a study (or a subgroup therein) might be necessary to evaluate the validity of the assumption of homogenous exposures in a job. These quantitative exposure measurements, in addition, can also be used to provide information on what levels of exposure are associated with excess health risks. However, because of the large number of people required for a study to statistically detect any small or medium increased risks, it is generally impossible to measure actual personal exposure to EMFs for all study participants; especially because this will have to be done for prolonged periods, long enough to cover the etiologically relevant time period of the health outcome of interest. Occupational studies to date have mostly been based on retrospective exposure assessment of the exposure, and it is very unlikely that prospective occupational studies will ever be done, given both the rarity of the health outcomes and the required follow-up time before these effects start occurring in the population [1]. Therefore, because exposure cannot be correctly measured for everyone for protracted periods, other exposure estimation methods have to be used.

As discussed above, most previous studies were characterized by crude exposure assessment, in that individual job titles or even groups of job titles were classified as "electrical occupations." The information on whether an individual was exposed to EMFs in these studies was often based on classification of a single job title as "electrical" or not [3].

A drawback of this approach is that it has been shown that often the relationship between occupational titles and actual EMF exposure is not very strong or predictable [46]. Nonetheless, rank correlations between the different metrics (arithmetic mean, geometric mean, median and higher and high and low percentiles of the exposure distributions) have been shown to be somewhat better [47], whereas also low and high exposed jobs can be separated [48] to identify these "electrical occupations." An additional complication of the use of JEMs is that, initially, JEMs focused on occupations predominantly held by men, and although more recently they have also included those jobs mostly held by women [3], nowadays women are more likely to have jobs once held exclusively by men [49] but their exposures may differ.

More so than in studies investigating occupational exposure to chemicals [50], an important problem for studies of occupational exposure to EMFs is that the job itself is often not the most important determinant of exposure [46,51]. Instead, the presence or absence of certain sources, the time spent near each of these, and the strength of individual sources may be more important. This can lead to large inter- but also intra-individual variability in exposure [52], that is, in fact, much larger than encountered by people in the general environment [45], and if only job titles are used will result in large opportunities for misclassification of exposure [53]. For example, it has been shown that the actual electrical wiring in a building can be an important factor that determines exposure levels [8]. To illustrate this issue of intraindividual variability (or for occupational exposures the "within-worker" variability), Figure 3.1 shows the results of personal static magnetic field exposure measurements from engineers building and testing MRI systems [54]. Exposure to static magnetic fields during MRI is supposed to be relatively straightforward to estimate because of homogeneity of the environment for similar systems as well as the temporal stability of the magnetic field; especially compared with exposures to RF-EMFs or ELF-EMFs. Nonetheless, large differences were still observed between tasks, workers, and days, and the MRI system could only explain 33% of the within-worker and 47% of the between-worker variability for time-weighted average static field exposure [54].

Depending on the ratio of within-worker and between-worker exposure variability, information on specific determinants of this exposure should be collected. For example, if within-worker variability (or day-to-day variability) is much larger than the variability between workers, then information on differences in work characteristics should be collected for a limited number of workers but for different working days, whereas if within-variability is small but the variability between workers is large, then it is more important to collect information on many different workers and obtaining data on different days for these workers is less important. The ratio of within- and between-worker variability depends on many factors and has been shown to differ between industries and occupations [6]. Improved exposure assessment methods have been developed aimed at accounting for between- and within-worker variability, and these methods, for example, include classification of exposure based on a combination of location and job [42] further extended on exposure times and information on locations relative to different sources of magnetic fields [55]. Another example is a study that included assessment of EMF exposure in the car manufacturing industry [56], in which the researchers relied on an industry-specific JEM developed for this study by experts, that was then further supplemented with individual expert assessment based on detailed developed job-specific questionnaires for a subsample of the workforce to address exposure variability. Nonetheless, although these methods may, to some extent, address problems of exposure variability, limitations remain because historic exposure databases of EMFs that could be used to quantitatively estimate

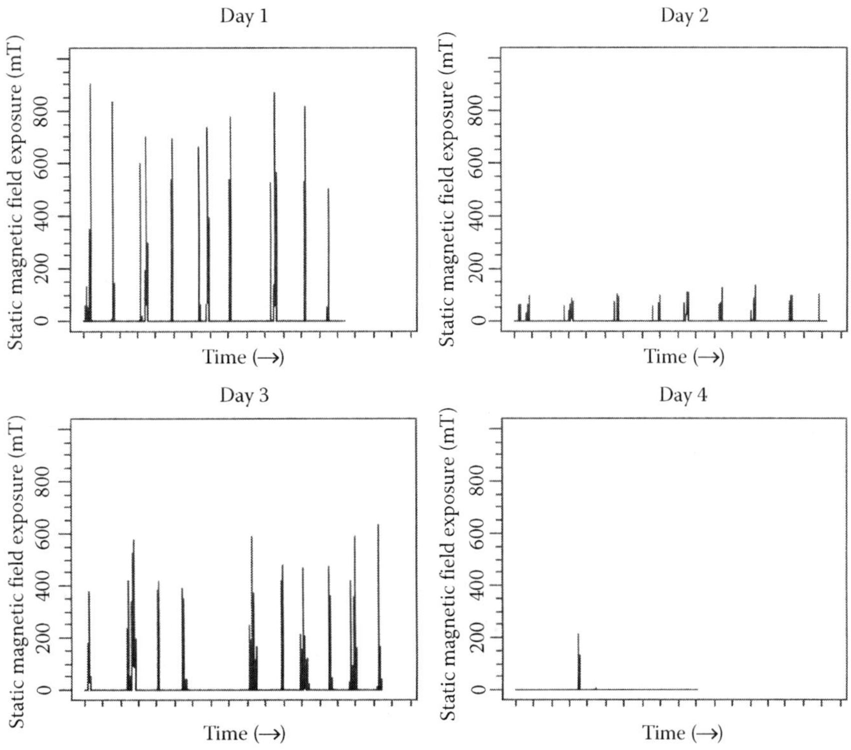

FIGURE 3.1
Illustration of measured personal exposure to static magnetic fields (in mT) during 4 days by four different magnetic resonance (MR) engineers. This figure illustrates that although all are measurements of MR engineers, personal exposure can be very different because the work is done by different engineers, who may conduct different tasks (system verification tests [day 1 and day 2], shimming [day 3], and turning of the magnet [day 4]) at different strengths of the magnet (3 T cylindrical whole-body scanner [day 1] or a 1-T whole-body open system [days 2–4]).

temporal variability are nonexistent [1]. In such cases, researchers rely on backcasting from existing measurement data or on experts' assessments to provide unbiased estimations for exposure situations, sometimes decades, in the past.

Nonetheless, even if researchers succeed in obtaining unbiased estimates of occupational EMF exposure, occupations with sole exposure to EMFs are rare or nonexistent [4], and confounding is of concern when comparing different occupations (see Chapter 6). For example, welders are also exposed to welding fumes [57]; electric utility workers to polychlorinated biphenyls [58]; chloralkali workers to mercury, chlorine, and asbestos [59]; and aluminum production workers to polycyclic aromatic hydrocarbons, sulfur dioxide, fluorides, quartz, and heat [4,60–62]. Seamstresses, or textile or garment workers, get exposed to magnetic fields [63] as well as cotton dust, endotoxin, formaldehyde, and dyes [64].

Diagnostic bias may also be of concern in some occupational studies if medical surveillance of specific occupations is more rigorous than in other occupations. A typical example may be railway drivers or pilots who have to do regular health checks. As a consequence, more health problems may be detected than in other occupational groups that could produce a false-positive study result.

As discussed above, occupational data may be often easily available, or large datasets may be generated by routine health screening in some occupations. The negative consequence of this may be publication bias, because unexpected interesting observations are more likely to be published than expected absences of associations.

Another limitation of occupational studies has, in fact, nothing to do with the occupational environment itself. Although exposures encountered in the occupational environment are generally higher than in the general environment, it has been pointed out [65] that by only including exposure from occupational sources, only about 6–8 hr (the working day) of the 24-hr exposures are accounted for, and that the remainder of exposures occur from nonoccupational household sources [66].

A Few Examples

Arguably, the era of contemporary research on the adverse health effects of occupational exposure to EMFs started with a study by Samuel Milham published in 1982 [11], which indicated increased leukemia risk for occupations with high exposure to electric or magnetic fields. This study was made possible because the cancer registry in Washington state had coded all male death records for 1950–1992 for occupation, and a judgment was made on whether certain occupations involved increased exposure to electric or magnetic fields. Milham published updated analyses after a further 3-year follow-up [67] that enabled more detailed analyses. In this study, the occupations considered to have electric or magnetic field exposures included electrical and electronic technicians, radio and telegraph operators, electricians, power and telephone linemen, television and radio repairmen, power station operators, aluminum workers, welders and flame-cutters, and motion-picture projectionists. Issues of coexposures were identified, in that arc welders were also exposed to ozone, nitrogen oxides, and metal fumes; power station workers to possible ozone exposure; aluminum potroom workers to polycyclic organic matter; and motion-picture projectionists to burning carbon electrodes and polycyclic organic matter. Excess risk, compared with the general population, was found for "electric occupations" for all malignant neoplasms, but further stratification indicated that whereas excess risk for cancers of the pancreas, lung, or brain were only observed for occupations with combined exposure to EMFs and other exposures, lymphatic and hematopoietic cancers—especially leukemia and acute leukemia—were associated with increased risk in occupations with EMFs only (most notably, for power station operators). However, there are some indications that the aforementioned healthy worker effect may have biased these results, because decreased risks were observed compared with the reference population for tuberculosis, diabetes mellitus, cerebral hemorrhage, other diseases of the heart, and cirrhosis of the liver without alcoholism. Nonetheless, increased leukemia risk in occupations with EMF exposure, also using cancer registry data, was corroborated in studies done in that same era [25–27].

One of the most recent publications using a similar study design as the one mentioned above and also assessing the carcinogenic risk of EMF-exposed workers, describes the update of a large cohort of electrical generation and transmission workers in the United Kingdom and was published in 2012 [5]. This cohort includes 81,842 employees of the former Central Electricity Generating Board of England and Wales between 1973 and 1982, with the updated results of this cohort not finding evidence of increased leukemia or brain cancer risk. Interestingly, the results of this study indicated that for leukemia the risk decreased with period from first employment, which seems inconsistent with occupational causation. Compared with the first study using job title as a proxy for exposure [11,67], this study described a cohort study in which all members were followed up over time and information was available on an individual level rather than relying on data only from cancer registers. Furthermore, this study looked at cancer incidence, which is a more precise health effect directly related to EMF exposure, rather than mortality, for which the actual rate depends on the specific cancer and may depend on other factors such as the quality of cancer care in different areas of countries. A limitation of this study was, similarly to the first study by Milham, that no information on ethnic origin, medical histories, smoking histories, or other lifestyle factors was available.

Overall, a meta-analysis conducted in 2008 [12] suggested a small increased risk of 14% for leukemia (95% confidence interval [CI], 7%–22%), as well as a 16% (95% CI, 11%–22%) overall increased risk for brain cancer associated with occupational electromagnetic field exposure. This summary study further indicated that from the first occupational studies in the early 1980s until 2007, the excess risk had slowly decreased with time, indicating that much (but not all) of the initially high excess risk could be ascribed to small studies (with a higher likelihood of "chance" findings) and exposure misclassification.

With the difficulty of looking specifically at health outcomes, additional research in occupational populations has been focused on intermediate, biological endpoints. For example, cytogenetic analyses were conducted among transformer and distribution line station workers in Turkey [68]. In this study, chromosomal aberrations and micronucleus test results were obtained from 55 exposed workers, including panelists and technicians, and 23 office workers at power generation and transmission systems, and compared with 17 controls. Spot measurements indicated workers were exposed to electric fields in the range from 130 to 15,000 V/m and electric fields between 0.25 and 17 A/m. Importantly, the exposed population in this study was reported to have no, or minimal exposure to other occupational (chemical) hazards. Smoking and alcohol intake rates, which have also been associated with these biological measures, were comparable in both groups. The results of this study are in broad agreement with older studies in linesmen [69] and substation workers [70], and indicated that chromosomal aberration and micronuclei frequencies were significantly higher in the highest exposed group compared with the control.

Interestingly, in addition to studies looking into occupational EMF exposure and direct adverse health effects (or intermediate biological markers of health effects), several studies have hypothesized that paternal or maternal exposures to EMFs at work may cause adverse health effects in their children. For example, increased risk of development of childhood cancers [71] has been shown to be associated with paternal and maternal [71,72] electrical occupations, whereas increased risk of birth defects was reported to be associated with parental occupational exposure [73]. Because there are no data to indicate that EMF exposure in the range up to RF-EMF directly damages DNA, this may indicate EMFs may cause epigenetic modifications, although no specific research has been done to date to investigate these associations. However, similar to direct health effects reported to be caused by occupational EMF exposure, these results are also not universally accepted, and other studies have shown lower increased cancer risks or even absence of any excess cancer risk, with parental [74] or maternal [75,76] exposures, or similarly for parental exposure and birth outcomes [77].

References

1. ICNIRP. *ICNIRP SCI Review: Review of the Epidemiological Literature on EMF and Health.* Oberschleissheim, Germany: ICNIRP; 2001.
2. Joseph W, Frei P, Röösli M, Thuroczy G, Gajsek P, Trcek T, et al. Comparison of personal radio frequency electromagnetic field exposure in different urban areas across Europe. *Environ Res.* 2010;110(7):658–63.
3. Kheifets L, Bowman JD, Checkoway H, Feychting M, Harrington JM, Kavet R, et al. Future needs of occupational epidemiology of extremely low frequency electric and magnetic fields: review and recommendations. *Occup Environ Med.* 2009;66(2):72–80.
4. Feychting M, Ahlbom A, Kheifets L. EMF and health. *Annu Rev Public Health.* 2005;26:165–89.
5. Sorahan T. Cancer incidence in UK electricity generation and transmission workers, 1973–2008. *Occup Med (Lond).* 2012;62(7):496–505.
6. Guenel P, Nicolau J, Imbernon E, Warret G, Goldberg M. Design of a job exposure matrix on electric and magnetic fields: selection of an efficient job classification for workers in thermoelectric power production plants. *Int J Epidemiol.* 1993;22(Suppl 2):S16–21.
7. Theriault G, Goldberg M, Miller AB, Armstrong B, Guenel P, Deadman J, et al. Cancer risks associated with occupational exposure to magnetic fields among electric utility workers in Ontario and Quebec, Canada, and France: 1970–1989. *Am J Epidemiol.* 1994;139(6):550–72.
8. McElroy JA, Egan KM, Titus-Ernstoff L, Anderson HA, Trentham-Dietz A, Hampton JM, et al. Occupational exposure to electromagnetic field and breast cancer risk in a large, population-based, case-control study in the United States. *J Occup Environ Med.* 2007;49(3):266–74.
9. Berg G, Spallek J, Schuz J, Schlehofer B, Bohler E, Schlaefer K, et al. Occupational exposure to radio frequency/microwave radiation and the risk of brain tumors: Interphone Study Group, Germany. *Am J Epidemiol.* 2006;164(6):538–48.
10. Asanova TP, Rakov AN. [Health conditions of workers exposed to electric fields of open switchboard installations of 400–500 kv. (Preliminary report)]. *Gig Tr Prof Zabol.* 1966;10(5):50–2.
11. Milham S, Jr. Mortality from leukemia in workers exposed to electrical and magnetic fields. *N Engl J Med.* 1982;307(4):249.
12. Kheifets L, Monroe J, Vergara X, Mezei G, Afifi AA. Occupational electromagnetic fields and leukemia and brain cancer: an update to two meta-analyses. *J Occup Environ Med.* 2008;50(6):677–88.
13. Santibanez M, Bolumar F, Garcia AM. Occupational risk factors in Alzheimer's disease: a review assessing the quality of published epidemiological studies. *Occup Environ Med.* 2007;64(11):723–32.
14. Wechsler LS, Checkoway H, Franklin GM, Costa LG. A pilot study of occupational and environmental risk factors for Parkinson's disease. *Neurotoxicology.* 1991;12(3):387–92.
15. Feychting M, Pedersen NL, Svedberg P, Floderus B, Gatz M. Dementia and occupational exposure to magnetic fields. *Scand J Work Environ Health.* 1998;24(1):46–53.
16. Johansen C. Electromagnetic fields and health effects—epidemiologic studies of cancer, diseases of the central nervous system and arrhythmia-related heart disease. *Scand J Work Environ Health.* 2004;30(Suppl 1):1–30.
17. Feychting M, Jonsson F, Pedersen NL, Ahlbom A. Occupational magnetic field exposure and neurodegenerative disease. *Epidemiology.* 2003;14(4):413–19; discussion 27–8.
18. Li CY, Sung FC. Association between occupational exposure to power frequency electromagnetic fields and amyotrophic lateral sclerosis: a review. *Am J Ind Med.* 2003;43(2):212–20.
19. de Vocht F, van Drooge H, Engels H, Kromhout H. Exposure, health complaints and cognitive performance among employees of an MRI scanners manufacturing department. *J Magn Reson Imaging.* 2006;23(2):197–204.
20. Wilen J, de Vocht F. Health complaints among nurses working near MRI scanners—a descriptive pilot study. *Eur J Radiol.* 2011;80(2):510–13.

21. Corbacio M, Brown S, Dubois S, Goulet D, Prato FS, Thomas AW, et al. Human cognitive performance in a 3 mT power-line frequency magnetic field. *Bioelectromagnetics*. 2011;32(8):620–33.
22. van Wijngaarden E, Savitz DA, Kleckner RC, Cai J, Loomis D. Exposure to electromagnetic fields and suicide among electric utility workers: a nested case-control study. *Occup Environ Med*. 2000;57(4):258–63.
23. van Wijngaarden E. An exploratory investigation of suicide and occupational exposure. *J Occup Environ Med*. 2003;45(1):96–101.
24. Johansson O. Disturbance of the immune system by electromagnetic fields-A potentially underlying cause for cellular damage and tissue repair reduction which could lead to disease and impairment. *Pathophysiology*. 2009;16(2–3):157–77.
25. Wright WE, Peters JM, Mack TM. Leukaemia in workers exposed to electrical and magnetic fields. *Lancet*. 1982;2(8308):1160–1.
26. McDowall ME. Leukemia mortality in electrical workers in England and Wales. *Lancet*. 1983;1(8318):246.
27. Coleman M, Bell J, Skeet R. Leukaemia incidence in electrical workers. *Lancet*. 1983;1(8331):982–3.
28. Knave B, Gamberale F, Bergstrom S, Birke E, Iregren A, Kolmodin-Hedman B, et al. Long-term exposure to electric fields. A cross-sectional epidemiologic investigation of occupationally exposed workers in high-voltage substations. *Scand J Work Environ Health*. 1979;5(2):115–25.
29. Johansen C, Raaschou Nielsen O, Olsen JH, Schuz J. Risk for leukaemia and brain and breast cancer among Danish utility workers: a second follow-up. *Occup Environ Med*. 2007;64(11):782–4.
30. Sahl JD, Kelsh MA, Greenland S. Cohort and nested case-control studies of hematopoietic cancers and brain cancer among electric utility workers. *Epidemiology*. 1993;4(2):104–14.
31. Savitz DA, Checkoway H, Loomis DP. Magnetic field exposure and neurodegenerative disease mortality among electric utility workers. *Epidemiology*. 1998;9(4):398–404.
32. Ferlay J, Shin HR, Bray F, Forman D, Mathers C, Parkin DM. *GLOBOCAN 2008 v2.0, Cancer Incidence and Mortality Worldwide: IARC CancerBase No. 10*. Lyon: International Agency for Research on Cancer; 2010.
33. Behrens T, Lynge E, Cree I, Sabroe S, Lutz JM, Afonso N, et al. Occupational exposure to electromagnetic fields and sex-differential risk of uveal melanoma. *Occup Environ Med*. 2010;67(11):751–9.
34. Chen C, Ma X, Zhong M, Yu Z. Extremely low-frequency electromagnetic fields exposure and female breast cancer risk: a meta-analysis based on 24,338 cases and 60,628 controls. *Breast Cancer Res Treat*. 2010;123(2):569–76.
35. Rothman KJ, Lash TL, Greenland S. *Modern Epidemiology*. 3rd ed. Philadelphia: Lippincott Williams and Wilkins; 2012, 816 p.
36. Bourdes V, Boffetta P, Pisani P. Environmental exposure to asbestos and risk of pleural mesothelioma: review and meta-analysis. *Eur J Epidemiol*. 2000;16(5):411–17.
37. Sahl JD, Kelsh MA, Smith RW, Aseltine DA. Exposure to 60 Hz magnetic fields in the electric utility work environment. *Bioelectromagnetics*. 1994;15(1):21–32.
38. Johansen C, Raaschou-Nielsen O, Skotte J, Thomsen BL, Olsen JH. Validation of a job-exposure matrix for assessment of utility worker exposure to magnetic fields. *Appl Occup Environ Hyg*. 2002;17(4):304–10.
39. Forssen UM, Mezei G, Nise G, Feychting M. Occupational magnetic field exposure among women in Stockholm County, Sweden. *Occup Environ Med*. 2004;61(7):594–602.
40. Bowman JD, Touchstone JA, Yost MG. A population-based job exposure matrix for power-frequency magnetic fields. *J Occup Environ Hyg*. 2007;4(9):715–28.
41. Huss A, Vermeulen R, Bowman JD, Kheifets L, Kromhout H. Electric shocks at work in Europe: development of a job exposure matrix. *Occup Environ Med*. 2013;70:261–7.
42. Miller RD, Neuberger JS, Gerald KB. Brain cancer and leukemia and exposure to power-frequency (50- to 60-Hz) electric and magnetic fields. *Epidemiol Rev*. 1997;19(2):273–93.

43. Renew DC, Cook RF, Ball MC. A method for assessing occupational exposure to power-frequency magnetic fields for electricity generation and transmission workers. *J Radiol Prot*. 2003;23(3):279–303.
44. Thygesen LC, Hvidtfeldt UA, Mikkelsen S, Bronnum-Hansen H. Quantification of the healthy worker effect: a nationwide cohort study among electricians in Denmark. *BMC Public Health*. 2011;11:571.
45. Ahlbom IC, Cardis E, Green A, Linet M, Savitz D, Swerdlow A. Review of the epidemiologic literature on EMF and Health. *Environ Health Perspect*. 2001;109(Suppl 6):911–33.
46. Kelsh MA, Kheifets L, Smith R. The impact of work environment, utility, and sampling design on occupational magnetic field exposure summaries. *AIHAJ*. 2000;61(2):174–82.
47. Savitz DA, Ohya T, Loomis DP, Senior RS, Bracken TD, Howard RL. Correlations among indices of electric and magnetic field exposure in electric utility workers. *Bioelectromagnetics*. 1994;15(3):193–204.
48. Bracken TD, Patterson RM. Variability and consistency of electric and magnetic field occupational exposure measurements. *J Expo Anal Environ Epidemiol*. 1996;6(3):355–74.
49. Zahm SH, Blair A. Occupational cancer among women: where have we been and where are we going? *Am J Ind Med*. 2003;44(6):565–75.
50. Kromhout H, Symanski E, Rappaport SM. A comprehensive evaluation of within- and between-worker components of occupational exposure to chemical agents. *Ann Occup Hyg*. 1993;37(3):253–70.
51. Floderus B. Is job title an adequate surrogate to measure magnetic field exposure? *Epidemiology*. 1996;7(2):115–16.
52. van der Woord MP, Kromhout H, Barregard L, Jonsson P. Within-day variability of magnetic fields among electric utility workers: consequences for measurement strategies. *AIHAJ*. 2000;61(1):31–8.
53. Man AK, Shahidan R. Variations in occupational exposure to magnetic fields among welders in Malaysia. *Radiat Prot Dosimetry*. 2008;128(4):444–8.
54. de Vocht F, Muller F, Engels H, Kromhout H. Personal exposure to static and time-varying magnetic fields during MRI system test procedures. *J Magn Reson Imaging*. 2009;30(5):1223–8.
55. Coble JB, Dosemeci M, Stewart PA, Blair A, Bowman J, Fine HA, et al. Occupational exposure to magnetic fields and the risk of brain tumors. *Neuro Oncol*. 2009;11(3):242–9.
56. Mester B, Schmeisser N, Lunzmann H, Pohlabeln H, Langner I, Behrens T, et al. Development and evaluation of a tool for retrospective exposure assessment of selected endocrine disrupting chemicals and EMF in the car manufacturing industry. *Ann Occup Hyg*. 2011;55(7):736–51.
57. Stern RM. Cancer incidence among welders: possible effects of exposure to extremely low frequency electromagnetic radiation (ELF) and to welding fumes. *Environ Health Perspect*. 1987;76:221–9.
58. Charles LE, Loomis D, Shy CM, Newman B, Millikan R, Nylander-French LA, et al. Electromagnetic fields, polychlorinated biphenyls, and prostate cancer mortality in electric utility workers. *Am J Epidemiol*. 2003;157(8):683–91.
59. Barregard L, Sallsten G, Jarvholm B. Mortality and cancer incidence in chloralkali workers exposed to inorganic mercury. *Br J Ind Med*. 1990;47(2):99–104.
60. Sim MR, Del Monaco A, Hoving JL, Macfarlane E, McKenzie D, Benke G, et al. Mortality and cancer incidence in workers in two Australian prebake aluminium smelters. *Occup Environ Med*. 2009;66(7):464–70.
61. Westberg HB, Bellander T. Epidemiological adaptation of quartz exposure modeling in Swedish aluminum foundries: nested case-control study on lung cancer. *Appl Occup Environ Hyg*. 2003;18(12):1006–13.
62. Logan PW, Bernard TE. Heat stress and strain in an aluminum smelter. *Am Ind Hyg Assoc J*. 1999;60(5):659–65.
63. Kelsh MA, Bracken TD, Sahl JD, Shum M, Ebi KL. Occupational magnetic field exposures of garment workers: results of personal and survey measurements. *Bioelectromagnetics*. 2003;24(5):316–26.

64. Wernli KJ, Astrakianakis G, Camp JE, Ray RM, Chang CK, Li GD, et al. Development of a job exposure matrix (JEM) for the textile industry in Shanghai, China. *J Occup Environ Hyg*. 2006;3(10):521–9.
65. Gobba F, Bravo G, Rossi P, Contessa GM, Scaringi M. Occupational and environmental exposure to extremely low frequency-magnetic fields: a personal monitoring study in a large group of workers in Italy. *J Expo Sci Environ Epidemiol*. 2011;21(6):634–45.
66. Behrens T, Terschuren C, Kaune WT, Hoffmann W. Quantification of lifetime accumulated ELF-EMF exposure from household appliances in the context of a retrospective epidemiological case-control study. *J Expo Anal Environ Epidemiol*. 2004;14(2):144–53.
67. Milham S, Jr. Mortality in workers exposed to electromagnetic fields. *Environ Health Perspect*. 1985;62:297–300.
68. Celikler S, Aydemir N, Vatan O, Kurtuldu S, Bilaloglu R. A biomonitoring study of genotoxic risk to workers of transformers and distribution line stations. *Int J Environ Health Res*. 2009;19(6):421–30.
69. Valjus J, Norppa H, Jarventaus H, Sorsa M, Nykyri E, Salomaa S, et al. Analysis of chromosomal aberrations, sister chromatid exchanges and micronuclei among power linesmen with long-term exposure to 50-Hz electromagnetic fields. *Radiat Environ Biophys*. 1993;32(4):325–36.
70. Nordenson I, Mild KH, Ostman U, Ljungberg H. Chromosomal effects in lymphocytes of 400 kV-substation workers. *Radiat Environ Biophys*. 1988;27(1):39–47.
71. Smulevich VB, Solionova LG, Belyakova SV. Parental occupation and other factors and cancer risk in children: II. Occupational factors. *Int J Cancer*. 1999;83(6):718–22.
72. Pearce MS, Hammal DM, Dorak MT, McNally RJ, Parker L. Paternal occupational exposure to electro-magnetic fields as a risk factor for cancer in children and young adults: a case-control study from the North of England. *Pediatr Blood Cancer*. 2007;49(3):280–6.
73. Blaasaas KG, Tynes T, Irgens A, Lie RT. Risk of birth defects by parental occupational exposure to 50 Hz electromagnetic fields: a population based study. *Occup Environ Med*. 2002;59(2):92–7.
74. Hug K, Grize L, Seidler A, Kaatsch P, Schuz J. Parental occupational exposure to extremely low frequency magnetic fields and childhood cancer: a German case-control study. *Am J Epidemiol*. 2010;171(1):27–35.
75. Sorahan T, Hamilton L, Gardiner K, Hodgson JT, Harrington JM. Maternal occupational exposure to electromagnetic fields before, during, and after pregnancy in relation to risks of childhood cancers: findings from the Oxford Survey of Childhood Cancers, 1953–1981 deaths. *Am J Ind Med*. 1999;35(4):348–57.
76. Feychting M, Floderus B, Ahlbom A. Parental occupational exposure to magnetic fields and childhood cancer (Sweden). *Cancer Causes Control*. 2000;11(2):151–6.
77. Tornqvist S. Paternal work in the power industry: effects on children at delivery. *J Occup Environ Med*. 1998;40(2):111–17.

4

Epidemiological Exposure Assessment

Martin Röösli and Danielle Vienneau

CONTENTS

Introduction

A primary objective of epidemiological investigations is to describe an exposure–response association that is unlikely to be explained by factors other than those under study. Thus, all epidemiological studies require estimation of exposure. The validity of an environmental epidemiology study is largely determined by the quality of the exposure measurements; likewise, the availability of appropriate exposure assessment methods determines the design and feasibility of a study. The exposure measure should be biologically relevant and should also show a range of levels in the study population. Exposure assessment in environmental epidemiology thus makes use of temporal variability, spatial variability, or both to optimize the exposure estimates and maximize the exposure contrasts. This principle is nicely described by Geoffrey Rose (1987): "If everyone in the country had smoked

20 cigarettes a day then clinical, case-control, and cohort studies alike would have led us to conclude that lung cancer was a genetic disease; and in one sense that would have been true, since if everyone is exposed to the necessary external agent then the distribution of cases becomes wholly determined by individual susceptibility. We reach then this paradox, that the more widespread is particular environmental hazard, the less it explains the distribution of cases. The cause that is universally present has no influence at all on the distribution of disease, and it may be quite unfindable by the traditional methods of clinical impression and case–control and cohort studies; for these all depend on heterogeneity of exposure."

Exposure Characterization

Exposure is defined as the contact between a substance or agent and a surface of the human body. Contact of agents with the body may be either by inhalation, ingestion, or by direct contact with the skin (i.e., dermal). The terms *agent* and *exposure* in epidemiological research can take on several meanings, and are sometimes used interchangeably. An agent may be chemical or physical, and may refer to environmental factors such as air pollution, noise, or nonionizing radiation (NIR). Exposure in an epidemiological study, however, is not restricted to chemical or physical agents. Exposure in epidemiology may also refer to lifestyle factors (e.g., physical activity, food choice, or quantity); medical procedures; and medications (e.g., X-ray, drugs), viruses, or pathogens. Even genetic variants can be conceptually considered as an exposure in an epidemiological study. By exposure, in this chapter and book as a whole, we are mostly referring to human contact with the physical agent electromagnetic fields (EMFs).

The course through the environment that an agent takes to reach a subject is often referred to as the *exposure pathway*, whereas the mode by which a substance enters the body is often referred to as the *exposure route*. The beginning of an exposure pathway is characterized by a source that emits a certain amount (i.e., *exposure intensity* or concentration) of the agent under study. During transport, the agent may be chemically transformed (e.g., formation of secondary air pollutants such as ozone) or may stay inert, as is the case for NIR. Nevertheless, the intensity generally decreases with increasing distance from the source as the agent disperses in the environment.

There are three dimensions that are relevant for determining the exposure, all of which need to be considered when designing an epidemiological study: (1) composition (e.g., frequency and signal modulation in the case of EMFs), (2) intensity or concentration, and (3) duration or temporal pattern. Together, these three aspects influence the exposure route and ultimately determine the uptake by the body (i.e., internal dose), and any or all of these dimensions may be used as an exposure index in an epidemiological study. In terms of exposure from NIR, induction is relevant for the extremely low-frequency electromagnetic field (ELF-EMF) range and heating is relevant for the microwave (MW) and radiofrequency (RF) range. Other interaction mechanisms have been postulated but not been established so far (see Chapter 17). NIR, however, interacts with the human body such that the exposure intensity at a given receptor point differs with the presence or absence of a human body. Furthermore, the physiology of the body is also relevant, which complicates the transition from the exposure pathway to the exposure route for this particular exposure agent. To simplify, most EMF epidemiologists refer to the exposure at the interface of the human body as measured without the presence of a human body.

Principles of Epidemiological Exposure Assessment

To a certain extent, every measured variable in an epidemiological study can only be considered a surrogate (or proxy) for some more appropriate measure of the underlying phenomenon. Let us take the example of cigarette smoking as a cause of lung cancer, outlined in the textbook *Modern Epidemiology* (Rothman 1998; pp. 141–143). Assume for discussion purposes that it is the inhaled amount of benzo[a]pyrene that best predicts lung cancer risk. In an epidemiological study, one cannot hope to measure the inhaled amount of benzo[a]pyrene for a large collective population over a long period. In place of a better measure, therefore, the daily consumption of cigarettes may be used as a proxy for the inhaled amount of benzo[a]pyrene. However, to accurately estimate the benzo[a]pyrene dose one also would need to know, among other things, the type of tobacco and whether there is a filter on the cigarette, how far down each cigarette is smoked, and how deeply the individual inhales. Generally, this level of detail cannot be determined with any reasonable accuracy. Even if it were possible to ascertain, the ideal measure of exposure must integrate this information over a period of time and allow for a reasonable, but often unknown, disease induction period. Because the relevant induction period is uncertain, in principle one needs accurate exposure information for many decades, including the details of how the exposure varied by time during this period (e.g., changes in individual smoking habits, or in the cigarette composition). Because historical information of such accuracy is not attainable, epidemiologists must use more simple proxies knowing that some misclassification of exposure is unavoidable. Nevertheless, taking a crude exposure measure, such as being a current smoker or not, as a surrogate (or a proxy) of the benzo[a]pyrene exposure would still reveal a substantial increase in lung cancer rate for smokers compared with nonsmokers. Furthermore, even if the benzo[a]pyrene exposure between smokers varies widely, the benzo[a]pyrene exposure of smokers would on average be much larger than that of nonsmokers. Thus, the exposure classification smoker versus nonsmoker is an appropriate, albeit far from perfect, proxy for benzo[a]pyrene exposure.

In epidemiological RF-EMF research, duration and intensity (e.g., number of calls per week) of mobile phone use is a widely used proxy for exposure to RF-EMFs. Similar to the cigarette smoking example, however, various additional factors are modifying the exposure, such as type of phone, type of network, physiology of the head, and many more factors. Some uncertainty in the exposure assessment is thus unavoidable, and its impact on the study results is discussed below (see section "Impact of Exposure Errors on Study Results"). Nevertheless, duration of use was found to be relatively well correlated ($R^2 = 0.57$) with absorbed energy (Erdreich et al. 2007).

In conclusion, the goal of each epidemiological exposure assessment is to find a good proxy or surrogate measure representative of the exposure of interest. To this end, the first priority for hazard identification is to accurately divide the study population into exposed or nonexposed groups, or into groups that are exposed to a varying degree, rather than to obtain an exact value for the past total exposure. After hazard identification, the second stage is hazard quantification and involves accurate determination of exposure levels for these groups.

Exposure Assessment Methods

Various techniques can be used for estimating exposures in environmental epidemiological studies. In general, there is a trade-off between the amount of information that can be collected from one individual and the sample size that can be reasonably addressed in a study. In the following examples, the advantages and disadvantages of various exposure assessment methods are described (Table 4.1). Where relevant, examples of these methods as used in EMF research to date are mentioned. More detailed examples are given in Chapter 8 on exposure to RF-EMFs in our everyday environment.

Fixed Site Monitoring

For various environmental factors, fixed site monitoring networks have been established and can be used for exposure assessment. Fixed site monitoring can be particularly useful to elucidate the temporal and spatial patterns of environmental factors, such as air pollution. Fixed monitoring sites are often operational for long periods, providing measurements at a fine temporal resolution. This means that a single site, located such that it is representative for the larger surrounding area, can be indicative for the day-to-day variation of air pollutants due to meteorology as well as the long-term average concentrations in the vicinity. The advantage of using fixed site monitoring data is the data availability and quality: such networks usually have established a quality program including processes for data verification and ratification. Given that they are expensive to set up and maintain, the density of such monitoring networks is often quite low, thus substantial extrapolation between the measurement sites is often needed to enable exposure estimation at unmonitored locations. This is particularly a problem for agents with a high spatial heterogeneity such as RF-EMFs or traffic noise. As a consequence, fixed site monitoring measurements are scarcely useful in this specific area of epidemiological research

TABLE 4.1

Overview of Advantages and Disadvantages of Exposure Assessment Methods

	Advantage	Disadvantage
Fixed site monitoring	Data availability, quality program, no cooperation with study participants needed	Low density, representativity (standard limits), change of techniques, etc.
Spot measurements	Objective	Elaborate, personal behavior and temporal pattern not considered, refusals for cooperation
Personal measurements	Objective, ~dose	Elaborate-impossible, compliance, manipulation
Biomarkers	Objective, internal dose	Elaborate-impossible, pharmacokinetic, reverse causality, refusals for cooperation
Routinely collected databases	Simple, cheap, no cooperation with study participants needed	Unspecific, data protection
Modeling	Objective, small marginal costs (large collectives long term), no cooperation with study participants needed	Expensive model construction, accuracy, not all input data accessible
Questionnaires	Designable to the study hypotheses	Accuracy, subjective data (recall bias), compliance

(Neubauer et al. 2007). Another issue is that measurement sites may not be selected to be representative of the population exposure. This is a known problem in air pollution epidemiology where monitoring networks were traditionally designed for compliance monitoring rather than exposure assessment; thus, sites are often located in highly polluted areas that are nonrepresentative for the population's exposure. A further challenge to use of fixed site monitoring is that equipment and measurement techniques change over time, and these changes must be carefully considered for long-term time trend analyses.

Spot Measurements

Spot measurements made with portable devices that can be set up in individuals home or work places provide an objective exposure surrogate at the place where people spend a lot of their time. As spot measurements are essentially one-off measures in a set of locations, exposure-relevant personal behavior of the study participants cannot be captured. A major advantage of this approach is that measurements can be conducted in a standardized way, thus minimizing systematic bias. This type of monitoring, however, is time consuming due to the logistics involved, and often requires trained personal to deploy and set up the measurement devices. It also requires the consent from the study participants, and if participation depends on the health status, selection bias may be introduced into the study (see Chapter 5). Unlike fixed monitoring, spot measurements can only capture a limited time period, with typical measurement duration for EMF spot measurements between a few minutes up to a few days. The temporal variability of the derived exposure is thus only partially captured, and it is not always clear how representative the measurement interval is for the long-term average. Given the benefits and limitations, spot measurements in the bedrooms of study participants have been used in epidemiological studies on childhood leukemia and exposure to ELF-EMFs (Michaelis et al. 1998; Green et al. 1999; UK Childhood Cancer Study Investigators 1999) and in studies on RF-EMFs (Hutter et al. 2006; Preece et al. 2007; Berg-Beckhoff et al. 2009; Tomitsch et al. 2010).

Personal Measurements

By means of personal measurements, exposure data are collected as study participants go about their daily lives. Personal measurements thus take into account the exposure-relevant behavior, and the collected data are proportional to the absorbed dose during the measurement period. Personal measurements are, however, very elaborate, and for an effective measurement campaign considerable compliance from the study participants is needed. Therefore, measurement duration cannot exceed a few days in general. If the agent of interest is highly variable and heavily depends on the activity of the subject, there is a risk that personal measurements are not representative for the typical exposure circumstances of the subject because carrying a device may interfere with typical activity. Study participants may also forget to wear or take the measurement meter with them, and there is even risk of manipulating the measurements by placing the device deliberately close or distant from the relevant exposure source.

Source-specific EMF measurements, which are of interest for epidemiology, are a challenge. For EMF research, commercially available meters remain relatively bulky and are somewhat inconvenient for study participants to wear or carry as they undertake their daily tasks and activities. Measurement accuracy is also a challenge for small devices, in particular for frequency-selective measurements (see Chapter 8). As mentioned in the section "Exposure Characterization," for EMF measurements the interaction of the body with the electric field adds additional uncertainty to the measurements. Nevertheless, studies using personal meters to quantify exposures in the ELF- and RF-EMF ranges have considerably improved the knowledge about the exposure situation of the general population (see Chapter 8).

Biomarkers

A biomarker is a constituent or metabolite that can be measured in a biological fluid or tissue. Biomarkers are supposed to best represent the biologically relevant dose from all available exposure measures because they are measured on the exposure route in the body. Biomarkers of exposure are typically determined in the blood, saliva, or urine. One may directly measure the substance of interest that is taken up by the body or a metabolite of it. A common biomarker for tobacco consumption, for example, is nicotine or its metabolite cotinine. The disadvantage of many biomarkers is their relatively short half-life. As a consequence, the measured concentration refers only to recent exposures, whereas long-term exposure is of most interest in many epidemiological study settings. Furthermore, to collect the biological samples, from which biomarkers are extracted, has ethical implications and needs consent from the study participants. If likelihood of consent depends on disease and exposure status, selection bias is introduced (see Chapter 5). A biomarker of EMFs, if one existed, must be the consequence of some biological interaction within the human body. So far, such a biological interaction has not been detected; thus, researchers continue to search for relevant EMF biomarkers.

Routinely Collected Databases

Routinely collected data may provide useful exposure indices. Most common in epidemiologic research is the use of occupation as a proxy to various types of exposure. In EMF research, job-exposure matrices (JEMs) have been established for exposure to ELF-MFs (Bowman et al. 2007; Vergara et al. 2012) or to electrical shocks (Huss et al. 2013). JEMs are usually based on expert judgment or retrieved from available spot measurements, personal measurements, or both (see Chapters 3 and 7). Unfortunately, in terms of measurement point selection or measurement devices, such measurements are not conducted in a standardized way for all occupations. This lack of standardization possibly introduces considerable uncertainty in the exposure ranking of various occupations. EMF exposure may also be quite different within the same occupational classification. In the RF-EMF range, JEMs are not yet widely used. Subscription data from mobile phone operators, however, have been occasionally used as exposure indices in epidemiologic studies (Dreyer et al. 1999; Schüz et al. 2009; Frei et al. 2011; Aydin et al. 2011c). The advantage of routinely collected data is their objectivity, low cost, and simple collection. It does not require contact with the study participants and thus minimizes selection bias (see Chapter 5). However, data protection may be a limitation for extensive access, and often the linkage with health data may be a challenge. The exposure indices may also be unspecific and not very detailed.

Modeling

Exposure modeling is appropriate for exposure agents where the emission and propagation pattern is known. The reliability of the modeled exposure depends to a large extent on the quality (accuracy) of the input data. Obtaining data of sufficient quality can be particularly challenging when attempting to construct retrospective exposures where historic data are required, yet often difficult or impossible to obtain. Past information about mobile phone base station traffic or current load for power lines, for example, may not be accessible, limiting the ability to model long-term averages. The main advantage of models is that they can provide an objective exposure measure and that the errors tend to be randomly distributed in the study population. Although the preparation of the input data and construction of a sophisticated exposure model is usually elaborate, the extra cost per additional calculation is very low, making this method appealing for a large study population.

Modeling in epidemiological EMF research is mainly used for exposure from broadcast transmitters (Ha et al. 2007; Merzenich et al. 2008) and mobile phone base stations (Bürgi et al. 2010; Elliott et al. 2010). From a physical point of view, EMF exposure at a given location is determined by the source emission characteristics; the topography; and shielding from buildings and other physical characteristics, such as the meteorology and the vegetation. In terms of modeling indoor exposure, detailed information about building characteristics including type of walls, roofs, and windows is desirable, but again is often not accessible. The computation capacity may be a limitation for purely deterministic models of large areas, but semi-empirical algorithms (see Chapter 8), to cut down calculation time, have shown to work relatively well for RF-EMF exposure (Bürgi et al. 2010). In ELF-MF research, the use of wire codes (see Chapter 8) can be considered as a simple exposure model (Wertheimer and Leeper 1979).

Questionnaires

Questionnaires are widely used in epidemiological research to obtain exposure information. Questionnaires are easily designed to fit the specific goals of an epidemiological study, assuming one is aware of relevant and suitable exposure indices, such as lifestyle factors or medical treatment. This exposure assessment approach allows for collection of individual level data that are otherwise not easily accessible. In EMF research, common exposure proxies inquired by questionnaires are use of mobile phones or electrical appliances. The accuracy of the responses is, however, difficult to evaluate, especially with regard to questions about past behaviors or exposure circumstances. A particular concern is recall bias, whereby the accuracy of the answers depends on the health status. For instance, in case–control studies, cases may have reflected more intensely on past exposure situations, and thus are more likely than controls to be classified as exposed. This is a serious problem because it creates a systematic bias in the study, resulting in either an overestimation or an underestimation of the true effect estimate depending on the direction of recall bias. Recall bias is prevented by a prospective study design because study participants are not aware of their future diseases status and this cannot influence their answers given on the questionnaire. Another disadvantage of questionnaire is that the willingness to fill out a questionnaire may be related to both the disease and the exposure status, a bias that is referred to as selection bias (see Chapter 5).

Biological Processes and Exposure Measures

One of the challenges in exposure assessment is to derive exposure measures that are biologically relevant to the disease under study. Knowledge of the underlying disease process and timing of such processes are vital to derive the most appropriate exposure measure.

Assume that you have perfect data of an exposure surrogate, over time, available for all study participants in an epidemiological study; for instance, the duration of each mobile phone call made in the entire lifetime. For a meaningful data analysis, this large amount of data has to be condensed to a single or a few exposure indices per person. In principle, there are an infinite number of possibilities to combine these data. Most straightforward would be the cumulative duration of calls or the average duration per week. From a biological point of view, however, other exposure indices may be more relevant. For instance, only exposure above a certain threshold may be linked to a biological effect, or the timing of exposure may be relevant in terms of age or even time of the day. One may also consider whether it is important to incorporate a time lag between exposure and disease (Figure 4.1). For instance, exposure shortly before the diagnosis of a brain tumor cannot be causally related to the disease because the development of a chronic disease needs some induction time (time between causal action and disease initiation) and latency time (time between disease initiation and detection) (Rothman and Greenland 1998).

The biological response to an exposure can be reversible or irreversible as well as proportional to the amount of exposure or discrete if a certain exposure threshold is exceeded. This results in four combinations of idealized time courses (Figure 4.2). A typical example of reversible, proportional response (Figure 4.2a) is an inflammation process or a cell damage. When a constant exposure level is applied, a biological response occurs until a steady state is reached. After elimination of the exposure, the biological response decreases and, ideally, complete recovery is achieved. Biological response may not occur until a steady-state condition is achieved after some time lag (Figure 4.2a). In the reversible discrete response (Figure 4.2b), the biological response occurs only if a certain threshold is exceeded. After exposure falls below this threshold, the biological response disappears as well. Typical examples are asthma attacks or dermatitis. Obviously, for reversible responses, cumulative lifetime exposure is not the most appropriate exposure index. The idealized time course for irreversible proportional diseases response is shown in Figure 4.2c. In this case, biological response is proportional to the cumulative exposure, and disease regeneration is negligible. Typical examples are silicosis from occupational dust exposure, central nervous system (CNS) damage in children due to lead exposure, or permanent hearing loss. Typical irreversible discrete responses are cancer and sensitization to an allergen (Figure 4.2d).

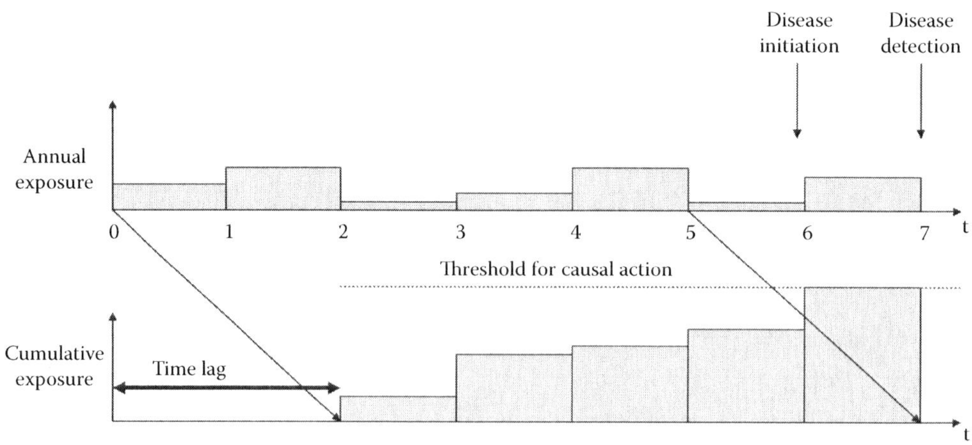

FIGURE 4.1
Illustration of a time lag between exposure and disease for a cumulative exposure index. After 5 years of exposure, cumulative exposure reached threshold for causal action and disease was initiated. After 7 years the disease was diagnosed. Exposure after fifth year is thus not relevant for disease development.

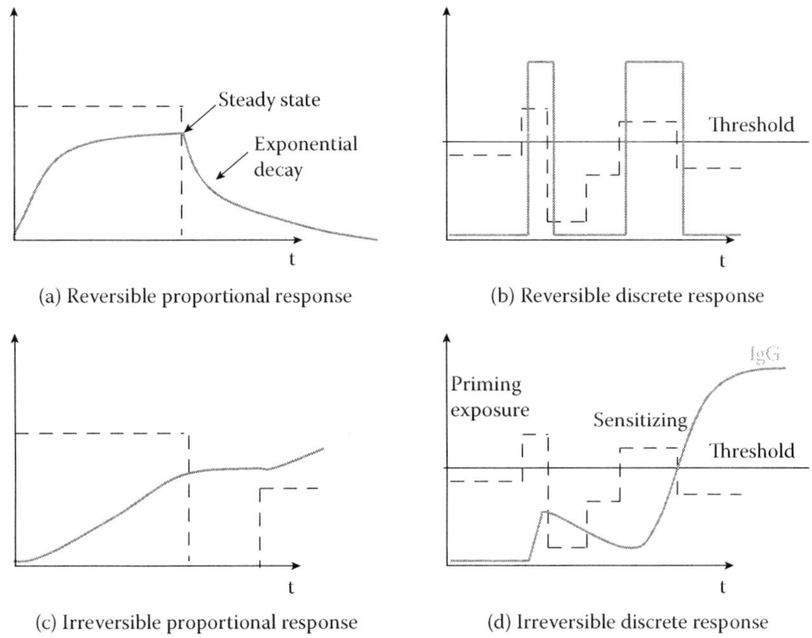

FIGURE 4.2
Illustration of idealized time courses for reversible and irreversible biological process. The biological reaction is either proportional to the amount of exposure (a and c) or discrete if a certain threshold is exceeded (b and d). The dashed line shows the exposure levels over time, and the solid gray line shows the biological response.

In reality, the disease processes are not that idealized, and the threshold for effects varies between individuals. As a result, the time course of all biological processes in a large collective tends to be proportional to the exposure; the higher the exposure, the higher the number of thresholds that are exceeded in the study population. For this reason, the selection of the correct exposure indices and time window may not be as crucial as first appears and, in many situations, cumulative exposure indices often work quite well. In practice, it is also important to realize that various exposure indices are highly correlated with each other. A person with high cumulative exposure is also more likely to have a higher maximum daily exposure value or to exceed an exposure threshold.

Because exposure data often show a lognormal distribution, it is sometimes not clear whether data are better summarized as arithmetic mean or geometric mean, for instance, when pooling daily or annual exposure measurements for an individual into a single exposure index. The arithmetic mean better represents the cumulative exposure and thus may better represent the underlying biological process for some diseases. In contrast, the geometric mean better describes the central tendency of lognormal data and thus may be statistically more appropriate. Again, from a practical point of view, the decision is rarely crucial because both measures are highly correlated.

Knowledge of the underlying disease process is helpful to determine the most appropriate exposure index for an exposure–response analysis. Conversely, if the underlying biological process is not known, the evaluation of various exposure indices may be useful to obtain hints about the relevant mechanism. In both cases, the high correlation between various exposure indices is a limitation for causal interpretation and has to be considered before drawing firm conclusions.

Exposure Errors and Misclassification

The term *error* is used differently in epidemiological exposure assessment than in everyday language. As outlined above, errors are unavoidable in exposure assessment because exposure indices are usually only a proxy or surrogate measure of the agent of interest. Error does not necessarily mean that something was done wrong, and thus the term *exposure* uncertainty may be preferred in many circumstances. If a binary or a group exposure variable is considered (e.g., occupational groups), the more common expression for error is exposure misclassification. If exposure is measured on a continuous scale (e.g., amount of mobile phone use), deviance of the observed exposure from the true exposure is usually called error.

Under certain circumstances, small errors may have a major impact on the study results, whereas under different circumstances seemingly large errors affect the study results only to a small degree. Thus, it is important to evaluate the effect of exposure errors on the study results.

Types of Exposure Errors

Three types of errors or exposure misclassification can be distinguished:

1. Nondifferential error/misclassification: The estimated exposure is randomly scattered around true value and the error is not correlated to health status. This is also referred to as random error, and an example is measurement uncertainty of a measurement device.
2. Systematic error: The estimated exposure is on average higher or lower than the true value. An example is an inaccurately calibrated measurement device.
3. Differential exposure error/misclassification: The extent of error depends on the disease status. This means that the direction or extent of random or systematic errors in recall differ between healthy and diseased study participants (or is correlated with the extent of health status in the case of a continuous health outcome). A typical example is that in a case–control study, cases may overestimate past exposure circumstances, or controls may underestimate their exposure status, or both. This happens because controls tend to think less about past exposures compared with an ill person who may search for a reason for their disease. Differential exposure error is also referred to as information bias or, in the case of self-reported exposure, as recall bias.

Measuring Exposure Errors

Exposure error is quantified in relation to the true exposure status or level. The true exposure value can be obtained from a gold standard method. This method may be more expensive and elaborate and only be applied to a subset of the study participants. For model evaluation, the gold standard is measurements that have not been used for model development.

In EMF research, several studies have addressed the relationship between self-reported mobile phone use and operator-recorded data, considering the latter as the gold standard (Parslow et al. 2003; Vrijheid et al. 2006a, 2009; Inyang et al. 2009; Aydin et al. 2011a, 2011b).

TABLE 4.2
Cross-Tabulation of the Four Possible Combinations of Exposure Classification with True Exposure Status

	True Exposure Status	
Exposure Classification	**Exposed**	**Not Exposed**
Exposed	a	b
Not Exposed	c	d

In ELF- and RF-EMF research, studies have compared spot measurements with personal measurements (Forssen et al. 2002, Frei et al. 2010). In practice, as described in Chapter 8, the gold standard may also be subject to uncertainty thus, exact error quantification is impossible from a theoretical point of view. Nevertheless, estimation of the extent of exposure assessment error is needed for a proper interpretation of epidemiological study results.

For exposure estimates on a continuous scale, various comparisons can be done, such as calculation of the correlation (Pearson or Spearman's rank correlation) or a summary statistics of the differences. Bland–Altman plots are useful for a graphical evaluation of the uncertainty. For the quantification of the agreement for two or more exposure categories, Kappa statistics are the method of choice. Weighted Kappa coefficients are useful for ordered exposure categories to take into account the seriousness of disagreement. In this case, disagreement between adjacent exposure categories is weighted less than categories that are further apart.

For a binary exposure classification "exposed versus nonexposed," the comparison of the estimated exposure in relation to the true exposure status results in four groups (Table 4.2). The reliability of the exposure assessment is then measured with the sensitivity and specificity. Sensitivity refers to the proportion of people being exposed and being (correctly) classified as exposed [= a/(a+c)]. Specificity refers to the proportion of people being unexposed and being (correctly) classified as unexposed [= d/(b+d)]. An exposure assessment with a sensitivity of 90% and a specificity of 80% means that 90% of the exposed people are correctly classified as exposed and 80% of the unexposed individuals are correctly classified as unexposed. The remaining study participants are assigned to the wrong exposure category.

Impact of Exposure Errors on Study Results

Random Exposure Misclassification

For the sake of simplicity, let us take a case–control study with a binary exposure classification and no true association between exposure and disease. In Figure 4.3, a calculation example is given for a hypothetical study with 3000 controls and 1500 cases. Exposure prevalence is 60% in controls and cases and the true odds ratio (OR) is 1.0 (no association). Let us further assume that exposure for 20% of cases and controls is wrongly estimated. (The same amount of misclassification for cases and controls implies that exposure misclassification is nondifferential.) Unfortunately, assuming exposure misclassification means that we cannot observe the "true" situation; instead, our data collection yields an erroneous table as shown on the right-hand side of Figure 4.3. Out of 900 exposed cases, 180 move to the unexposed group and 120 of the unexposed cases move into the exposed group. The resulting net

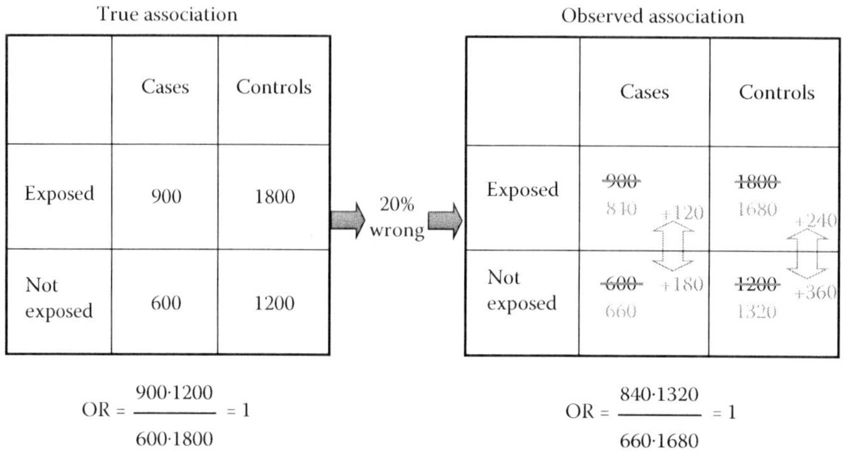

FIGURE 4.3
Effect of nondifferential exposure misclassification in a case–control study with binary exposure categorization and no relation between outcome and exposure (true OR: 1.0).

observed distribution would be 840 exposed cases and 660 unexposed cases. The same type of misclassification among controls results in 1680 observed exposed and 1320 observed unexposed controls. Strikingly, the observed OR remains at 1.0; thus, the study correctly shows an absence of risk. No bias occurs. The only problem in practice is that random data fluctuation increases according to the increasing imprecision of the exposure assessment. Thus, in real studies the observed ORs scatter randomly around 1. In a given study, therefore, an over- or underestimated OR may occur due to chance.

The situation looks different if there is a real association between exposure and outcome. An exposure prevalence of 75% in cases and of 60% in controls results in an OR of 2.0 (Figure 4.4). Exposure misclassification of 20% results in an observed 975 exposed and 525 unexposed cases. For controls, the number of exposed individuals is 1680 while unexposed individuals is 1320. The resulting OR is 1.46. These examples illustrate a general principle: Nondifferential exposure misclassification

- … is unlikely to create spurious associations if there is no true association.
- … results in underestimation of the association if there is a true association.

Importantly, the magnitude of underestimation for nondifferential exposure misclassification depends heavily on the number of exposed individuals. In Figure 4.5, the calculation example of Figure 4.4 is repeated, but true exposure prevalence in controls is reduced from 60% to 10%. Accordingly, the true exposure prevalence in cases was set to 18% to obtain an OR of 2.0. As can be seen in Figure 4.5, the observed OR is now reduced to 1.27 if exposure misclassification is 20% in cases and controls. The reason for the different result compared with the example in Figure 4.4 is that the dilution of the exposed control group by nonexposed controls is larger than for cases where exposure prevalence is somewhat higher. In conclusion, if only a small proportion of the collective is exposed, a small decrease of specificity results in a considerable dilution of the exposed group (Figure 4.6). A low sensitivity, however, does not have a noticeable impact on the proportion of nonexposed persons and underestimation of risk. Conversely, if the exposure is common, that is, around 50%, both specificity and sensitivity are important.

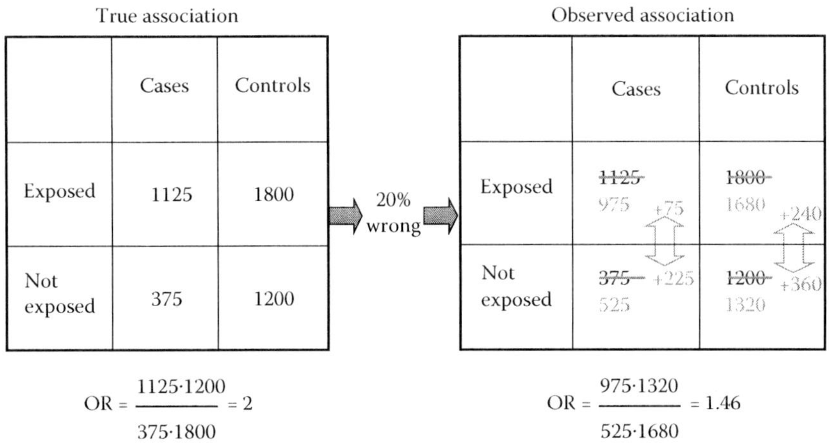

FIGURE 4.4
Effect of nondifferential exposure misclassification in a case–control study with binary exposure categorization and a true odds ratio of 2.

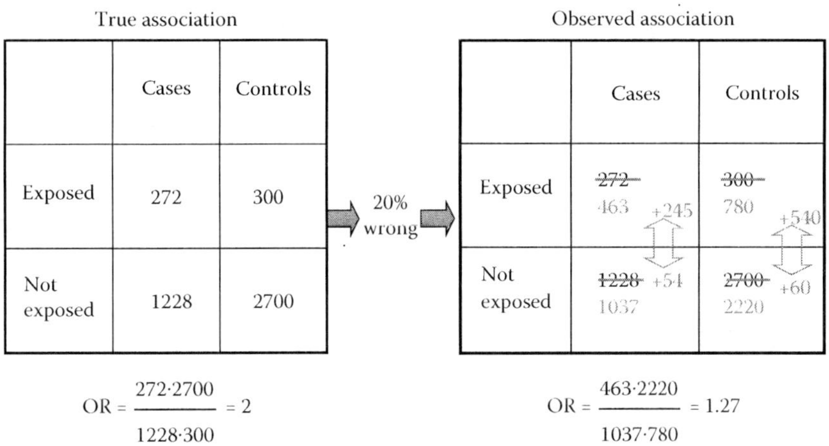

FIGURE 4.5
Effect of nondifferential exposure misclassification in a case–control study with binary exposure categorization and a true OR of 2. True exposure prevalence is rare, that is, only 10% in controls.

It is somewhat counterintuitive that under certain circumstances it is more relevant to correctly estimate nonexposed study participants than exposed study participants. A typical example is the Danish cohort study on mobile phone use (Frei et al. 2011) where cancer rates of early mobile phone subscribers (1982–1995) were compared with the rest of the Danish population. This study was criticized (Söderqvist et al. 2011) because from 720,000 mobile phone subscriptions only 360,000 could be attributed to an individual, corresponding to a sensitivity of 50%. The main reason for nonlinkage was the high number of business subscriptions. This means that the unexposed group was somewhat diluted. Because the proportion of nonexposed is much larger, the impact on the study result is relatively small if we assume a specificity of 100%, that is, that those identified without a subscription indeed did not have a subscription. This can be shown with the following

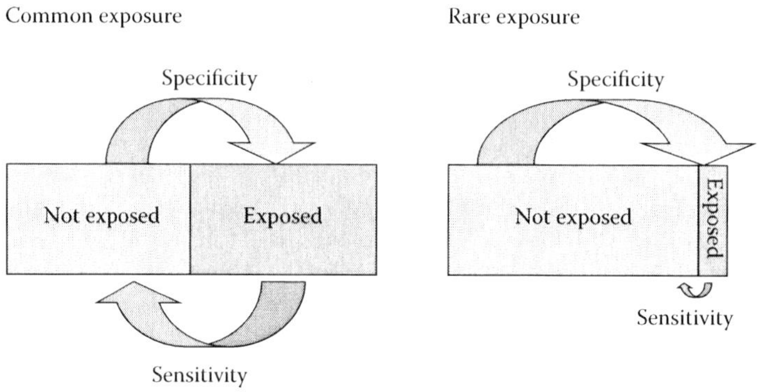

FIGURE 4.6
Effect of specificity and sensitivity on exposure misclassification for a common and rare exposure.

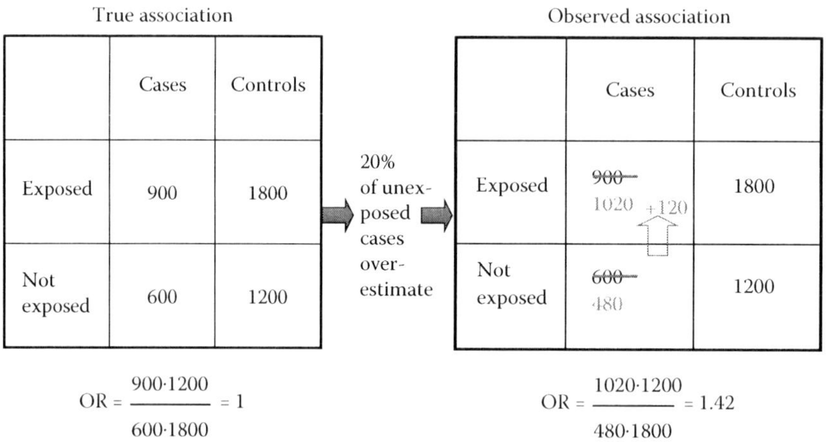

FIGURE 4.7
Effect of differential exposure misclassification in a case–control study with binary exposure categorization and a true OR of 1. It is assumed that 20% of the nonexposed cases are classified as exposed and the exposure assessment is otherwise correct.

calculation based on the assumption of a relative risk (RR) for subscribers of 1.5. Adding 360,000 subscribers with an RR of 1.5 to the 4,100,000 Danes without a subscription would result in an observed RR of 1.44 [= 1.5/((360,000 × 1.5 + 4,100,000 × 1.0)/4,460,000)]. Thus, the observed glioma risk for male (RR = 1.08; 95% confidence interval [CI], 0.96–1.22) and female (RR = 0.98; 95% CI, 0.69–1.40) subscribers is unlikely indicative for a major risk related to early mobile phone subscription. Because access to the amount of use was not possible in this cohort, corresponding exposure misclassification is expected to be higher and the potential risk corresponding to the amount of use cannot be evaluated (see Chapter 12).

Differential Exposure Misclassification

It is intuitive that differential exposure misclassification leads to biased study results. If cases overestimate exposure compared with controls or if controls underestimate exposure compared with cases, the observed RR will be biased upward (Figure 4.7).

TABLE 4.3

Effect of Nondifferential and Differential Exposure Misclassification in a Case–Control Study with Either a True Risk of 2.0 or 1.0 Assuming an Exposure Prevalence in Controls of 60%

	Sensitivity		Specificity		OR	
Row	Cases	Controls	Cases	Controls	True	Observed
1	0.9	0.8	0.6	0.8	2	2.7
2	0.8	0.8	0.6	0.8	2	1.8
3	0.9	0.8	0.5	0.8	2	3.1
4	0.8	0.8	0.5	0.8	2	2.1
5	0.9	0.7	0.6	0.7	2	2.9
6	0.8	0.7	0.6	0.7	2	2.0
7	0.9	0.7	0.5	0.7	2	3.4
8	0.8	0.7	0.5	0.7	2	2.2
9	0.9	0.8	0.6	0.8	1	1.8
10	0.8	0.8	0.6	0.8	1	1.4
11	0.9	0.8	0.5	0.8	1	2.2
12	0.8	0.8	0.5	0.8	1	1.7
13	0.9	0.7	0.6	0.7	1	2.0
14	0.8	0.7	0.6	0.7	1	1.5
15	0.9	0.7	0.5	0.7	1	2.4
16	0.8	0.7	0.5	0.7	1	1.8

Conversely, if cases underestimate exposure and controls overestimate exposure, the RR is downward biased.

In reality, differential exposure misclassification is usually accompanied with a nondifferential exposure misclassification; thus, sensitivity, specificity, or both are different for cases and controls but <1 for both. For a case–control study with some recall bias, sensitivity for cases is expected to be higher than for controls, whereas specificity would be higher in controls compared with cases. For this assumption, Table 4.3 shows how the observed OR changes in a case–control study with an exposure prevalence of 60% among controls (3000 controls, 1500 cases). If a true risk of 2.0 is assumed, the (erroneously) observed OR is larger than the true exposure–disease association in most of the examples shown. That means that overestimation of the OR from differential exposure misclassification dominates the dilution effect from nondifferential exposure misclassification. In one example, however, the true effect is underestimated (second row) and in another example one effect cancels the other out and the true OR of 2 is observed (sixth row). If there is no real risk (true OR = 1), the observed OR is always overestimated. Interestingly, an observed risk of 1.8 is simulated from a true risk of 2.0 (second row) or a true risk of 1.0 (ninth row) by only a slight change in the sensitivity for cases.

Table 4.3 illustrates that the differential exposure misclassification becomes more relevant when the risk is small, whereas for large risk some compensation between overestimation of differential exposure misclassification and underestimation of unavoidable random misclassification can be expected. Similarly, for an epidemiological study on mobile phone use and brain tumor risk (Vrijheid et al. 2006b), it was shown that the presence of random recall errors of plausible levels leads to a large underestimation of the risk. On the other hand, differential errors in recall had little additional impact on the effect estimates (see Chapter 13).

Ecological versus Individual Exposure Estimates

Exposure of the study population can be assessed using an individual or a group approach. Individual exposure assessment means that exposure for each study participant is individually obtained, for instance, from a measurement or from a questionnaire. In the group approach, the study population is split into exposure groups, for example, on the basis of the presence of or distance to an exposure source. In EMF epidemiology, distance to power lines or occupational groups have been repeatedly used as exposure proxies (Röösli et al. 2007; Huss et al. 2009). In the former it is, for example, assumed that people living in the vicinity of power lines have, on average, a higher exposure than people farther away. With regard to occupation, exposure in certain occupations is expected to be higher than in other occupations although variability within one occupational group may be high.

Intuitively, one expects that individual exposure estimates are superior to group assignments. This, however, is not necessarily the case as explained by the Classical and Berkson-type error models (Armstrong 1998). Observed exposure is always the combination of true exposure and possible bias and random error (Figure 4.8). For the sake of simplicity, bias is ignored in this example. If the exposure of interest shows a high temporal variability and only a few samples or measurements are available for assessing individual exposure, the random error becomes large. As a consequence, high exposure values are more likely due to a random overestimation, whereas low exposure values are more likely to somewhat underestimate the truth (Figure 4.8). For this reason, exposure difference in the sample is overestimated on average and the exposure–response association would be underestimated as outlined above, in the example of nondifferential misclassification for a binary exposure variable.

If exposure is assessed on a group level, there is random exposure error within the group. If, however, the group mean is measured correctly, the difference between groups would not be underestimated. For this Berkson-type error model, an exposure–response association would not be biased to the null association, which is counterintuitive. The larger the random exposure error within the groups the lower the statistical precision and

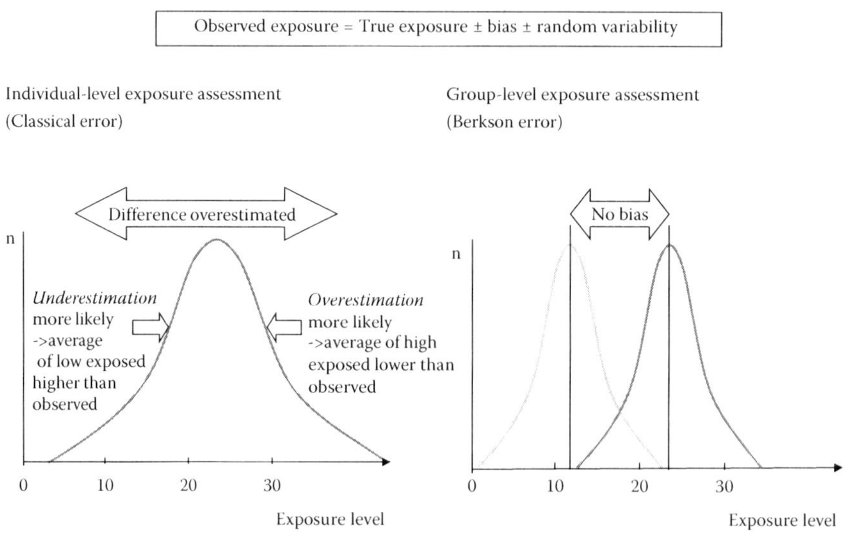

FIGURE 4.8
Illustration of Classical and Berkson-type error models resulting from individual- or group-level exposure assessment.

the higher the SE and CI of the effect estimate. The choice between individual-based and group-based exposure assessment thus also implies a choice between retaining power and reducing error. The use of prediction models for assessing exposure of each study participant is conceptually equivalent to the group-based approach, because exposure values are assigned for a set of predictor variables. Thus, the resulting error is mainly of Berkson type. Some classical error is, however, introduced if the predictor variables are measured with some error.

Exposure–Response Modeling

In environmental epidemiology, the crude binary variable exposed versus nonexposed is often not adequate because it does not consider nor capture the range of exposures that may, in fact, be large. Quantification on a continuous scale of exposure will, in general, provide a higher statistical power (Royston et al. 2006) and more insight in to the form of the exposure–response association.

The form of the exposure–response association is often not clear for an environmental factor. Although not necessarily true, a common approach is to assume a linear association without a threshold. There are, however, examples where exposure–response associations were found to be nonlinear. For instance, exposure to relative low concentration of tobacco smoke from passive smoking was found to increase the myocardial infarction risk considerably, more than what would be expected from linear interpolation of active smoking studies (Pechacek and Babb 2004; Pope et al. 2011). For regulatory purposes, it is also of interest to evaluate whether a threshold without any adverse effect exists. Exposure–response modeling is particularly relevant if both exposure and outcome are measured on a continuous scale.

If exposure is assessed on a continuous scale, the form of the exposure–response relationship may, and should, be evaluated if the statistical power of the study is sufficient. Various approaches can be applied to estimate exposure–response relationship, and the goodness of fit will inform about the quality of the statistical model. A linear exposure–response model will be the first choice in most circumstances, unless there is a strong prior knowledge for assuming a different relationship.

In a second step, the exposure variable may be categorized into quantiles appropriate for the study sample size. For each exposure category, the median of the respective quantile may be assigned or the exposure group may simply be considered in order. In case of a skewed exposure distribution and a sufficient sample size, irregular quantile intervals may be selected, such as using a cutoff at the 50th, 75th, and 90th percentiles to evaluate the risk for high exposure values. Criteria for exposure categorization should be determined beforehand based on data distribution and biological plausibility and not be data driven. The advantage of a categorical analysis is that the form of the exposure–response association is not restricted and the results are relatively easily to communicate. Furthermore, extreme values have a lower impact on the effect estimates, which is desirable if these values have a high uncertainty. The disadvantage of a categorical analysis is that the statistical power of the study is somewhat decreased.

Using quadratic or polynomial functions is a further alternative for nonlinear exposure–response associations, in particular if the outcome is also measured on a continuous scale. It does not introduce (arbitrary) jumps in the exposure–response association as would occur in a categorical analysis and, thus, the statistical power is higher.

Transformation of the exposure variable is another option for modeling nonlinear exposure–response functions. Log transformation is often used for skewed exposure distributions because the assumptions of regression models are generally better fulfilled and the back-transformed effect estimates, corresponding to a percentage increase in risk per unit of the exposure variable, are simple to communicate.

In some situations, nonparametric functions such as splines may be used for elucidating the form of the exposure–response association. This approach is highly flexible but may induce random fluctuations in the exposure–response association if the power of study is not sufficient. The effect estimates, however, are difficult to communicate or pooled with other studies and thus the benefit of this approach is mainly for illustrative purposes and hypotheses generation. In EMF research, this approach has been applied with respect to the association between childhood leukemia and ELF-MFs (Greenland et al. 2000).

Conclusions

Sound exposure assessment is crucial for the validity of an epidemiological study. The relevance of exposure assessment in occupational epidemiology has been expressed by Blair et al. (2007) and their sentiment is expected to also be true for epidemiological research on EMF:

> We believe of the two of the major methodological issues raised in epidemiological studies of occupational exposures, that is, confounding and exposure misclassification, the latter is of far greater concern. [...] It is rare to find substantial confounding in occupation studies (or in other epidemiologic studies for that matter), even by risk factors that are strongly related to the outcome of interest. On the other hand, exposure misclassification probably occurs in nearly every epidemiologic study.

The selection of the most appropriate exposure assessment method is often a trade-off between the accuracy and the costs or determined by the sample size that can feasibly be assessed. Very elaborate methods that need high compliance from the study participants may produce selective study participation leading to a biased study even if exposure errors are minimized. Thus, under certain circumstances crude exposure surrogates that can be easily obtained for large study populations may be more valuable.

The effect of errors in exposure assessment on the study results is often counterintuitive and should therefore be carefully evaluated. Small errors can have a large impact, in particular, if the true risk is small. Alternatively, seemingly large exposure assessment errors have little impact on the study results. When epidemiological studies are evaluated, one has to consider both potential systematic and random exposure misclassification and the following rules can be applied for interpretation of study results:

- Nondifferential exposure misclassification is of particular concern in studies that show no association between exposure and disease. Here, the differentiation between "no true association" and substantial underestimation of the true exposure response association due to random exposure misclassification is crucial.
- Differential exposure misclassification is of particular concern in studies that show an association between exposure and disease. In this case, the observed risk can either be "a true association" or "a biased association" due to differential errors in the exposure assessment.

In practice, however, it is not easy to determine, whether and, to what extent systematic misclassification, random exposure misclassification, or both have actually occurred. Moreover, random data fluctuation, confounding or other types of bias, can superpose the effects from exposure errors.

In EMF epidemiology, a crude binary variable is often not adequate because exposure levels vary across a large range and, for many EMF sources, people who are completely unexposed do not exist. Quantification on a continuous scale not only, often, produces results with greater sensitivity, but also allows for evaluating the shape of the exposure–response association, an evaluation that can provide the basis for regulatory decision making.

References

Armstrong, B. G. 1998. Effect of measurement error on epidemiological studies of environmental and occupational exposures. *Occup Environ Med* 55(10): 651–6.

Aydin, D., Feychting, M., Schüz, J., Andersen, T. V., Poulsen, A. H., Prochazka, M., et al. 2011a. Impact of random and systematic recall errors and selection bias in case—control studies on mobile phone use and brain tumors in adolescents (CEFALO study). *Bioelectromagnetics* 32(5): 396–407.

Aydin, D., Feychting, M., Schüz, J., Klaeboe, L. and Röösli, M. 2011b. Predictors of Mobile Phone Use and Overestimation of Recalled Mobile Phone use among Children and Adolescents in a Case-Control Study (CEFALO-Study). Paper read at European Bioelectromagnetics Association, Rome, 22–24 February, 2011.

Aydin, D., Feychting, M., Schüz, J., Tynes, T., Andersen, T. V., Schmidt, L. S., et al. 2011c. Mobile phone use and brain tumors in children and adolescents: a multicenter case-control study. *J Natl Cancer Inst* 103(16): 1264–76.

Berg-Beckhoff, G., Blettner, M., Kowall, B., Breckenkamp, J., Schlehofer, B., Schmiedel, S., et al. 2009. Mobile phone base stations and adverse health effects: phase 2 of a cross-sectional study with measured radio frequency electromagnetic fields. *Occup Environ Med* 66(2): 124–30.

Blair A., Stewart P., Lubin J.H., Forastiere F. 2007. Methodological issues regarding confounding and exposure misclassification in epidemiological studies of occupational exposures. *Am J Ind Med* 50(3): 199–207.

Bowman, J. D., Touchstone, J. A. and Yost, M. G. 2007. A population-based job exposure matrix for power-frequency magnetic fields. *J Occup Environ Hyg* 4(9): 715–28.

Bürgi, A., Frei, P., Theis, G., Mohler, E., Braun-Fahrländer, C., Fröhlich, J., et al. 2010. A model for radiofrequency electromagnetic field predictions at outdoor and indoor locations in the context of epidemiological research. *Bioelectromagnetics* 31(3): 226–36.

Dreyer, N. A., Loughlin, J. E. and Rothman, K. J. 1999. Cause-specific mortality in cellular telephone users. *JAMA* 282(19): 1814–16.

Elliott, P., Toledano, M. B., Bennett, J., Beale, L., de Hoogh, K., Best, N., et al. 2010. Mobile phone base stations and early childhood cancers: case-control study. *BMJ* 340: c3077.

Erdreich, L. S., Van Kerkhove, M. D., Scrafford, C. G., Barraj, L., McNeely, M., Shum, M., et al. 2007. Factors that influence the radiofrequency power output of GSM mobile phones. *Radiat Res* 168(2): 253–61.

Forssen, U. M., Ahlbom, A. and Feychting, M. 2002. Relative contribution of residential and occupational magnetic field exposure over twenty-four hours among people living close to and far from a power line. *Bioelectromagnetics* 23(3): 239–44.

Frei, P., Mohler, E., Bürgi, A., Fröhlich, J., Neubauer, G., Braun-Fahrländer, C., et al. 2010. Classification of personal exposure to radio frequency electromagnetic fields (RF-EMF) for epidemiological research: evaluation of different exposure assessment methods. *Environ Int* 36(7): 714–20.

Frei, P., Poulsen, A. H., Johansen, C., Olsen, J. H., Steding-Jessen, M. and Schuz, J. 2011. Use of mobile phones and risk of brain tumours: update of Danish cohort study. *BMJ* 343: d6387.

Green, L. M., Miller, A. B., Villeneuve, P. J., Agnew, D. A., Greenberg, M. L., Li, J., et al. 1999. A case-control study of childhood leukemia in southern Ontario, Canada, and exposure to magnetic fields in residences. *Int J Cancer* 82(2): 161–70.

Greenland, S., Sheppard, A. R., Kaune, W. T., Poole, C. and Kelsh, M. A. 2000. A pooled analysis of magnetic fields, wire codes, and childhood leukemia. Childhood Leukemia-EMF Study Group. *Epidemiology* 11(6): 624–34.

Ha, M., Im, H., Lee, M., Kim, H. J., Kim, B. C., Gimm, Y. M., et al. 2007. Radio-frequency radiation exposure from AM radio transmitters and childhood leukemia and brain cancer. *Am J Epidemiol* 166(3): 270–9.

Huss, A., Spoerri, A., Egger, M. and Röösli, M. 2009. Residence near power lines and mortality from neurodegenerative diseases: longitudinal study of the Swiss population. *Am J Epidemiol* 169(2): 167–75.

Huss, A., Vermeulen, R., Bowman, J. D., Kheifets, L. and Kromhout, H. 2013. Electric shocks at work in Europe: development of a job exposure matrix. *Occup Environ Med* 70(4): 261–7.

Hutter, H. P., Moshammer, H., Wallner, P. and Kundi, M. 2006. Subjective symptoms, sleeping problems, and cognitive performance in subjects living near mobile phone base stations. *Occup Environ Med* 63(5): 307–13.

Inyang, I., Benke, G., Morrissey, J., McKenzie, R. and Abramson, M. 2009. How well do adolescents recall use of mobile telephones? Results of a validation study. *BMC Med Res Methodol* 9: 36.

Merzenich, H., Schmiedel, S., Bennack, S., Bruggemeyer, H., Philipp, J., Blettner, M., et al. 2008. Childhood leukemia in relation to radio frequency electromagnetic fields in the vicinity of TV and radio broadcast transmitters. *Am J Epidemiol* 168(10): 1169–78.

Michaelis, J., Schuz, J., Meinert, R., Zemann, E., Grigat, J. P., Kaatsch, P., et al. 1998. Combined risk estimates for two German population-based case-control studies on residential magnetic fields and childhood acute leukemia. *Epidemiology* 9(1): 92–4.

Neubauer, G., Feychting, M., Hamnerius, Y., Kheifets, L., Kuster, N., Ruiz, I., et al. 2007. Feasibility of future epidemiological studies on possible health effects of mobile phone base stations. *Bioelectromagnetics* 28(3): 224–30.

Parslow, R. C., Hepworth, S. J. and McKinney, P. A. 2003. Recall of past use of mobile phone handsets. *Radiat Prot Dosimetry* 106(3): 233–40.

Pechacek, T. F. and Babb, S. 2004. How acute and reversible are the cardiovascular risks of second-hand smoke? *BMJ* 328(7446): 980–3.

Pope, C. A., 3rd, Burnett, R. T., Turner, M. C., Cohen, A., Krewski, D., Jerrett, M., et al. 2011. Lung cancer and cardiovascular disease mortality associated with ambient air pollution and cigarette smoke: shape of the exposure-response relationships. *Environ Health Perspect* 119(11): 1616–21.

Preece, A. W., Georgiou, A. G., Dunn, E. J. and Farrow, S. C. 2007. Health response of two communities to military antennae in Cyprus. *Occup Environ Med* 64(6): 402–8.

Röösli, M., Lörtscher, M., Egger, M., Pfluger, D., Schreier, N., Lörtscher, E., et al. 2007. Mortality from neurodegenerative disease and exposure to extremely low-frequency magnetic fields: 31 years of observations on Swiss railway employees. *Neuroepidemiology* 28(4): 197–206.

Rose, G. 1987. Environmental factors and disease: the man made environment. *Br Med J (Clin Res Ed)* 294(6577): 963–5.

Rothman, K. J. and Greenland, S. 1998. *Modern Epidemiology*. 2nd ed. Philadelphia: Lippincott Williams & Wilkins.

Royston, P., Altman, D. G. and Sauerbrei, W. 2006. Dichotomizing continuous predictors in multiple regression: a bad idea. *Stat Med* 25(1): 127–41.

Schüz, J., Waldemar, G., Olsen, J. H. and Johansen, C. 2009. Risks for central nervous system diseases among mobile phone subscribers: a Danish retrospective cohort study. *PLoS One* 4(2): e4389.

Söderqvist, F., Carlberg, M., Hansson Mild, K. and Hardell, L. 2011. Childhood brain tumour risk and its association with wireless phones: a commentary. *Environ Health* 10(1): 106.

Tomitsch, J., Dechant, E. and Frank, W. 2010. Survey of electromagnetic field exposure in bedrooms of residences in lower Austria. *Bioelectromagnetics* 31(3): 200–8.

UK Childhood Cancer Study Investigators. 1999. Exposure to power-frequency magnetic fields and the risk of childhood cancer. *Lancet* 354(9194): 1925–31.

Vergara, X. P., Kheifets, L., Silva, M., Bracken, T. D. and Yost, M. 2012. New electric-shock job exposure matrix. *Am J Ind Med* 55(3): 232–40.

Vrijheid, M., Armstrong, B. K., Bedard, D., Brown, J., Deltour, I., Iavarone, I., et al. 2009. Recall bias in the assessment of exposure to mobile phones. *J Expo Sci Environ Epidemiol* 19(4): 369–81.

Vrijheid, M., Cardis, E., Armstrong, B. K., Auvinen, A., Berg, G., Blaasaas, K. G., et al. 2006a. Validation of short term recall of mobile phone use for the Interphone study. *Occup Environ Med* 63(4): 237–43.

Vrijheid, M., Deltour, I., Krewski, D., Sanchez, M. and Cardis E. 2006b. The effects of recall errors and of selection bias in epidemiologic studies of mobile phone use and cancer risk. *J Expo Sci Environ Epidemiol* 16(4): 371–84.

Wertheimer, N. and Leeper, E. 1979. Electrical wiring configurations and childhood cancer. *Am J Epidemiol* 109(3): 273–84.

5

Selection and Detection Bias

Maria Feychting

CONTENTS

Introduction

In epidemiological studies of rare diseases, prospective collection of detailed exposure information for each individual in the entire study base may be very time consuming and expensive, because the study base must be very large to achieve sufficient statistical power. A cohort design may therefore not be feasible, unless exposure assessment can be made through registry-based sources independent of participating cohort members. Therefore, a more efficient and less expensive alternative is to perform a case–control study that identifies cases that occur over a certain time period and where a random sample of the cohort is used to gain information about the exposure distribution in the study base, instead of collecting this information from the entire study population (see Chapter 2).

If population-based disease registries, for example, cancer registries, and registries of the population are available, enabling a complete follow-up of the cohort, a cohort study is unlikely to be affected by selection bias, because participation in the study is determined before the observation period, that is, before any disease occurrence. Therefore, the remainder of this chapter focuses on the case–control design, although selection bias is also relevant for cross-sectional studies (see Chapter 2) that are often applied to investigate the association between symptom prevalence and electromagnetic field (EMF) exposure (see Chapter 16). In this chapter, reverse causality and detection bias are also briefly discussed from the perspective of selection bias. Additional considerations on reverse causality are made in Chapter 6 from the perspective of confounding.

Potential Sources of Selection Bias

Control Selection

The majority of epidemiological studies on cancer risk from EMF exposure have used a case–control design. Cases of the studied disease have been identified over a certain time period, and individuals free from the disease (controls) are selected during the same time period. The purpose of the controls is to provide information about the exposure distribution in the population that has generated the cases. It is therefore essential to be able to identify the population that constitutes the basis for the study (study base), to allow random selection of controls from the study base and evaluation of the representativeness of the controls. The preferable control selection method is population based, that is, a random selection from a population register throughout the study period, or from a corresponding list that covers the whole population from which the cases have come, for example, electoral rolls. Sometimes, the study base is not available for random sampling, for example, when no population registers are available. In this situation, other methods for control selection are used, for example, random digit dialing, hospital controls, friends, or neighborhood controls.

Random digit dialing (random dialing of phone numbers until a suitable control agrees to participate) attempts to achieve a population-based control sampling, but the actual number of potential controls available for sampling is unknown, and it is not possible to calculate an adequate participation rate. Hospital controls are usually other patients at the same hospital as where the cases are identified, with diseases that are thought to be unrelated to the studied exposure. Participation rates are often higher among hospital controls than controls from the general population, but a limitation is that, it is not known whether the patients used as controls actually represent the population from which the cases came, in terms of the exposure distribution; it is not a random sample of the population who ends up as patients in a hospital. Using friend or neighborhood controls is also associated with the problem of not knowing whether they actually are representative of the study base.

Case Identification

Identification of cases should ideally also be population based, that is, all new cases of the disease that occur during the study period in the population that constitutes the basis for the study should be identified. These are the eligible cases. If no population-based health data registries are available, for example, a cancer registry, sometimes a hospital-based, or even clinic-based, recruitment of cases is used instead. With hospital-based case recruitment, it is difficult or even impossible to make the study base available for random sampling of controls, because the theoretical study base is all persons who would end up in the same hospital as where the cases are identified should they get the studied disease. If the hospital has a well-defined catchment area and there are no other alternative hospitals or clinics in the area, hospital-based case recruitment could essentially be population based. This is, however, rarely the case, because there may be several hospitals and clinics in the area, and some people may travel outside the area to get to a well-reputed hospital. Several different factors may influence a patient's choice of hospital of clinic. These may be factors associated with socioeconomic status, income, education, health insurance, and others, as well as distance to the treatment center. These factors may also be related to the exposure. Sometimes, not all hospitals or clinics in an area are willing to collaborate with the study investigators, leading to a nonrandom selection of cases that can be included in the study. Hospital-based case recruitment is often accompanied by hospital-based control recruitment, but there are also examples of hospital-based case recruitment and population-based controls (and vice versa).

Nonparticipation

Once cases and controls have been identified, information about the exposure and other factors that can influence the studied associations, so-called confounding factors, are collected from cases and controls, usually retrospectively, unless register-based information is available. Many studies of extremely low-frequency (ELF) magnetic fields and childhood leukemia have taken measurements inside the children's homes, which require participation of the eligible cases and controls. Case–control studies of mobile phone use have usually collected retrospective self-reported information about mobile phone use history and potential confounding factors, either through personal interviews or mailed questionnaires. Not all cases and controls, however, are willing or able to participate in scientific studies. Sometimes, it is not possible to contact all identified cases and controls; some may have died before contact could be made, or the responsible physician may not allow contact. For population-based controls, a physician is usually not involved, but contact information may not be available for all eligible persons. For those who are contacted, some decline participation for various reasons, for example, some patients may be too ill to be able to participate, or persons may refuse for other reasons. Nonparticipation may be a source of selection bias because it may be related to the exposure, either directly or indirectly.

Another potential source of selection bias is missing information on specific items in the interview or questionnaire, which makes individuals drop out from some, but not all, analyses.

Selection Bias in EMF Studies

As mentioned above, within EMF research on cancer outcomes, selection bias is primarily a potential problem in the case–control studies, because the cohort studies have had access to population-based registers for complete follow-up of cancer occurrence. Selection bias occurs if the likelihood of participation is related to both the disease and the exposure, that is, the association between the exposure and disease differ among participants and nonparticipants (Rothman et al. 2008). A low participation rate, by itself, does not necessarily lead to selection bias, not even if the participation rates differ between cases and controls (Greenland and Criqui 1981; Hartge 2006). Participation rates are often higher among cases, who usually want to contribute to increased knowledge about the causes of their disease, and are more motivated to participate and answer numerous questions in interviews or questionnaires. Healthy controls are often less motivated, and usually participation rates are lower. However, if the likelihood of participation is independent of the exposure it does not matter that participation rates are higher among cases than controls, this will not introduce selection bias. Correspondingly, similar response rates among cases and controls are not a guarantee against selection bias (Hartge 2006); the reasons for nonparticipation may not be the same for cases and controls. Nevertheless, high participation rates are always preferable because low participation gives more room for potential selection bias.

When the participation probability is related to both exposure and disease, selection bias is introduced. The magnitude and direction of the bias depends on the bias ratio, that is, the ratio between exposed and unexposed participation probabilities among cases to that among controls (Hartge 2006).

ELF Magnetic Fields and Childhood Leukemia

It has long been speculated that the observed association between exposure to residential ELF magnetic fields and childhood leukemia (see Chapter 9) may have been affected by selection bias in the case–control studies, primarily originating from nonparticipation among controls (e.g., Hatch et al. 2000; Ahlbom et al. 2001; Mezei and Kheifets 2006; Schuz and Ahlbom 2008). Mezei and Kheifets (2006) have provided a thorough review and discussion of this potential bias in the studies of ELF magnetic fields and childhood leukemia. The hypothesis is that low socioeconomic status is associated with higher levels of residential magnetic fields and that participation in case–control studies is often lower among persons from lower socioeconomic groups. Because participation rates generally are higher among cases, socioeconomic status will have less influence on participation among cases than among controls. Thus, the bias caused by nonparticipation would tend to bias risk estimates upward, because the participating controls would have a lower prevalence of exposure to ELF magnetic fields than the prevalence in the study base, that is, the population that generated the cases. Support for this hypothesis was presented by Hatch et al. (2000) who found that exclusion of subjects for whom only wire codes were available or who only allowed measurements taken outside the front door, but not inside the home (partial participants), tended to produce higher odds ratios related to very high-current configuration wire codes than when all subjects were included in the analyses. They also observed that partial participants tended to be characterized by lower socioeconomic status than subjects who participated in all aspects of the study. A few other studies have also found higher estimated magnetic fields through wire coding among nonparticipants (Mezei and Kheifets 2006). This reasoning is, however, only applicable to studies where direct contact with study subjects is necessary, for example, when measurements are taken inside the homes of cases and controls. The studies from the Nordic countries used exposure assessment methods that did not require contact with the study subjects (Feychting and Ahlbom 1993; Olsen et al. 1993; Verkasalo et al. 1993; Tynes and Haldorsen 1997), by estimating exposure through calculations of the magnetic field levels generated by transmission lines situated in the vicinity of the children's homes. The calculations were based on detailed information available in various registries; therefore, investigators did not need access to the subjects' homes, and participation was almost complete for both cases and controls. In the large pooled analysis published by Ahlbom and collaborators, results from pooling of studies using calculated fields to estimate exposure were very similar to the results from pooling the studies that had used 24hr measurements (Ahlbom et al. 2000). In addition, adjustment for socioeconomic status had little effect on the risk estimates, which is contrary to what would have been expected if selection bias caused by nondifferential participation related to socioeconomic status was the explanation for the observed findings. It is likely, however, that selection bias may explain some, but not the entire, risk increase observed for childhood leukemia (Mezei and Kheifets 2006; Schuz and Ahlbom 2008).

Radiofrequency Fields and Cancer

The Interphone study found risk estimates associated with regular mobile phone use that were generally below unity (Interphone Study Group 2010, 2011), for example, odds ratios for ever regular mobile phone use was 0.81 (95% confidence interval [CI], 0.70–0.94), 0.79 (95% CI, 0.68–0.91), and 0.85 (95% CI, 0.69–1.04) for glioma, meningioma, and acoustic neuroma, respectively (see Chapter 13). Regular mobile phone use was defined as making or receiving at least one call per week on average during last 6 months. Most risk estimates in various categories of amount and duration of use were also below unity, sometimes considerably reduced, for example, 1–2 years of mobile phone use was associated with a risk estimate of 0.62 (95% CI, 0.46–0.81) for glioma; and for acoustic neuroma, the ninth decile of cumulative hours of use was associated with an odds ratio of 0.48 (95% CI, 0.30–0.78). It has generally been regarded as implausible that radiofrequency fields associated with mobile phone use would protect against brain tumor development, although the biological plausibility for an increased risk also lacks scientific support. Selection bias from nonparticipation among controls in the case–control studies has been discussed as a potential explanation for the reduced risk estimates observed.

As discussed above, cases are often more willing to participate in scientific studies aiming at increased knowledge about causes of their disease, whereas controls are usually less motivated. It is also conceivable that persons who themselves are mobile phone users are more willing to participate in a study where the purpose is to investigate the potential carcinogenicity of radiofrequency fields from mobile phone use. Within Interphone, a validation study was performed to estimate the magnitude of potential bias caused by nonparticipation (Vrijheid et al. 2009). Cases and controls who refused to participate were asked a few questions in a nonresponder questionnaire. From this questionnaire, it was evident that nonparticipants were less frequently mobile phone users than participants, both among cases and among controls. Among participants, 66% of cases and 69% of controls were regular mobile phone users, compared with 50% of cases and 56% of controls who answered the nonresponder questionnaire. Even though this pattern was seen for both cases and controls, selection bias may occur because of the higher response rates among cases than among controls. Reasons for nonparticipation also differed between cases and controls, and refusal was a more common reason for controls than for cases (30% compared with 11%). The magnitude of the bias estimated in the validation study was an underestimation of odds ratios by approximately 10% (bias ratios for the most plausible scenarios varied between 0.87 and 0.92). This means that selection bias is likely to explain some of the apparently reduced risk observed in the Interphone study, but not the entire risk reduction. A limitation in the validation study is that the nonresponder questionnaire was only answered by 41% of case refusers and 4% of other case nonparticipants, and 57% of control refusers and 2% of other control nonparticipants. Thus, potential selection bias cannot be fully evaluated.

Other means to evaluate potential selection bias is to compare risk estimates in subsets of the study with higher participation rates to subsets where participation was lower (Hartge 2006). In the Interphone study, reduced risk estimates were found also in studies with high response rates, and there was no correlation between level of risk reduction and participation rates (see Appendix 1 in Interphone Study Group 2010), which also supports the notion that selection bias cannot explain the entire risk reduction observed. Interestingly enough, risk estimates below unity were observed both in centers that presented the study as an investigation of health effects of mobile phone use, and in centers where mobile phones were not mentioned in the introduction letter.

An attempt to reduce the impact of potential selection bias was made by the Interphone study investigators by restricting the analyses to regular users of mobile phones (see Appendix 2 in Interphone Study Group 2010), and using the lowest exposure category as the reference category instead of nonregular mobile phone users. This method is based on the assumption that selection bias from nonparticipation among controls is the only reason for the downward bias of the risk estimates. If this assumption is incorrect, however, new bias may be introduced and any gain in validity is questionable, because the odds ratios can be distorted in any direction, and the results will be difficult or impossible to interpret. There is strong evidence that selection bias cannot be the only reason for the reduced risk estimates, both from the validation study (Vrijheid et al. 2009) and from internal comparisons of results from different centers, as discussed above, and from comparisons of results for the different tumor types (Interphone Study Group 2010, 2011). For glioma, and to some extent acoustic neuroma, risk estimates in the lowest categories of duration and magnitude of use were reduced below unity far more than would have been expected from selection bias alone, whereas this was not the case for meningioma. There is no reason to believe that nonparticipation among meningioma controls would be less prone to introduce selection bias than nonparticipation among controls for glioma or acoustic neuroma. In addition, several study centers used the same controls for all tumor types.

An alternative explanation for the strong risk reductions in the lowest exposure categories is prodromal symptoms (see Chapter 6) from the tumor that would make not yet diagnosed cases less likely to take on a new habit like mobile phone use (reverse causation). If prodromal symptoms explain some of the risk reduction, restriction to regular users would introduce upward bias in the risk estimates. The validity of analyses restricted to regular users can also be questioned when inspecting the results; for glioma, a statistically significant 70% increased risk was observed already after 2–4 years of mobile phone use, an observation that is incompatible with national glioma incidence trends (de Vocht et al. 2011; Deltour et al. 2012; Little et al. 2012). Also, the raised glioma risks for longer durations of use are incompatible with incidence trends (Deltour et al. 2012). In addition, for meningioma odds ratios were still considerably reduced below unity. Thus, the results from the analyses restricted to regular users would imply that radiofrequency fields from mobile phone use increase the risk of glioma and at the same time protect against meningioma, which sounds biologically implausible. One cannot choose without indisputable reasons to believe some of the results from the analyses restricted to regular users (e.g., for glioma) but not others (e.g., for meningioma), unless convincing explanations for the conflicting results can be presented.

If prodromal symptoms in not yet diagnosed glioma cases affect the likelihood that the case becomes a new mobile phone user (reverse causation), risk estimates below unity for short durations of use would be expected also in other studies of mobile phone use and glioma, especially in the early studies conducted when mobile phone use was still uncommon in the general population, but gaining in popularity. Apart from being found in 11 of the 14 Interphone study centers for which center-specific odds ratios were presented, risk estimates below unity for glioma for the shortest duration of mobile phone use was also reported by the two U.S. hospital-based studies and the Danish cohort study (Muscat et al. 2000; Inskip et al. 2001; Schuz et al. 2006). Selection bias is of no concern in the Danish cohort study, where a population-based cancer registry was available for complete follow-up of the cohort. The hospital-based case–control studies from the United States had high response rates, but selection bias originating from potential noncomparability between patients used as controls and the population that has generated the brain tumor cases is difficult to assess. In addition, the magnitude and direction of such bias, if any, is impossible to predict. Attempts were made to ensure that control patients came from the same population as brain tumor cases by frequency matching on distance to the hospital and adjusting for this variable in the analyses (Inskip et al. 2001). There are also a few studies that did not find reduced risk estimates for the shortest duration of use; a Finnish register-based case–control study (Auvinen et al. 2002) and the studies by the Hardell group (Hardell et al. 2006). Like the Danish cohort study, the Finnish study is unlikely to be affected by selection bias but is based on very small numbers of exposed cases. The Hardell studies have several methodological limitations discussed in detail in a review by an expert group commissioned by the British Health Protection Agency (AGNIR 2012). Thus, there is some support for apparently reduced risk estimates for glioma associated with short duration of mobile phone use, which would be compatible with an effect of prodromal symptoms, although the evidence is not entirely consistent.

Selection bias associated with case recruitment has been discussed to a lesser degree. This bias would primarily potentially affect studies using hospital-based case recruitment and population-based controls and studies with incomplete and nonrandom recruitment of cases. A few study centers within the Interphone study used hospital-based case recruitment, but exclusion of these studies from the pooled analyses did not change the risk estimates (Interphone Study Group 2010). The studies by the Hardell group included only living subjects (Hardell et al. 2006), and they did not use a rapid case recruitment procedure. This resulted in a substantial loss of malignant brain tumors due to death of the case before contact had been made, likely to be the most severe cases. The effect on the risk estimates from nonparticipation of the most severe cases is impossible to predict. To address this potential selection problem, Hardell and colleagues contacted relatives of cases that had died before recruitment started in the original study, and they selected controls among persons who had died from other causes at the same time as the brain tumor case (Hardell et al. 2010). This was, however, done several years later, meaning that a relative had to report about their deceased relative's mobile phone habits many years back in time. From the considerably higher amounts of mobile phone use reported by relatives to both deceased cases and deceased controls compared with the original study, despite the fact that they were supposed to cover the same time period, it is obvious that exposure misclassification (see Chapter 4) must be substantial, and the validity of this study is questionable.

Detection Bias

Detection bias may also be classified as a type of selection bias, or it may be defined as dependent disease misclassification. If exposed cases of the studied disease are more likely to be diagnosed than the unexposed cases, detection bias occurs. Essentially, this means that in a case–control study, exposure is associated with the likelihood of being included as a case in the study. This may be the situation in studies of acoustic neuroma, because the most common early symptom of acoustic neuroma is unilateral hearing loss. Acoustic neuroma is a slow growing tumor that may be present many years before diagnosis (Thomsen and Tos 1990). It is possible that mobile phone use (or any phone use) increase the likelihood that the tumor is detected as the person may easily become aware of a one-sided impaired hearing while talking on the phone. This would give rise to an apparently (noncausal) increased risk of acoustic neuroma after a short latency period.

Conclusions

Selection bias is a potential source of error in case–control studies of electromagnetic fields and health effects that needs to be thoroughly evaluated and taken into consideration in health risk assessment. The evidence so far from studies of ELF magnetic fields and childhood leukemia risk speak in favor of selection bias as a partial explanation for the increased risk observed, but it is unlikely that selection bias explains the entire risk increase.

For studies of radiofrequency fields from mobile phone use, selection bias has been identified as a likely explanation for some of the observed risk reductions, primarily in the Interphone study. There is, however, strong evidence that speaks against the hypothesis that selection bias is the only cause of the apparently reduced risk estimates. Another potential explanation discussed is prodromal symptoms from the not yet diagnosed tumor, making the case less likely to take up a new habit such as mobile phone use. Causality as an explanation for the reduced risks observed has been dismissed as biologically implausible. It is, however, noteworthy that biological plausibility is equally weak for an increased risk of cancer related to the low level of radiofrequency fields associated with mobile phone use.

References

AGNIR. 2012. Health effects from radiofrequency electromagnetic fields. Report from the Independent Advisory Group on Non-Ionising Radiation. R., *Documents of the Health Protection Agency, Chemical and Environmental Hazards*. RCE 20. London, UK: Health Protection Agency.

Ahlbom, A., Cardis, E., Green, A., Linet, M., Savitz, D. and Swerdlow, A. 2001. Review of the epidemiologic literature on EMF and Health. *Environ Health Perspect* 109(Suppl 6): 911–33.

Ahlbom, A., Day, N., Feychting, M., Roman, E., Skinner, J., Dockerty, J., et al. 2000. A pooled analysis of magnetic fields and childhood leukaemia. *Br J Cancer* 83(5): 692–8.

Auvinen, A., Hietanen, M., Luukkonen, R. and Koskela, R. S. 2002. Brain tumors and salivary gland cancers among cellular telephone users. *Epidemiology* 13(3): 356–9.

Deltour, I., Auvinen, A., Feychting, M., Johansen, C., Klaeboe, L., Sankila, R., et al. 2012. Mobile phone use and incidence of glioma in the Nordic countries 1979–2008: consistency check. *Epidemiology* 23(2): 301–7.

de Vocht, F., Burstyn, I. and Cherrie, J. W. 2011. Time trends (1998–2007) in brain cancer incidence rates in relation to mobile phone use in England. *Bioelectromagnetics* 32(5): 334–9.

Feychting, M. and Ahlbom, A. 1993. Magnetic fields and cancer in children residing near Swedish high-voltage power lines. *Am J Epidemiol* 138(7): 467–81.

Greenland, S. and Criqui, M. H. 1981. Are case-control studies more vulnerable to response bias? *Am J Epidemiol* 114(2): 175–7.

Hardell, L., Carlberg, M. and Hansson Mild, K. 2006. Pooled analysis of two case-control studies on use of cellular and cordless telephones and the risk for malignant brain tumours diagnosed in 1997–2003. *Int Arch Occup Environ Health* 79(8): 630–9.

Hardell, L., Carlberg, M. and Hansson Mild, K. 2010. Mobile phone use and the risk for malignant brain tumors: a case-control study on deceased cases and controls. *Neuroepidemiology* 35(2): 109–14.

Hartge, P. 2006. Participation in population studies. *Epidemiology* 17(3): 252–4.

Hatch, E. E., Kleinerman, R. A., Linet, M. S., Tarone, R. E., Kaune, W. T., Auvinen, A., et al. 2000. Do confounding or selection factors of residential wiring codes and magnetic fields distort findings of electromagnetic fields studies? *Epidemiology* 11(2): 189–98.

Inskip, P. D., Tarone, R. E., Hatch, E. E., Wilcosky, T. C., Shapiro, W. R., Selker, R. G., et al. 2001. Cellular-telephone use and brain tumors. *N Engl J Med* 344(2): 79–86.

INTERPHONE Study Group. 2010. Brain tumour risk in relation to mobile telephone use: results of the INTERPHONE international case-control study. *Int J Epidemiol* 39(3): 675–94.

INTERPHONE Study Group. 2011. Acoustic neuroma risk in relation to mobile telephone use: results of the INTERPHONE international case-control study. *Cancer Epidemiol* 35(5): 453–64.

Little, M. P., Rajaraman, P., Curtis, R. E., Devesa, S. S., Inskip, P. D., Check, D. P., et al. 2012. Mobile phone use and glioma risk: comparison of epidemiological study results with incidence trends in the United States. *BMJ* 344: e1147.

Mezei, G. and Kheifets, L. 2006. Selection bias and its implications for case-control studies: a case study of magnetic field exposure and childhood leukaemia. *Int J Epidemiol* 35(2): 397–406.

Muscat, J. E., Malkin, M. G., Thompson, S., Shore, R. E., Stellman, S. D., McRee, D., et al. 2000. Handheld cellular telephone use and risk of brain cancer. *JAMA* 284(23): 3001–7.

Olsen, J. H., Nielsen, A. and Schulgen, G. 1993. Residence near high voltage facilities and risk of cancer in children. *BMJ* 307(6909): 891–5.

Rothman, K. J., Greenland, S. and Lash, T. L., eds. 2008. *Modern Epidemiology*, 3rd ed. Philadelphia, PA: Lippincott Williams & Wilkins.

Schuz, J. and Ahlbom, A. 2008. Exposure to electromagnetic fields and the risk of childhood leukaemia: a review. *Radiat Prot Dosimetry* 132(2): 202–11.

Schuz, J., Jacobsen, R., Olsen, J. H., Boice, J. D., Jr., McLaughlin, J. K. and Johansen, C. 2006. Cellular telephone use and cancer risk: update of a nationwide Danish cohort. *J Natl Cancer Inst* 98(23): 1707–13.

Thomsen, J. and Tos, M. 1990. Acoustic neuroma: clinical aspects, audiovestibular assessment, diagnostic delay, and growth rate. *Am J Otol* 11(1): 12–19.

Tynes, T. and Haldorsen, T. 1997. Electromagnetic fields and cancer in children residing near Norwegian high-voltage power lines. *Am J Epidemiol* 145(3): 219–26.

Verkasalo, P. K., Pukkala, E., Hongisto, M. Y., Valjus, J. E., Jarvinen, P. J., Heikkila, K. V., et al. 1993. Risk of cancer in Finnish children living close to power lines. *BMJ* 307(6909): 895–9.

Vrijheid, M., Richardson, L., Armstrong, B. K., Auvinen, A., Berg, G., Carroll, M., et al. 2009. Quantifying the impact of selection bias caused by nonparticipation in a case-control study of mobile phone use. *Ann Epidemiol* 19(1): 33–41.

6

Confounding, Reverse Causation, and Ecological Fallacy

Susanna Lagorio

CONTENTS

Introduction

To evaluate the possible beneficial or detrimental effect of a given agent, the ideal error-free study design would be one that compares the outcome experiences of a single group of individuals under the opposite conditions of having and having not been exposed to the agent of interest. Such an unattainable comparison may be simulated in animal research. In cancer bioassays, for example, groups of rodents of the same strain, reared under identical conditions (environment, housing, and food and beverage) are randomly and blindly assigned to one or more exposed and unexposed subgroups and followed up life-long for the outcomes of interest. In experimental studies on humans, however, the assumptions of genetic homogeneity and habitat identity are violated. Furthermore, when ethical constraints prevent human experiments, the exposure–disease relationship may only be assessed by observational studies.

Epidemiological studies are inherently susceptible to bias from multiple sources, as effectively shown by Maclure and Schneeweiss [1] in their "episcope" (Figure 6.1).

Three validity issues are explored in separate sections of this chapter: confounding, reverse causality, and interpretative problems of ecological studies. These seemingly

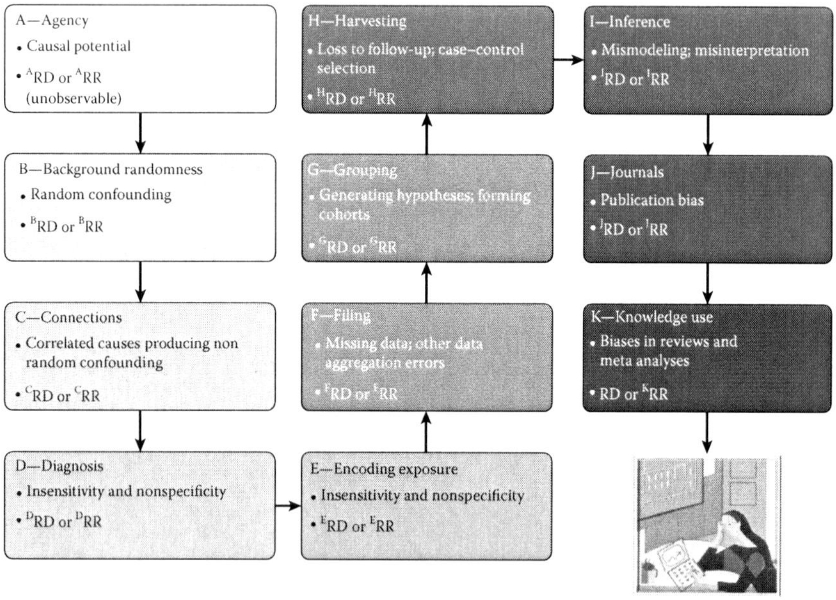

FIGURE 6.1
Domains of bias causation in epidemiology. (Adapted from Maclure M., Schneeweiss S., *Epidemiology* 12, 114–122, 2001.)

unrelated topics are closely connected, in reality, because reverse causality can be regarded as a particular form of confounding, and confounding and temporal bias are two components of the complex distortion commonly referred to as ecological fallacy.

Confounding

The word "confounding" derives from the Latin term *confundere* that means "to mix together." A few definitions of confounding, drawn from a nonexhaustive series of major epidemiological textbooks, are shown in Box 6.1 [2–7].

In a nutshell, confounding

- is an error in the empirical measure of association between an exposure (denoted X) and a disease (denoted Y), whereas the true association may be positive, negative, or null;
- may arise when the disease occurrence in the comparison populations differ for reasons other than the exposure to the risk factor of interest, and the key difference is in the distribution of exposure to another risk factor for the disease under study (denoted C), which is also associated with the agent of interest; and
- is not an "all-or-none" phenomenon; when present, it occurs in degrees, depending on the strength of the associations between X and C, and between C and Y.

A confounding variable, therefore, is a risk (or preventive) factor for the disease of interest, also associated with the exposure under study. A confounder, however, is neither

BOX 6.1 DEFINITIONS OF CONFOUNDING FROM MAJOR EPIDEMIOLOGY TEXTBOOKS OR DICTIONARIES

- On the simplest level, confounding is a confusion of effects: the apparent effect of the exposure of interest is distorted because the effect of an extraneous factor is mistaken for or mixed with the actual exposure effect (which can be null). A more precise definition of confounding begins by considering the manner in which effects are estimated. When we wish to estimate the degree to which an exposure has changed the frequency of a disease in an exposed cohort, we must estimate what the frequency of disease would have been in the exposed cohort had the exposure been absent. To accomplish this task, we observe the disease frequency in an unexposed cohort. But rarely could we take this unexposed frequency as fairly representing what the frequency of disease would have been in the exposed cohort had the exposure been absent, because the unexposed cohort would differ from the exposed cohort on many factors that affect disease frequency besides exposure. … We say that the comparison of the exposed and the unexposed is confounded because the difference in disease frequency between exposed and unexposed results from a mixture of several effects, included (but not limited to) any exposure effect [2 (pp. 120–125)].
- The term confounding refers to a situation in which a noncausal association between a given exposure and an outcome is observed as a result of the influence of a third variable (or group of variables), usually designated as *confounding variable*, or merely as *confounder* [3 (pp. 177–210)].
- Confounding is one of the fundamental methodological concerns in epidemiology. … The ideal comparison group for the exposed group is the exposure group itself but under the condition of not having been exposed, an experience that did not, in fact, occur (thus it is counterfactual). If we could observe this experience, we would be able to compare the disease occurrence under the situation in which exposure has occurred to the counterfactual one in which everything else is the same except exposure was not present. Instead, we choose some other group, sometimes one that has been randomized not to receive exposure, to provide an estimate of what the experience of the exposed group would have been absent the exposure. Ignoring various forms of selection, measurement errors, and random processes, the reason that comparing the exposed to the unexposed group would fail to accurately measure the causal effect of exposure is confounding [4 (pp. 137–161)].
- In epidemiology, confounding is the error in the estimate of the measure of association between a risk factor and a disease, which may arise when there are differences in the comparison populations other than the risk factor under study. In this circumstance confounding may occur but is not inevitable. These differences must include factors that are associated both with the disease and the risk factor under study for confounding to be present. Confounding is a major problem in epidemiology, and probably the most difficult one to understand, show and counteract. The potential for it to occur is there whenever the cardinal rule "compare like-with-like" is broken. […] The degree of confounding is the difference between the measure of risk

when the study group is compared with a counterfactual population (which can never be known) and that seen in real data [5, pp. 93–97].

- Bias of the estimated effect of an exposure on an outcome due to the presence of a common cause of the exposure and the outcome [6, pp. 49–50].
- In an outcome's empirical association with an antecedent (descriptively valid association, possibly nil in magnitude, with an antecedent that in principle could be causal), confounding is the possible explanation, partial or full, other than the antecedent's degree of role (when present) in the outcome's etiology/etiogenesis [7, pp. 110–111].

an intermediate in the causal pathway from X to Y nor a variable affected by both X and Y (the latter is called a collider).

Are There Epidemiological Designs Especially Susceptible to Confounding Bias?

Because confounding is an error originating in the target population (a domain that precedes the phases of subject selection and information retrieval), no epidemiological study design is truly free from potential confounding bias [1].

Random confounding, arising in the domain of "background randomness" (Figure 6.1), is characterized by the fact that X and C (both risk factors for Y) are not associated in the source population, but they may appear correlated in the study sample just by chance (Figure 6.2A).

Nonrandom confounding, instead, arises in the domain of "connections" (Figure 6.1) and is due to correlated causes of the disease under study (Figure 6.2B1 and 6.2B2).

What Effects May Confounding Exert on the Empirical Measures of Association?

Confounding may bias the empirical measures of association in either direction, toward or away from the null. Therefore, confounding is an issue in the interpretation of any epidemiological study, irrespective of its finding (i.e., positive, negative, or null associations between the exposure and the disease of interest) [3–5].

The direction of the distortion will depend on (1) the direction of the association between X and C and (2) the direction of the association between C and Y. In the simplified

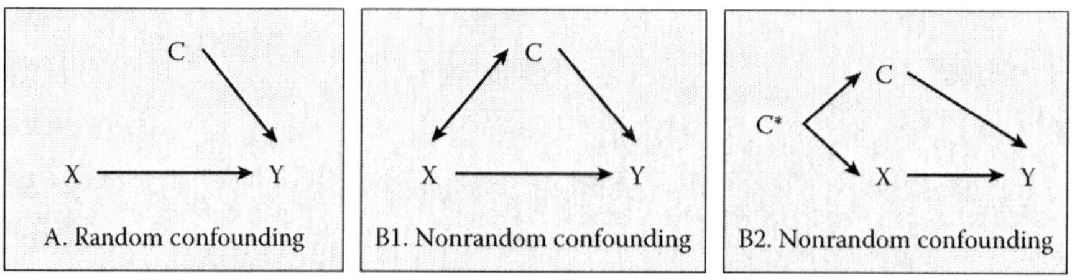

FIGURE 6.2
Schematic representation of random and nonrandom confounding. X, exposure of interest; Y, disease under study; C, confounder factor. The arrows indicate that the event at the arrow tail can produce the event at the arrowhead. In diagram A, X and C, both risk factors for D, are not associated, but they may appear correlated in the study sample just by chance. In diagrams B1 and B2, X and C are associated, directly in B1, or through a common cause (C*) in B2, and both are independent risk factors for Y.

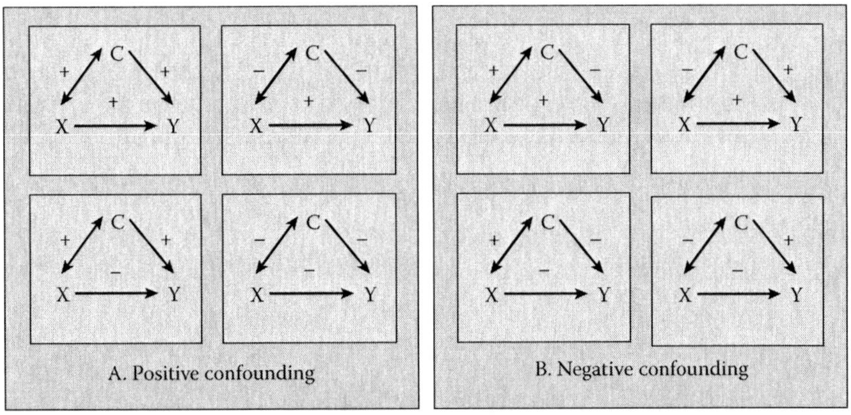

FIGURE 6.3
Schematic representation of positive and negative confounding. X, exposure of interest; Y, disease under study; C, confounder factor. The arrows indicate that the event at the arrow tail can produce the event at the arrowhead, and the signed edges represent monotonically increasing (+) or decreasing (–) relationships.

situation where a causal path consists of X, Y, and C, and there is a monotonically increasing or decreasing relationship between any pair of these variables, the output of the different combinations may be depicted as shown in Figure 6.3. Positive confounding is when the absolute value of a crude estimate (i.e., unadjusted for the level of C) is greater than the adjusted estimate, and negative confounding is when the reverse is true. Thus, independently from the direction of the relation between X and Y, positive confounding biases the risk estimate away from the null hypothesis (diagrams in Figure 6.3A), and negative confounding biases the risk estimate toward the null (diagrams in Figure 6.3B).

Intuition concerning the sign of confounding may fail in more complex situations; when there are multiple unmeasured confounding variables, conclusions can be drawn only if these variables are independent of one another, conditional on the measured covariates [8].

How to Address Confounding?

Confounding is an issue to consider at the design stage.

A comprehensive understanding of the etiology and natural history of the disease under study is essential to identify potential confounders of the exposure–disease relationship of interest [9].

Randomization (i.e., casual assignment of subgroups of the study population to different levels of the exposure of interest), when ethically possible and acceptable to the study subjects, is a strategy to control *random confounding*, and it is effective in very large provocation (or intervention) studies.

Eligibility restriction and matching are design options suitable to control *nonrandom confounding* in observational etiological studies, although both have disadvantages [2,5].

Confounding by a known risk predictor can be controlled by restricting enrolment to groups with comparable levels of the confounder (e.g., a particular age group, men or women, nonsmokers only). Restriction, however, limits the possibility to generalize the study findings, prevents the assessment of the X–Y relationship at different levels of C, and can lead to erroneous conclusions when there is interaction between X and C. Moreover, residual confounding may persist if the restriction categories are not sufficiently narrow.

Pair- or frequency-matching (e.g., on age, sex, residence, and other personal characteristics) is common in case–control studies. If the matching variables are true confounders, however, the resulting study sample is no more representative of the source population, and proper analytical techniques (matched pair analysis or conditional logistic regression) are needed to obtain reliable measures of effect. Matching should be used with caution, because the expected gain in precision may be diminished or nullified by the exclusion of cases without suitable controls, as well as by unnecessary matching on weak or too many confounders [10].

Confounding can also be controlled in the analysis of data, provided the information on exposure to potential confounder(s) has been collected. The relevant statistical techniques include Mantel–Haenszel estimators (odds ratios [ORs], risk ratios, and rate ratios) of adjusted effects from stratified data (a suitable option to control for one or two confounders), and multivariable regression models (logistic regression in case–control studies, Poisson regression in cohort studies) [2]. Specific methods have been developed for studies including time-varying exposures (or treatments) affected by time-dependent confounders [11,12].

However, the amount and direction of actual confounding in the data set should always be assessed before any attempt to control it in the statistical analysis. There are three data-based methods to accomplish this, and they are to be used as integrated rather than alternative tools [3].

- Assessment of the C*–X* and C*–Y* relationships, where Y*, X*, and C* denote the operational variables for the disease under study, the exposure of interest, and the potential confounder of their relationship, respectively. Confounding from the extraneous factor is not a likely candidate explanation for the observed association (or lack of) between X* and Y* if the two criteria for confounding (C* is associated with both X* and Y*) are not concurrently satisfied within the study population. However, the true relationships between C and Y, and between C and X, could be missed if the operational variable C* is a poor proxy of exposure to C [2,4].
- Stratification (assessment of the X*–Y* associations in separate strata of C*). Confounding in the data is unlikely when the stratum-specific estimates do not differ from the crude (unadjusted) overall measure of effect. When confounding is present, the measures of association across strata of C* are of similar magnitude, but all differ from the crude estimate.
- Comparison between crude and adjusted estimates of association, for example, using a test for homogeneity [13]. In the presence of many potential confounders, this allows one to estimate the amount of confounding introduced by C_i^* and to apply a threshold criterion for "substantial" confounding (e.g., a 10% change-in-estimate criterion) to decide whether to adjust for C_i^* in the analyses [14].

Two caveats are worth mentioning here. First, variables affected by the exposure of interest should not be controlled for in the analysis (either by matching or through statistical adjustment), because this introduces selection bias [10,15]. Second, stratification on a collider variable can induce confounding even if there is no confounding in the crude estimate [16].

Residual Confounding and Unmeasured Confounding

Confounding can persist after adjustment (residual confounding), due to imperfectly measured confounders, improperly treated confounding variables (i.e., categorized continuous variables), mismodeled confounding effects, or unnecessary adjustment [4,10,17].

Unmeasured confounding may be due to unavailable information on exposure to known risk factors for the disease under study. Confounding from unknown factors, due to poor knowledge about the disease etiology (as it is true for many cancers and other chronic diseases), is also a possibility.

Spurious increases in the relative risk of disease per unit increase in exposure from 1.1 to 3 can be generated by residual and/or unmeasured confounding alone [18].

Unmeasured confounding can be dealt with through quantitative bias analyses that may enable

- estimating the impact of an unmeasured, but known, confounder on the study findings;
- identifying the combination of bias parameters that would wholly explain the observed association if there were no true effect of the exposure of interest on the disease under study; and
- assessing what combination of bias parameters would be necessary to reverse the direction of a true effect to what was observed [19].

In the simplest situation of a dichotomous exposure (X = 0,1), a dichotomous outcome (absence or presence of the disease Y = 0,1) and a dichotomous confounder (C = 0,1), the required bias parameters consist of (a) the association between C and Y among the unexposed to X; (b) the association between C and X in the source population; and (c) the prevalence of exposure to C in the source population.

The availability of internal or external sources of information on the bias parameters is a prerequisite for carrying out bias analyses for unmeasured known confounders. Such information may derive from side-studies on subsets of the study population, or literature sources concerning similar populations.

Confounding from unknown confounders can also be addressed by simulations; in this case, however, the bias parameters can only be estimated based on educated guesses [19].

Confounding bias analyses are relatively straightforward assuming that unmeasured confounding is the only source of distortion of the study findings, that there is no effect modification in subjects coexposed to X and C and that the bias parameters are known without error. When the latter assumption does not hold, different approaches are needed (i.e., multidimensional or probabilistic bias analyses) [19].

The epidemiological evidence on the relation between childhood leukemia (CL) and exposure to extremely low-frequency magnetic fields (ELF-MFs), briefly summarized below, may be considered a paradigmatic example of inferential challenges related to possible uncontrolled confounding.

Confounding in Studies of CL and ELF-MF Exposure

Confounding was the main alternative explanation discussed by the authors of the first study suggesting an association between exposure to childhood leukemia [20]. It was also one source of bias that subsequent epidemiological studies on the same topic tried to explore and control.

The relevant epidemiological evidence is discussed at length elsewhere in this book (see Chapter 9). The focus herein is on confounding factors (Table 6.1) considered in the 15 original studies [20–34] included in two nonoverlapping pooled analyses [35,36].

TABLE 6.1

Confounders Examined in Studies of CL and ELF-MF Exposure Included in the Nonoverlapping Pooled Analyses by Ahlbom et al. [35] and Kheifets et al. [36]

						Potential Confounder							
			Matching Variable			Common		Study Specific (No. of Categories)					
								SES Indices					
Pooled Analysis	Study	Country	Sex	Year of Birth	Area of Diagnosis	Detached or Other House	Mobility Quintile	Social Group	Mother's Education	Family Income	Urbanization	Vehicle Exhausts	East or West Germany
Ahlbom [35]	Feychting and Ahlbom [20]	Sweden	✓	✓	✓	✓	✓	4			2	3	
	Olsen et al. [21]	Denmark	✓	✓			✓	5			4		
	Verkasalo et al. [22]	Finland	✓	✓							2		
	Tyne and Haldorsen [23]	Norway	✓	✓	✓	✓	✓	6			2		
	Linet et al. [24]	United States	✓	✓	✓	✓	✓			6	4		
	Dockerty et al. [25]	New Zealand	✓	✓			✓		5		2		
	Michaelis et al. [26]	Germany 1	✓	✓	✓	✓	✓	2			3	2	✓
	McBride et al. [27]	Canada	✓	✓	✓	✓	✓		3		2		
	UKCCSI [28]	United Kingdom 1	✓	✓	✓			7					
Kheifets [36]	Bianchi et al. [29]	Italy 1	✓	✓	✓				3			✓	
	Kroll et al. [30]	United Kingdom 2	✓	✓	✓				3		✓		
	Malagoli et al. [31]	Italy 2	✓	✓	✓		✓		3				
	Schüz et al. [32]	Germany 2	✓	✓	✓	✓	✓		3		✓		
	Kabuto et al. [33]	Japan	✓	✓	✓		✓		3		✓		
	Wünsch-Filho et al. [34]	Brazil	✓	✓	✓		✓		3				

The main results of the pooled analysis by Ahlbom et al. (2000) were adjusted for age, sex, and socioeconomic status (SES), plus East/West in Germany; further control of other confounding variables (urbanization, type of dwelling, residential mobility, and vehicle exhausts), carried out on subsets of studies with available information, implied irrelevant changes in the combined measures of effect [35].

The findings reported by Kheifets et al. (2010) were adjusted for age, sex, and SES; the estimates also adjusted for residential mobility (stable residents vs. two or more change of residence between birth and diagnosis) did not differ from those obtained controlling only for the three baseline confounders [36].

Age and sex clearly meet the criteria for a variable to be considered a potential confounder: both incidence of CL and residential exposure to ELF-MF (because of different patterns of time spent at home) vary by age and sex.

The role of SES as a confounder of the relationship between ELF-MF and CL risk is less clear; inverse (negative) associations of CL risk with individual-level measures of family income, mother's education, and father's education (rescaled from the original score to make increasing values correspond to increasing SES) have been consistently reported in interview-based case–control studies conducted in the United States and Europe, whereas positive associations of CL with father's occupational class in record-based case–control studies, and with average occupational class in ecological studies, have been observed [37]. Connections of SES indexes to CL are likely to vary with place and time, and they are difficult to interpret because they might also be due to selection or participation bias [38]. It must be stressed that the majority of studies on CL and exposure to ELF-MF have a large potential for selection bias [39], likely arising through an SES-associated differential participation of cases and controls, and an inverse gradient of exposure to ELF-MF by SES for which there is limited empirical evidence [40,41].

A positive correlation between a composite area-level index of SES and distance of the dwellings from high-voltage power lines (the higher the SES level, the shorter the distance) has also been observed [42]. This finding needs to be viewed with caution, however, because distance from overhead power lines is a very poor predictor of the in-home MF level [43].

Notwithstanding its relevance, we leave aside the association between CL and SES resulting from selection bias to concentrate on the extent to which indicators of exposure to ELF-MFs act as nonspecific markers of social class, or as proxies for leukemogenic environmental exposures with a social class gradient.

The single available empirical estimate of the amount of confounding exerted by SES on the observed association between risk of CL and residential exposure to ELF-MF, in addition to the impact of participation bias, comes from a U.S. case–control study [24,40]. There were many "partial participants" in this study, that is, cases and controls who, although refusing the full measurement protocol, permitted a front-door spot measurement of MF induction. The study investigators first recalculated the OR of disease among all participants (complete plus partial, controlling for age and sex) as their best estimate of the measure of association net of participation bias, and then compared the "corrected" estimates adjusted for age and sex only to those adjusted for age, sex, type of residence, type of area, and home ownership. Adjustment for these SES indicators resulted in additional reductions of the risk estimates of −65% and −25%, depending on the cutoff used to define the upper exposure category (Figure 6.4) [24,40].

According to a recent update of the epidemiological evidence on CL and ELF-MF exposure, the new pooled analysis reinforces the consistency of the observation and

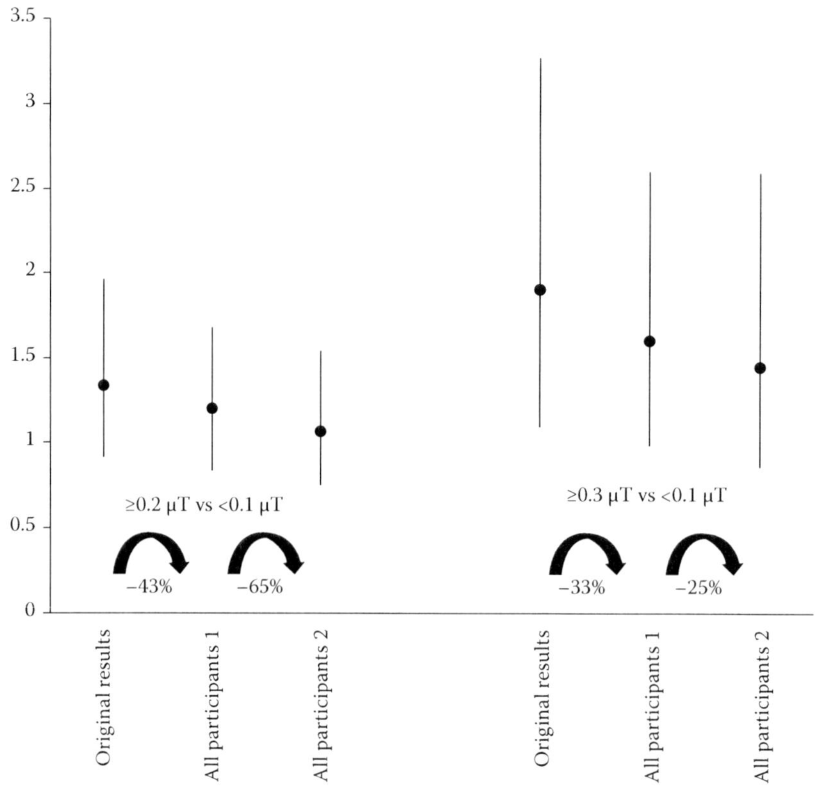

FIGURE 6.4
Estimated impact of participation bias and confounding by indicators of SES in the U.S. study of CL and exposure to ELF-MFs. (From Hatch EE et al., *Epidemiology* 11, 189–198, 2000.) Original results: based on complete participants, as reported in the published paper by Linet MS et al., *New Engl J Med* 337, 1–7, 1997. All participant 1: based on complete and partial participants and adjusted for age and sex; All participant 2: based on complete and partial participants, and adjusted for age and sex, type of residence, type of area, and home ownership.

lessens the likelihood of a chance finding, but a common bias or confounding cannot still be excluded [44].

It cannot be a single confounding factor because, to explain the observed association between ELF-MF and CL in full, it should be a strong risk factor for CL highly correlated with all sources of elevated MFs [45]. Confounding may persist after adjustment, however, and several unmeasured confounders with small or moderate effects may be able to produce the same effects of a single strong confounder [4,18].

A relevant issue may be which potential confounders should be considered by new original studies of the relation between CL and ELF-MFs, or by updated pooled analyses of available studies (assuming the availability of information in the latter), to gain further insight into the nature of the observed association.

The first key question is what we know today about the etiology of CL. This disease has an incidence rate of 30–50 per million children per year and is actually a family of biologically heterogeneous neoplasms [46]. The major subtypes in children are acute lymphoblastic leukemia (ALL) and acute myeloid leukemia (AML) accounting, respectively, for about 80% and 15% of CL cases in white populations aged 0 to 14 years; ALL has a peak incidence around the age of 2–5 years, whereas the incidence of AML increases with increasing age. Such differences make a single causal exposure or mechanism unlikely [47].

Both subtypes are thought to develop through a first initiating genetic event *in utero* followed by further postnatal genetic changes [48–50]. The most common chromosomal abnormalities observed in ALL are hyperdiploidy (35% of cases) and the TEL-AML1 gene fusion (25%). Chromosomal translocations occur *in utero*; the prevalence of TEL-AML1 fusion gene in newborns has been estimated at around 1%, but only 1% of these children will develop ALL with a TEL-AML1 gene fusion before the age of 15 years [48]. Thus, a "second hit" is a critical necessary event that might consist of additional genetic changes arising either before or after the formation of the chromosomal translocation, as well as of exposures to biological, chemical, or physical agents [47–50].

Few causative environmental agents are known for AML (high doses of ionizing radiation and benzene, alkylator, and topoisomerase II inhibitors), whereas a delayed or dysregulated response to common infections seems to be implied in the pathogenesis of pre-B ALL [48–50].

The role of environmental exposures to ionizing radiation (e.g., indoor levels of radon and gamma radiation) in the etiology of CL is unclear. A meta-analysis of seven case–control studies of CL and measured domestic radon concentration gave mixed results, with some indication of a weak exposure–disease association (OR < 2) [51]. Indoor radon levels were measured in 3 of the more than 30 case–control studies of CL and ELF-MF [52–54], whereas exposure to gamma radiation was assessed in one study only [55]. None of these studies, however, reported any association between environmental exposure to ionizing radiation and CL risk.

In adults, benzene exposure and acute nonlymphocytic leukemia (AnLL) are causally related, whereas there is limited evidence for increased risks of acute or chronic lymphocytic leukemia, multiple myeloma, and non-Hodgkin lymphoma among those exposed to this chemical [56]. Exposure to benzene would increase the risk of AnLL at ≥40 ppm-years of occupational cumulative exposure, equivalent to a lifetime (76 years) environmental exposure of ≥120 ppb [57]. Thus, it seems unlikely that benzene is a major cause of leukemia in the general population exposed in the ppb range. That notwithstanding, children may represent a subpopulation with increased susceptibility. Findings from available studies of benzene and CL are inconsistent, possibly due the use of indirect estimates of exposure (e.g., traffic density, residential proximity to gas stations, parental smoking) and lack of analyses by leukemia subtype [58]. For these reasons, studies of pediatric cancers including quantitative estimates of environmental exposure to benzene, rather than surrogate exposure indicators, have been recommended [59].

Both the time window of exposures to EMF-MF and to other known or putative risk factors for the disease, and the leukemia subtype may be important parameters to take into account, as shown in a recent meta-analysis of 13 case–control studies of CL and residential exposure to pesticides published between 1987 and 2009. The strongest associations were observed for exposure during pregnancy (OR 2.19 [95% CI, 1.92–2.50]), and for AnLL (OR 2.30 [95% CI, 1.53–3.45]) [60].

The results of a meta-analysis of 14 case–control studies indicated that day-care attendance is associated with a reduced risk of ALL (OR = 0.76 [95% CI, 0.67–0.87]), supporting the hypothesis that childhood ALL might be a rare response to common infection(s) [61]. Early socialization is likely to vary across countries, within countries, and by SES (depending, *inter alia*, on the proportion of working women in childbearing age, on the availability and accessibility of public and private crèches, and on sociocultural differences in the support provided to the nuclear family by older family members).

Population mixing in rural areas might also been a mechanism through which a biological agent (probably a virus) might be responsible of leukemia clusters among children living in rural communities, after settlement of migrants from populations characterized by a high degree of herd immunity toward this biological agent [62]. The association between population mixing and CL was examined in the UK Childhood Cancer Study (UKCCS) Research Centre, and it was actually found associated with a decreased risk of the disease [63]. It has been argued, however, that the proxy variable used in the latter study (diversity of origin of migrants at one census in the mainly urban census wards where the study children lived) was unrelated to the original conceptual construct, so that the study findings do not hinder the possible validity of this hypothesis [62].

The second key question is what we know about the mechanisms possibly involved in a leukemogenic effect of ELF-MF. The answer is quite a lot, but with important knowledge gaps. Long-term carcinogenicity assays have provided overwhelming negative results: overall, there is no evidence that ELF exposure alone causes tumors (including lymphoma and leukemia) in rodents, whereas there is inadequate evidence that ELF field exposure can enhance tumor development in combination with carcinogens [64]. One open issue is the extrapolability of laboratory results to humans, in that the available animal models do not closely resemble childhood ALL. A first study using a chemically induced rat model of acute pro-B lymphoblastic leukemia did not observe any difference in cumulative incidence and type of leukemia between exposed to 50 Hz MF (100 μT) and unexposed groups [65], but independent replications of the finding are needed [64].

One hypothesis raised to reconcile the epidemiological and experimental evidence concerning the possible carcinogenicity of ELF-MFs, as well as the heterogeneity of results across original studies of CL and measured or calculated ELF-MFs [66,67] was that unmeasured EMF characteristics could act as effect modifiers or confounders.

One such characteristic would be contact currents linked to residential electrical wiring safety practices in the United States, in particular to conductive residential plumbing lines connected to the municipal water main in the street [68]. Contact current exposure would occur when a child contacts a bathtub's water fixtures, usually contiguous with a residence's electrical ground, and when the drainpipe is conductive. The association between MF level and contact voltage observed in a pooled analysis of measurements made in about 700 U.S. residences was indeed of sufficient magnitude for contact current exposure to confound the ELF-MF–CL association [69]. That notwithstanding, in the Northern California Childhood Leukemia Study no association between CL and contact currents was observed; the disease was not associated with ELF-MF measurements either, and the two agents were not correlated [70].

The third key question is what we know about the joint distribution of exposures to ELF-MFs and known or putative environmental risk factors for CL in the population. The answer, in this case, is in fact very little.

In an Italian pilot study of CL (mainly pre-B ALL) and benzene exposure (assessed by repeated seasonal weekly measurements in breathing zone air samples and outside the children's dwellings), a positive correlation between outdoor benzene concentration (yearly average) and ELF-MF level in the child's bedroom (48-hr time-weighted average) was observed; the overall association was driven by children participating in the personal monitoring survey, rather than by those who only accepted outdoor measurements, suggesting that the finding might be in part attributable to participation bias [71].

In conclusion, the association between CL and exposure to ELF-MFs is consistent and apparently specific, but its causality is still questionable. In the epidemiological domain, new insight can only be provided by novel studies less prone to bias by design, with greater proportions of highly exposed children, devoting similar efforts to collect accurate and biologically relevant data on both ELF-MF exposure and exposures to potential confounders. Side-investigations aimed at evaluating the amount and direction of distortions from multiple sources, and exploring gene–environment interactions, would also be beneficial.

To be realistic, it is very unlikely that all these requirements can be concurrently met by any single study. A more plausible scenario would include a limited number of multicenter studies affording one or few uncertainties in the available epidemiological evidence. It has therefore been suggested that the status of uncertainty might remain for further decades, if ever fully resolved [72].

Regardless, among the candidate potential environmental confounders to consider in further epidemiological studies of morphology-specific types of CL and ELF-MFs, we would suggest personal exposure to benzene in investigations focused on AML and exposure to pesticides, early socialization, and substantial rises of population density in the child's place of residence in studies of pre-B ALL.

Reverse Causality

According to the Hill's clues to causal inference, temporality (i.e., exposure always precedes the outcome) is the only necessary property of a cause [73]. Notwithstanding its obviousness, assessment of the exposure–outcome time order may not be straightforward in observational epidemiological studies [5]. Assuming absence of bias and confounding, an observed exposure–disease association (either positive or negative) may reflect either an effect of the exposure on the occurrence of the outcome or an effect of the disease on the frequency, intensity, and/or duration of the exposure. Reverse causality occurs when the second option is true.

Reverse causation bias may be thought of as a special form of nonrandom confounding arising when, in the causal path between the exposure and disease of interest, the disease itself acts as a confounding variable [1].

Susceptibility to temporal bias varies by study design: whereas clinical trials (or provocation studies) are practically immune, the degree of vulnerability increases moving from cohort and case–control studies to cross-sectional and ecological studies [3, pp. 159–161; 5, table 5.9, p. 157].

The latter two study designs are particularly prone to reverse causation bias because the exposure(s) and the disease(s) are simultaneously assessed; for this reason, such studies are given little or no weight in health risk assessments.

By contrast, historical information on the exposure of interest (start and stop dates of exposure, and changing patterns in exposure over time) are generally available in cohort and case–control studies, so that the exposure–disease relationship can be investigated in the purported correct time order.

Cohort studies are less prone to temporal bias than case–control studies because information on the exposure of interest is collected before the occurrence of the outcome(s), whereas the retrospective exposure estimates (either self-reported or measured) in case–control studies may be affected by the current exposure situation.

Nevertheless, temporally biased measures of effect may be obtained in both cohort and case–control studies because of inadequate allowance for a sufficiently long induction-latency period.

Alternatively, information on the endpoint onset may be difficult or impossible to obtain in studies of long-term outcomes such as cancer or neurodegenerative diseases. In addition, for postulated (as opposed to established) exposure–outcome associations, the average length of the induction-latency period is obviously unknown.

Analyses of the risk of disease at different lag times (i.e., censoring the exposure histories of the study subjects to varying time points before the diagnosis) are often used to address temporal bias. Such analyses, however, are ineffective against reverse causation stemming from retrospective exposure assessment, or due to a too short period of observation.

Epidemiological research on possible detrimental effects of mobile phone use is an area abounding in opportunities for reverse causality bias. The underlying reason is that mobile phones may be either particularly useful, or completely unsuitable, to persons affected by different diseases or disorders.

Because these communicating devices are always at hand, permitting quick contact with family members or the attending physician, mobile phone use might be more prevalent among persons affected by severe asthma or epilepsy, elderly in poor health, or children with physical or psychological problems compared with persons of the same age in a given population. Other health impairments or prodromal symptoms of a not-yet-diagnosed disease (such as brain tumors, neurodegenerative diseases, hearing losses, or idiopathic environmental intolerance attributed to EMFs), instead, could dissuade sufferers from becoming mobile phone users, or substantially reduce their phone use.

Thus, studies of possible health effects from mobile phone use are susceptible to both upward and downward reverse causation bias, depending on the outcome at hand. A few examples are provided below.

Mobile Phone Use and "Electrical Hypersensitivity"

Persons who perceive themselves as electrical hypersensitive will avoid any source of EMF exposure, and cross-sectional studies (based on objective exposure index, such as operator data) might therefore observe a reduced risk of idiopathic environmental intolerance attributed to electromagnetic fields (IEI-EMFs) in relation to mobile phone use. That is why most studies of this topic have used an experimental design to test whether single- or double-blind exposures to handset-related radiofrequency (RF) fields, compared with sham exposures, trigger higher levels of symptoms or changes in measured physiological or cognitive endpoints, as well as to assess the ability to detect the presence of an RF field [74,75].

Mobile Phone Use and Behavioral Problems in Children

Reverse causation bias might also affect findings from studies of behavioral problems in children and adolescents, for example, if parents of children in psychotherapy for concentration difficulties or other problems provided a mobile phone to the child to help him or her keep appointments.

In a cross-sectional German study based on personal exposure measurements and mental behavior assessed by the Strengths and Difficulties Questionnaire (SDQ), an association between high levels of RF exposure (highest quartile) and overall behavioral problems was observed in adolescents (OR 2.2 [95% CI, 1.1–4.5]) [76]. The authors addressed the issue of reverse causality by introducing an interaction term between measured RF exposure and mobile phone use in the multivariate models controlling for potential confounders (age, sex, education level, self-reported environmental worries, self-reported general RF exposure, and study place); because the interaction term did not reach the level of statistical significance, they concluded that "the observed association was true independent of mobile phone use" [76]. Whereas the lack of statistical significance might be due to the very high prevalence of mobile phone owners in the study population (92% among adolescents), the cross-sectional design *de facto* impeded examination of the time order between exposure and outcome.

Longitudinal studies are expected to provide more reliable information on this topic.

The Danish National Birth Cohort (DNBC) recruited nearly 100,000 pregnant women during 1996–2002 to follow them and their offspring in a life-course perspective. The data were used to explore the relationships between prenatal maternal mobile phone use and/or postnatal use of mobile phones by the child, and behavioral problems in children at 7 years of age [77,78]. An association between prenatal and postnatal exposure to mobile phones and overall behavioral problems was observed (OR 1.5 [95% CI, 1.4–1.7]); the OR for prenatal exposure only was 1.4 (1.2–1.5) and that for postnatal exposure only was 1.2 (1.0–1.3) [78]. However, with reference to the relationship between postnatal exposure and behavioral problems, this study is actually a cross-sectional investigation, because both phone usage by the child and the outcome were concurrently assessed at one point in time. Therefore, reverse causality is one possible noncausal explanation of the findings.

Mobile Phone Use and Intracranial Tumors

Reverse causality is one contributing noncausal explanation for the apparent protective effect of mobile phone use observed in the Interphone international case–control studies of glioma, meningioma, and acoustic neuroma (ORs and 95% CI, 0.81 [0.70–0.94], 0.79 [0.68–0.91], and 0.85 [0.69–1.04], respectively (see Chapter 5) [79,80].

An Interphone side-study, carried out to the purpose of assessing whether mobile phone use was eventually associated with both participation in the study and case–control status (so that selection bias from differential participation would arise), found a lower prevalence of regular mobile phone users in nonparticipant compared with participant subjects (both cases and controls), and the authors anticipated a 10% (5%–15%) spurious reduction in the ORs for regular use [81].

The observed deficits of risk were greater than expected based on the estimated size of the participation bias. There were further arguments in support of the idea that participation bias was unlikely to fully account for the findings. First, the ratio between the participation rates of cases and controls, quite variable across the local Interphone centers, was not associated to the country-specific users versus nonusers ORs for glioma or meningioma, and it did not sensibly vary by study (glioma vs. meningioma) within each local center [79 (Appendix 1, Table 6)].

Second, the analyses restricted to mobile phone users only, carried out to "correct" for the impact of participation bias, provided diverging results for the glioma and meningioma studies. The ORs for glioma were mostly above the null in all categories of times since start of use, cumulative call time, and cumulative number of calls, while persisting widespread deficits of meningioma risk were observed [79 (Appendix 2)].

Reverse causation was one of other possible sources of downward bias discussed by the Interphone authors. A temporal bias could occur if prodromal symptoms of a brain tumor dissuaded subjects from becoming phone users or reduced their use before diagnosis. An accentuated risk reduction among recent users would support such a hypothesis. Actually, accentuated deficits of risk in the "1–1.9 year" category of time since start of use were observed for glioma (OR 0.62 [0.46–0.81]) and acoustic neuroma (0.73 [0.49–1.09]), but not for meningioma (0.90 [0.68–1.18]) [79,80].

Nevertheless, if glioma were not the short-latency and fast-growing disease that is currently believed, reverse causation might contribute to explain the risk reduction in regular mobile phone users. Moreover, if prodromal symptoms were a better and earlier predictor of a subsequent diagnosis of glioma than of meningioma, a temporal bias would also accommodate the differences between the glioma and meningioma studies in the amount of risk reduction observed among recent users. In fact, there is limited support to these hypotheses: in a longitudinal study of the association between previous hospitalization for epilepsy and brain tumors incidence, prodromal symptoms several years before a glioma diagnosis (up to 8 years) were more frequently observed among low-grade and high-grade glioma cases than in controls; a similar association, but considerably weaker, was also seen with meningioma [82].

The main analyses of the Interphone study of acoustic neuroma were carried out with all exposure variables censored at 1 year before the reference date (as in the brain tumor study), but analyses based on a 5-year lag were also performed to allow for a longer latency period. Although the ORs from the 5-year lag analyses were generally increased compared with those using a 1-year lag, there was no major difference in the pattern of findings [80]. Censoring the exposure variables at 5 years before the reference date is a way to disregard exposures occurring after the etiologically relevant time window, but it is unclear how effective they are to assess the occurrence of reverse causality, or to mitigate its impact on the results. Furthermore, the kind of recall bias observed in the Interphone exposure validation study (increasing overestimation of use by cases, but not by controls, with increasing time between the period of reported use and the date of interview) [83], may conceivably have a greater impact on the 5-year lag analyses compared with those based on 1-year censoring of the exposure variables.

Mobile Phone Use and Neurodegenerative Diseases

In the cohort study of Danish mobile phone subscribers followed up through 2003 for hospital contacts due to central nervous systems (CNS) disorders, decreased standardized hospitalization ratios (by 30%–40%) were seen for Alzheimer's disease or other dementia, Parkinson's disease, and epilepsy in men [84]. These findings, in the opinion of the authors, were attributable to "an interplay of a healthy cohort effect and reverse causation bias due to prodromal symptoms."

In summary, the key information required to detect or avoid a reverse causation bias lies in the dates of disease onset and start of exposure. The date of disease onset is rarely available in observational studies of long-latency diseases, whereas the accuracy of the start of exposure is often suboptimal. The use of self-reported exposure proxies, collected after diagnosis, is especially challenging, and may nullify the attempt to control or mitigate a possible temporal bias by introducing information bias.

Ecological Fallacy and Other Bias in Ecological Studies

The term "ecological study" is commonly used in epidemiology to address investigations of exposure–disease relationships where "the units of analysis are populations or groups of people, rather than individuals" [6].

However, the potential value of "ecological variables" should not be underestimated. In fact, such variables may be used as substitute for individualized data that would be better but may not exist or that may have no equivalent at the individual level, while having intrinsic importance in a public health perspective (e.g., the weather or indicators of political structure). Ecological variables can be studied in their own (e.g., description of time trends or variation between countries), correlated with each other, or correlated with health data collected on individuals but summarized by period of time or by place [5].

The surveys covered in this chapter are analyses of aggregated data on disease occurrence in individuals (such as prevalence or incidence rates) by aggregated data on places or time trend comparisons of exposure and disease data.

Such ecological studies are prone to two sorts of error, within-group bias and across-level bias, each of which can have different sources.

Within-group bias is not specific to ecological analyses and may consist of selection bias (from migration across groups); outcome ascertainment and numerator and denominator bias; temporal ambiguity (induction-latency bias and misspecification of the exposure–disease model); exposure inaccuracy bias (within-group misclassification, collinearity, errors in variable modeling); spatial dependency of the outcomes; and inappropriate significance tests [2 (pp. 469–480)].

Moreover, because a relation between two variables at the population level does not necessarily imply the existence of such an association at the individual level, incorrect inferences may be drawn from these studies due to one or more of the following components of *across-level bias* (or *ecological fallacy*): specification bias, confounding by group, and effect modification by group [2 pp. 469–480, 5 pp. 321–322, 85].

Specification bias (also called homogeneity fallacy) may arise from the use of area-level exposure proxies that, although accurate per se, have poor validity as individual exposure markers because the between-subject exposure variability within each study area is almost as great as (or even greater than) the across area exposure variability [5,85].

Confounding and effect modification by group can occur when

- extraneous risk factors for the disease of interest (at the individual levels) are differentially distributed across groups;
- disease risk (at the individual levels) depends on the disease prevalence among other members of the group, as it is true for many infectious diseases;
- the ecological level exposure variable has a contextual effect on risk different from the biological effect of its individual-level analog, typically exemplified by the contrasting findings one may obtain in studies of the relationship between suicide and religion, whereas living in mainly catholic (Protestant, Jewish, Hindu, Muslim, … or atheist) areas is something conceptually different from being of that religion (or atheist).

Occurrence and amount of bias in ecological estimates are difficult to detect and predict [2,85].

Ecological studies have been, and continue to be, very useful to describe differences in disease rates across populations or over time [5]. When it comes to evaluating biological etiological hypotheses, however, their results can be very difficult to interpret, when not misleading.

These contrasting properties are illustrated in two examples drawn from the EMF epidemiology area.

Radio and TV Transmitters and CL

Inverse trends of CL risk with increasing distance from radio and TV transmitters have been reported by several spatial epidemiology studies, often carried out in response to a perceived cluster (Figure 6.5) [86–90]. In a few cases, these studies were followed by further investigations aimed at checking whether the local findings could be accounted for by within-group or across-level bias [91], or if they could be replicated on a larger geographical scale [92].

An excess of CL over the period 1972–1990 was observed in three local governmental areas (LGAs) of Sydney, Australia, in proximity of three TV broadcast antennas, compared with seven other LGAs within the same health service area [87]. This finding was given a "second look" by another research team [91]. The study area was extended to additional districts not considered in the first survey, although equidistant from the antennas compared with those included in the "unexposed" zone. Such an improvement, however, had no major impact on the findings, suggesting that the original results were not attributable to a "Texas sharpshooter" bias. A possible exposure inaccuracy bias in the original study was addressed by evaluating the risk of CL in relation to the predicted level of RF signal as a continuous variable; but this resulted in only a small, downward, change of the effect estimates. Finally, the possibility of an effect modification by group was assessed by sensitivity analyses that excluded one district at a time from the "exposed" category. These analyses did show that the excess risk was limited to one specific district out of the three forming the "inner zone" (Figure 6.5) and that it was driven by an increased incidence of CL during 1972–1978, whereas 24-hr TV transmission was only introduced toward the end of such period [91].

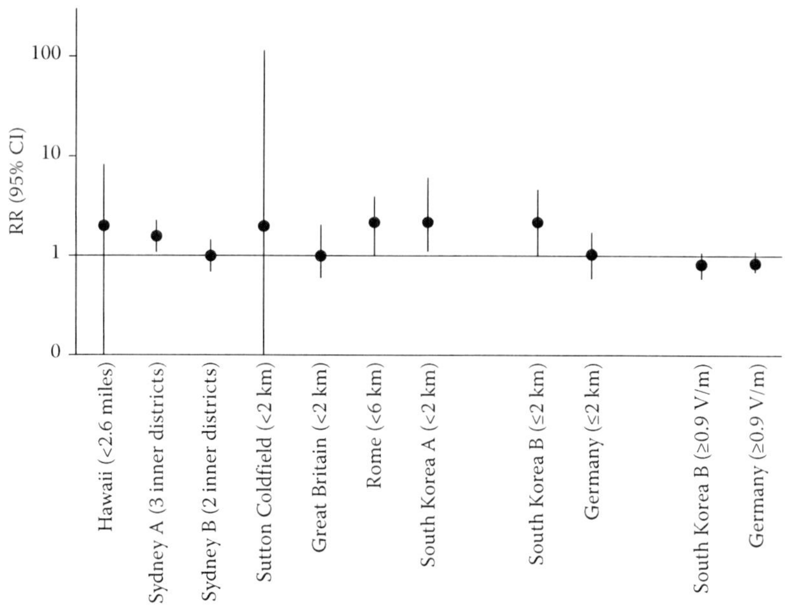

Design	Ecological							Case–control			
Exposed cases	12	33	16	2	10	8	11	36	25	494	293
Reference #	86	87	91	88	92	89	90	93	95	94	95

FIGURE 6.5
Ecological and case–control studies of CL and residential proximity to radio and TV transmitters.

The observation of an excess number of CL cases (two vs. one expected, over the period 1974–1986) in proximity to the Sutton Coldfield radio–TV transmitter [88] could not be replicated in a larger ecological study including all high-power transmitters in Great Britain [92].

More recently, two large case–control studies of CL and residential proximity to radio–TV transmitters were carried out [93–95], both using the predicted RF level at the child's dwelling as exposure proxy, validated by *in situ* measurements [96]. These studies were consistent in not showing any exposure–disease association, and their results weaken findings from earlier reports on leukemia clusters around radio and TV broadcast antennas [97]. Longitudinal studies based on personal RF exposure assessment would likely provide further and more reliable insight on this topic.

Mobile Phone Use and Time Trends of Brain Tumor Incidence Rates

In spite of the previously outlined inferential challenges, there are research topics on which ecological studies could provide valuable insight. The impact of mobile phone use on risk of intracranial tumors is one of these issues.

Monitoring of brain tumor incidence trends through well-established population-based cancer registries, possibly combined with population exposure data, has been identified by World Health Organization (WHO) as a high-priority research need [98]. The reasons underlying this recommendation include the almost universal use of mobile phones, the availability of complete cancer registry data in several countries, the few known environmental risk factors for these neoplasms, along with the unavoidable shortcomings of individual-based studies of intracranial tumors and mobile phone use (mainly exposure measurements errors and susceptibility to selection and participation bias).

Two simulation studies, based on time trend data of glioma incidence, have been published. These studies are consistent in showing that the increased or decreased glioma risks observed in some epidemiological studies are not compatible with the incidence rates of glioma recorded in the Nordic countries and the United States in middle-aged adults (40–59 years) during the past 20 years (up to 2007–2008), assuming induction periods up to 10 years, although a modest risk increase among heavy users cannot yet be excluded [99,100].

The possibility that exposure to mobile phone-related RF fields started in childhood might entail an increased risk of brain tumors in the short term and later in life is also of concern. The issue, so far, has been addressed by one multicenter case–control study of childhood brain tumors [101] and by several time trend analyses [102–105]. Although these data do not suggest a link between mobile phone use and brain tumor risk in persons below 20 years of age, some uncertainties remain (e.g., concerning the relationship between heavy mobile phone use and risk of rare histological subtypes of intracranial tumors). Thus, further monitoring of incidence rates of brain tumors in children and adolescents is warranted.

References

1. Maclure M, Schneeweiss S. Causation of bias: the episcope. *Epidemiology* 2001;12:114–122.
2. Rothman KJ, Greenland S. *Modern epidemiology*, 2nd ed. Philadelphia PA: Lippincott Williams & Wilkins, 1998.

3. Szklo M, Nieto RJ. *Epidemiology beyond the basics*. Gaithersburg, MD: Aspen Publishers, 2000.
4. Savitz DA. *Interpreting epidemiologic evidence*. New York, NY: Oxford University Press, 2003.
5. Bhopal. R. *Concepts of epidemiology*, 2nd ed. New York, NY: Oxford University Press, 2008.
6. Porta M (Ed), Greenland S, Last JM (Ass Eds). *A dictionary of epidemiology*, 5th ed. New York, NY: Oxford University Press, 2008.
7. Miettinen OS. *Epidemiological research: terms and concepts*. New York, NY: Springer, 2011.
8. VanderWeele TJ, Hernán MA, Robins JM. Causal directed acyclic graphs and the direction of unmeasured confounding bias. *Epidemiology* 2008;19:720–728.
9. Hernán MA, Hernández-Díaz S, Werler MM, Mitchell AA. Causal knowledge as a prerequisite for confounding evaluation: an application to birth defects epidemiology. *Am J Epidemiol* 2002;155:176–184.
10. Schisterman EF, Cole SR, Platt RW. Overadjustment bias and unnecessary adjustment in epidemiologic studies. *Epidemiology* 2009;20:488–495.
11. Robins JM, Blevins D, Ritter G, Wulfsohn M. G-estimation of the effect of prophylaxis therapy for Pneumocystis carinii pneumonia on the survival of AIDS patients. *Epidemiology* 1992;3:319–336.
12. Robins JM, Hernán MA, Brumback B. Marginal structural models and causal inference in epidemiology. *Epidemiology* 2000;11:550–560.
13. Higgins JPT, Thompson SG, Deeks JJ, Altman DG. Measuring inconsistency in meta-analyses. *BMJ* 2003;327:557–560.
14. Greenland S, Mickey RM. The impact of confounder selection criteria on effect estimation. *Am J Epidemiol* 1989;130:1066.
15. Hernán M. Fallibility in estimating direct effects. *Int J Epidemiol* 2002;31:163–165.
16. Greenland S. Quantifying biases in causal models: classical confounding vs collider-stratification bias. *Epidemiology* 2003;14:300–306.
17. Brenner H, Blettner M. Controlling for continuous confounders in epidemiologic research. *Epidemiology* 1997;8:429–434.
18. Fewell Z, Davey Smith G, Sterne JA. The impact of residual and unmeasured confounding in epidemiologic studies: a simulation study. *Am J Epidemiol* 2007;166:646–655.
19. Lash TL, Fox MP, Fink AK. *Applying quantitative bias analysis to epidemiologic data*. New York, NY: Springer, 2009.
20. Feychting M, Ahlbom A. Magnetic fields and cancer in children residing near Swedish high voltage power lines. *Am J Epidemiol* 1993;138:467–481.
21. Olsen JH, Nielsen A, Schulgen G. Residence near high voltage facilities and risk of cancer in children. *BMJ* 1993;307:891–895.
22. Verkasalo PK, Pukkala E, Hongisto MY, Valjus JE, Järvinen PJ, Heikkilä PV, et al. Risk of cancer in Finnish children living close to power lines. *BMJ* 1993;307:895–899.
23. Tynes T, Haldorsen T. Electromagnetic fields and cancer in children residing near Norwegian high-voltage power lines. *Am J Epidemiol* 1997;145:219–226.
24. Linet MS, Hatch EE, Kleinerman RA, Robison LL, Kaune WT, Friedman DR, et al. Residential exposure to magnetic fields and acute lymphoblastic leukemia in children. *N Engl J Med* 1997;337:1–7.
25. Michaelis J, Schüz J, Meinert R, Zemann E, Grigat J-P, Kaatsch P, et al. Combined risk estimates for two German population-based case-control studies on residential magnetic fields and childhood acute leukemia. *Epidemiology* 1998;9:92–94.
26. Dockerty JD, Elwood JM, Skegg DCG, Herbison GP. Electromagnetic field exposures and childhood cancers in New Zealand. *Cancer Causes Control* 1998;9:209–309. Erratum in 1999;10:641.
27. McBride ML, Gallagher RP, Theriault G, Armstrong BG, Tamaro S, Spinelli JJ, et al. Power frequency electric and magnetic fields and risk of childhood leukemia in Canada. *Am J Epidemiol* 1999;149:831–842.
28. UK Childhood Cancer Study Investigators. Exposure to power frequency magnetic fields and the risk of childhood cancer: a case/control study. *Lancet* 1999;354:1925–1931.
29. Bianchi N, Crosignani P, Rovelli A, Tittarelli A, Carnelli CA, Rossitto F, et al. Overhead electricity power lines and childhood leukemia: a registry-based, case-control study. *Tumori* 2000;86:195–198.

30. Kroll ME, Swanson J, Vincent TJ, Draper GJ. Childhood cancer and magnetic fields from high voltage power lines in England and Wales: a case-control study. *Br J Cancer* 2010;103:1122–1127.
31. Malagoli C, Fabbi S, Teggi S, Calzari M, Poli M, Ballotti E, et al. Risk of hematological malignancies associated with magnetic fields exposure from power lines: a case-control study in two municipalities of northern Italy. *Environ Health* 2010;9:16.
32. Schüz J, Grigat JP, Brinkmann K, Michaelis J. Residential magnetic fields as a risk factor for childhood acute leukaemia: results from a German population-based case-control study. *Int J Cancer* 2001;91:728–735.
33. Kabuto M, Nitta H, Yamamoto S, Yamaguchi N, Akib S, Honda Y, et al. Childhood leukemia and magnetic fields in Japan: a case-control study of childhood leukemia and residential power-frequency magnetic fields in Japan. *Int J Cancer* 2006;119:643–650.
34. Wünsch-Filho V, Pelissari DM, Barbieri FE, Sant'Anna L, de Oliveira CT, de Mata JF, et al. Exposure to magnetic fields and childhood acute lymphocytic leukemia in São Paulo, Brazil. *Cancer Epidemiol* 2011;35:534–539.
35. Ahlbom A, Day N, Feychting M, Roman E, Skinner J, Dockerty J, et al. A pooled analysis of magnetic fields and childhood leukaemia. *Br J Cancer* 2000;83:692–698.
36. Kheifets L, Ahlbom A, Crespi CM, Draper G, Hagihara J, Lowenthal RM, et al. Pooled analysis of recent studies on magnetic fields and childhood leukemia. *Br J Cancer* 2010;103:1128–1135.
37. Poole C, Greenland S, Luetters C, Kelsey JL, Mezei G. Socioeconomic status and childhood leukaemia: a review. *Int J Epidemiol* 2006;35:370–384.
38. Savitz DA, Poole C. Do studies of wire code and childhood leukemia point towards or away from magnetic fields as the causal agent? *Bioelectromagnetics* 2001;(Suppl 5):S69–S85.
39. Mezei G, Kheifets L. Selection bias and its implications for case–control studies: a case study of magnetic field exposure and childhood leukemia. *Int J Epidemiol* 2006;35:397–406.
40. Hatch EE, Kleinerman RA, Linet MS, Tarone RE, Kaune WT, Auvinen A, et al. Do confounding or selection factors of residential wiring codes and magnetic fields distort findings of electromagnetic fields studies? *Epidemiology* 2000;11:189–198.
41. Schüz J. Implications from epidemiologic studies on magnetic fields and the risk of childhood leukemia on protection guidelines. *Health Phys* 2007;92:642–648.
42. Wartenberg D, Greenberg MR, Harris G. Environmental justice: a contrary finding for the case of high-voltage electric power transmission lines. *J Expo Sci Environ Epidemiol* 2010;20:237–244.
43. Maslanyj M, Simpson J, Roman E, Schüz J. Power frequency magnetic fields and risk of childhood leukaemia: misclassification of exposure from the use of the 'distance from power line' exposure surrogate. *Bioelectromagnetics* 2009;30:183–188.
44. Schüz J. Exposure to extremely low-frequency magnetic fields and the risk of childhood cancer: update of the epidemiological evidence. *Prog Biophys Mol Biol* 2011;107:339–342.
45. Langholz B. Factors that explain the power line configuration wiring code-childhood leukemia association: what would they look like? *Bioelectromagnetics* 2001;5:S19–S31.
46. Steliarova-Foucher E, Stiller C, Kaatsch P, Berrino F, Coebergh JW, Lacour B, et al. Geographical patterns and time trends of cancer incidence and survival among children and adolescents in Europe since the 1970s (the ACCIS project): an epidemiological study. *Lancet* 2004;364:2097–2105.
47. Schüz J, Lightfoot T. Interpreting studies on magnetic fields and the risk of childhood leukaemia: an ongoing challenge. In: Obe G, Vijayalaxmi (Eds.), *Chromosomal alterations: methods, results and importance in human health*. Berlin: Springer-Verlag, 2007, pp. 411–426.
48. Greaves M. Infection, immune responses and the aetiology of childhood leukemia. *Rev Cancer* 2006;6:193–203.
49. Eden T. Aetiology of childhood leukemia. *Cancer Treat Rev* 2010;36:286–297.
50. Wiemels J. Perspectives on the causes of childhood leukemia. *Chem Biol Interact* 2012;196:59–67.
51. Raaschou-Nielsen O. Indoor radon and childhood leukaemia. *Radiat Prot Dosimetry* 2008;132:175–181.
52. Lubin JH, Linet MS, Boice JD, Jr, Buckley J, Conrath SM, Hatch EE, et al. Case-control study of childhood acute lymphoblastic leukemia and residential radon exposure. *J Natl Cancer Inst* 1998;90:294–300.

53. Kaletsch U, Kaatsch P, Meinert R, Schüz J, Czarwinski R, Michaelis J. Childhood cancer and residential radon exposure—results of a population-based case-control study in Lower Saxony (Germany). *Radiat Environ Biophys* 1999;38:211–215.
54. UK Childhood Cancer Study Investigators. The United Kingdom Childhood Cancer Study of exposure to domestic sources of ionising radiation: I: radon gas. *Br J Cancer* 2002;86:1721–1726.
55. UK Childhood Cancer Study Investigators. The United Kingdom Childhood Cancer Study of exposure to domestic sources of ionising radiation: II: gamma radiation. *Br J Cancer* 2002;86:1727–1731.
56. IARC (International Agency for Research on Cancer). *A review of human carcinogens. F. Chemical agents and related occupations*. IARC Monographs on the Evaluation of Carcinogenic Risks to Humans, Vol. 100. Lyon, France: IARC Press, 2012.
57. U.S. EPA. *Carcinogenic effects of benzene: an update*. Prepared by the National Center for Environmental Health, Office of Research and Development. Washington, DC: U.S. EPA, 1998.
58. Pyatt D, Hays S. A review of the potential association between childhood leukemia and benzene. *Chem Biol Interact* 2010;184:151–164.
59. Smith MT, Zhang L, McHale CM, Skibola CF, Rappaport SM. Benzene, the exposome and future investigations of leukemia etiology. *Chem Biol Interact* 2011;192:155–159.
60. Van Maele-Fabry G, Lantin A-C, Hoet P, Lison D. Residential exposure to pesticides and childhood leukaemia: a systematic review and meta-analysis. *Environ Int* 2011;37:280–291.
61. Urayama KY, Buffler PA, Gallagher ER, Ayoob JM, Ma X. A meta-analysis of the association between day-care attendance and childhood acute lymphoblastic leukaemia. *Int J Epidemiol* 2010;39:718–732.
62. Kinlen LJ. An examination, with a meta-analysis, of studies of childhood leukemia in relation to population mixing. *Br J Cancer* 2012;107:1162–1168.
63. Law GR, Parslow RG, Roman E, UK Childhood Cancer Study Investigators. Childhood cancer and population mixing. *Am J Epidemiol* 2003;158:328–336.
64. Poulletier de Gannes F, Lagroye I, Veyret B. *D3—report on the analysis of risks associated to exposure to EMF:* in vitro *and* in vivo *(animals) studies*. EFHRAN (European Health Risk Assessment Network on Electromagnetic Fields Exposure), 2010. Available from: http://efhran.polimi.it/docs/IMS-EFHRAN_09072010.pdf
65. Bernard N, Alberdi AJ, Tanguy ML, Brugere H, Helissey P, Hubert C, et al. Assessing the potential leukemogenic effects of 50 Hz magnetic fields and their harmonics using an animal leukemia model. *J Radiat Res (Tokyo)* 2008;49:565–577.
66. Loomis D, Lagorio S, Salvan A, Comba P. Update of evidence on the association of childhood leukemia and 50/60 Hz magnetic field exposure. *J Expo Anal Environ Epidemiol* 1999;9:99–105.
67. Lagorio S, Salvan A. Infantile leukemia and exposure to 50/60 Hz magnetic fields: review of epidemiologic evidence in 2000 [Italian]. *Ann Ist Super Sanita* 2001;37:213–224.
68. Kavet R, Zaffanella LE, Daigle JP, Ebi KL. The possible role of contact current in cancer risk associated with residential magnetic fields. *Bioelectromagnetics* 2000;21:538–553.
69. Kavet R, Hooper C, Buffler P, Does M. The relationship between residential magnetic fields and contact voltage: a pooled analysis. *Radiat Res* 2011;176:807–815.
70. Does M, Scélo G, Metayer C, Selvin S, Kavet R, Buffler P. Exposure to electrical contact currents and the risk of childhood leukemia. *Radiat Res* 2011;175:390–396.
71. Lagorio S, Ferrante D, Ranucci A, Negri S, Sacco P, Rondelli R, et al. Exposure to benzene and childhood leukemia: a pilot study. *BMJ Open* 2013;3:e002275.
72. Maslanyj M, Lightfoot T, Schüz J, Sienkiewicz Z, McKinlay A. A precautionary public health protection strategy for the possible risk of childhood leukaemia from exposure to power frequency magnetic fields. *BMC Public Health* 2010;10:673.
73. Hill AB. The environment and the disease: association or causation. *Proc R Soc Med* 1965;58:295–300.
74. Rubin GJ, Hillert L, Nieto-Hernandez R, van Rongen E, Oftedal G. Do people with idiopathic environmental intolerance attributed to electromagnetic fields display physiological effects when exposed to electromagnetic fields? *Bioelectromagnetics* 2011;32:593–609.
75. AGNIR (Independent Advisory Group on Non-Ionising Radiation). *Health effects from radiofrequency electromagnetic fields*. Chilton, UK: Health Protection Agency (HPA), RCE-20, 2012.

76. Thomas S, Heinrich S, von Kries R, Radon K. Exposure to radio-frequency electromagnetic fields and behavioural problems in Bavarian children and adolescents. *Eur J Epidemiol* 2010;25:135–141.
77. Divan HA, Kheifets L, Obel C, Olsen J. Prenatal and postnatal exposure to cell phone use and behavioural problems in children. *Epidemiology* 2008;19:523–529.
78. Divan HA, Kheifets L, Obel C, Olsen J. Cell phone use and behavioural problems in young children. *J Epidemiol Community Health* 2012;66:524–529.
79. INTERPHONE Study Group. Brain tumour risk in relation to mobile telephone use: results of the INTERPHONE international case–control study. *Int J Epidemiol* 2010;39:675–694.
80. INTERPHONE Study Group. Acoustic neuroma risk in relation to mobile telephone use: results of the INTERPHONE international case–control study. *Cancer Epidemiol* 2011;35:453–464.
81. Vrijheid M, Richardson L, Armstrong BK, Auvinen A, Berg G, Carroll M, et al. Quantifying the impact of selection bias caused by nonparticipation in a case-control study of mobile phone use. *Ann Epidemiol* 2009;19:33–41.
82. Schwartzbaum J, Jonsson F, Ahlbom A, Preston-Martin S, Malmer B, Lönn S, et al. Prior hospitalization for epilepsy, diabetes, and stroke and subsequent glioma and meningioma risk. *Cancer Epidemiol Biomarkers Prev* 2005;14:643–650.
83. Vrijheid M, Armstrong BK, Bédard D, Brown J, Deltour I, Iavarone I, et al. Recall bias in the assessment of exposure to mobile phones. *J Expo Sci Environ Epidemiol* 2009;19:369–381.
84. Schüz J, Waldemar G, Olsen JH, Johansen C. Risks of central nervous system diseases among mobile phone subscribers: a Danish retrospective cohort study. *PLoS One* 2009;4:e4389.
85. Elliott P, Wakefield JC. Bias and confounding in spatial epidemiology. In: Elliott P, Wakefield J, Best N, Briggs D (Eds.), *Spatial epidemiology: methods and applications.* Oxford, UK: Oxford University Press, 2000; pp. 68–84.
86. Maskarinec G, Cooper J, Swygert L. Investigation of increased incidence in childhood leukemia near radio towers in Hawaii: preliminary observations. *J Environ Pathol Toxicol Oncol* 1994;13:33–37.
87. Hocking B, Gordon IR, Grain HL, Hatfield GE. Cancer incidence and mortality and proximity to TV towers. *Med J Aust* 1996;165:601–605.
88. Dolk H, Shaddick G, Walls P, Grundy C, Thakrar B, Kleinschmidt I, et al. Cancer incidence near radio and television transmitters in Great Britain. I. Sutton Coldfield transmitter. *Am J Epidemiol* 1997;145:1–9.
89. Michelozzi P, Capon A, Kirchmayer U, Forastiere F, Biggeri A, Barca A, et al. Adult and childhood leukemia near a high-power radio station in Rome, Italy. *Am J Epidemiol* 2002;155:1096–1103.
90. Park SK, Ha M, Im HJ. Ecological study on residences in the vicinity of AM radio broadcasting towers and cancer death: preliminary observation in Korea. *Int Arch Occup Environ Health* 2004;77:387–394.
91. McKenzie DR, Yin Y, Morrell S. Childhood incidence of acute lymphoblastic leukemia and exposure to broadcast radiation in Sydney—a second look. *Aust N Z J Public Health* 1998;22(Suppl 3):360–367.
92. Dolk H, Elliott P, Shaddick G, Walls P, Thakrar B. Cancer incidence near radio and television transmitters in Great Britain. II. All high power transmitters. *Am J Epidemiol* 1997;145:10–17.
93. Ha M, Im H, Lee M, Kim HJ, Kim BC, Gimm YM, et al. Radio-frequency radiation exposure from AM radio transmitters and childhood leukemia and brain cancer. *Am J Epidemiol* 2007;166:270–279.
94. Ha M, Im H, Kim BC, Gimm YM, Pack JK. Five authors reply [letter]. *Am J Epidemiol* 2008;167:884–885.
95. Merzenich H, Schmiedel S, Bennack S, Brüggemeyer H, Philipp J, Blettner M, et al. Childhood leukemia in relation to radio frequency electromagnetic fields in the vicinity of television and radio broadcast transmitters. *Am J Epidemiol* 2008;168:1169–1179.
96. Schmiedel S, Brüggemeyer H, Philipp J, Wendler J, Merzenich H, Schüz J. An evaluation of exposure metrics in an epidemiologic study on radio and television broadcast transmitters and the risk of childhood leukemia. *Bioelectromagnetics* 2009;30:81–91.
97. Schüz J, Ahlbom A. Exposure to electromagnetic fields and the risk of childhood leukaemia: a review. *Radiat Prot Dosimetry* 2008;132:202–211.

98. WHO (World Health Organization). *WHO research agenda for radiofrequency fields*. 2010. Available from: http://www.who.int/peh-emf/research/agenda/en/index.html
99. Deltour I, Auvinen A, Feychting M, Johansen C, Klaeboe L, Sankila R, et al. Mobile phone use and incidence of glioma in the Nordic countries 1979–2008: consistency check. *Epidemiology* 2012;23:301–307.
100. Little MP, Rajaraman P, Curtis RE, Devesa SS, Inskip PD, Check DP, et al. Mobile phone use and glioma risk: comparison of epidemiological study results with incidence trends in the United States. *BMJ* 2012;344:e1147.
101. Aydin D, Feychting M, Schüz J, Tynes T, Andersen TV, Schmidt LS, et al. Mobile phone use and brain tumors in children and adolescents: a multicenter case–control study. *J Natl Cancer Inst* 2011;103:1264–1276.
102. Inskip PD, Hoover RN, Devesa SS. Brain cancer incidence trends in relation to cellular telephone use in the United States. *Neuro Oncol* 2010;12:1147–1151.
103. de Vocht F, Burstyn I, Cherrie JW. Time trends (1998–2007) in brain cancer incidence rates in relation to mobile phone use in England. *Bioelectromagnetics* 2011;32:334–339.
104. Schmidt LS, Schmiegelow K, Lahteenmaki P, Träger C, Stokland T, Grell K, et al. Incidence of childhood central nervous system tumors in the Nordic Countries. *Pediatr Blood Cancer* 2011;56:65–69.
105. Aydin D, Feychting M, Schüz J, Röösli M, CEFALO Study Team. Childhood brain tumours and use of mobile phones: comparison of a case–control study with incidence data. *Environ Health* 2012;11:35.

Part II

State of Scientific Knowledge

7

Exposures to ELF-EMF in Everyday Environments

Joseph D. Bowman

CONTENTS

Introduction

A crucial step in epidemiology involves understanding the exposures that may affect human health. This chapter covers the basic elements that determine exposures to electric and magnetic fields (EMFs) at extremely low frequencies (ELFs ≡ >0–3000 Hz) in everyday environments, particularly electric utilities, residences, schools, offices, transportation, and manufacturing. The chapter starts with an overview of basic concepts. For each environment, important sources of ELF-EMFs are described, followed by a brief survey of exposure measurements.

Basic ELF-EMF Concepts

The objectives of this section are to introduce the terminology, notation, and physical concepts used in the assessment of ELF-EMF exposures. The focus is on the physical characteristics of EMF that have been hypothesized to affect human health. Important terms are denoted by italics. Additional details on these terms can usually be found in *Wikipedia*®, whereas instructive illustrations can be found by searching with *Google*® *Images*.

Definitions and Units

From an epidemiologic perspective, electric fields (EFs) and magnetic fields (MFs) are *force fields* that electricity exerts on charged particles in the body, such as electrons, ions, and polarized molecules. In everyday environments, an EF (ε) is determined by an electric circuit's *voltage* and is measured in volts per meter (V/m). An MF is determined by the circuit's *current*. In the ELF range, MF exposures are measured as the *magnetic flux density* (B) in units of microtesla (μT). (In the United States, the MF unit is often reported in milligauss (mG), whose conversion factor is 1 μT = 10 mG.) More rigorous definitions of EMF are given in electromagnetism textbooks such as Jackson (1999) and Griffiths (2013).

The *electromagnetic spectrum* divides EMF into categories, such as ultraviolet (UV) and radio frequency (RF) radiation, as a function of *frequency* (f) measured in hertz (Hz = cycles per second). The spectrum ranges from static EMF at 0 Hz to gamma rays with frequencies above 10^{19} Hz. Frequency is a key indicator of EMF toxicity because it determines the severity of molecular changes (ionization, photochemistry, heating) that result from exposure to *electromagnetic radiation*.

In everyday environments, ELF-EMFs are unsynchronized fields (known technically as *near fields*) rather than radiation, yet their frequency can still determine their effects on the body, especially through *magnetic induction*. This relationship between frequency and magnetic induction is so important to ELF-EMF health effects that its mathematical basis should be understood by all practitioners. According to *Faraday's law*, a time-varying MF $B(t)$ induces EFs inside the body whose strength is proportional to the derivative dB/dt. For MFs emitted by *alternating current* (AC) electricity, $B(t)$ has the same sinusoidal time dependence as the AC current with f = 50 or 60 Hz. Therefore, the induced electric field $E_{in}(t)$ is related to frequency by

$$E_{\text{in}}(t) \propto \frac{d}{dt} B_{\text{pk}} \sin(2\pi ft) \\ \propto 2\pi f B_{\text{pk}} \cos(2\pi ft) \tag{7.1}$$

In other words, both the MFs magnitude B_{pk} and the frequency f determine the induced field. Due to this frequency dependence of magnetic induction, the ELF frequency band has served as a useful indicator of toxicity in health effects studies.

Of the many definitions for *extremely low frequencies* (SCENIHR 1997; WHO 2007; ITU 2008), this chapter uses ELF ≡ >0–3000 Hz as the range best representing time-varying EMF exposures in everyday environments where AC electricity is the primary source. The 3000 Hz upper limit originated in the International Telecommunications Union's system for radio bands (ITU 2008) and is currently used with the ELF band for atmospheric radio waves (Barr et al. 2000). The criterion for the lower end of the ELF band is that an MF exposure varies with time ($f > 0$ Hz) so that the *magnetic induction* mechanism can operate. This leads to the consideration of the time-varying exposures from a person's motion in the earth's MF, fields that are called *motion gradient magnetic fields* (MG-MFs).

Occasionally, sources of ELF-EMF also emit fields in other frequency bands: static MFs (0 Hz), intermediate frequencies (3000 Hz–10 MHz), and radio frequencies (10 MHz–300 GHz). In such cases, this chapter mentions a source's other frequencies but focuses on its ELF emissions.

Summary of Definitions and Units

- The EF is determined by the voltage of electric circuits and is measured in volts per meter (V/m).
- An ELF-MF is determined by the circuit's current and is measured as the magnetic flux density (B) in units of microtesla (μT).
- The ELF band encompasses time-varying EMF in everyday environments where AC electricity is the primary source, and magnetic induction is an important biophysical mechanism.

EMF Physical Characteristics

EMFs are vectors, denoted **E** and **B**, with a magnitude and a direction in space. **E**'s direction at a given location is determined by the voltage's sign (+ or –), and the relative positions of the wires and nearby objects, particularly the earth. Likewise, the current's direction in a wire along with its geometry determines **B**'s direction (Figure 7.1). *Direct currents* (DCs) therefore generate a static MF $\mathbf{B}_0$ (Figure 7.1a), whereas AC currents generate an oscillating field **B**(t) with the same frequency as the current (Figure 7.1c).

In everyday environments, a person is surrounded by many wires whose EMFs add vectorially to give the person's net EMF exposure. A simple example is an electric line with only "hot" and neutral return wires and equal currents running in opposite directions, therefore generating opposing MF vectors (Figure 7.1b and 7.1c). When two insulated wires in a cable are touching, their MF vectors effectively cancel each other. With separated electric lines, the cancellation is incomplete, leaving a net MF > 0 (Figure 7.1b–7.1d).

As Figure 7.1 shows, the *polarization* or shape traced by the MF vector depends on the currents' *phases* or timing of their peaks. Single-phase currents where the two wires have

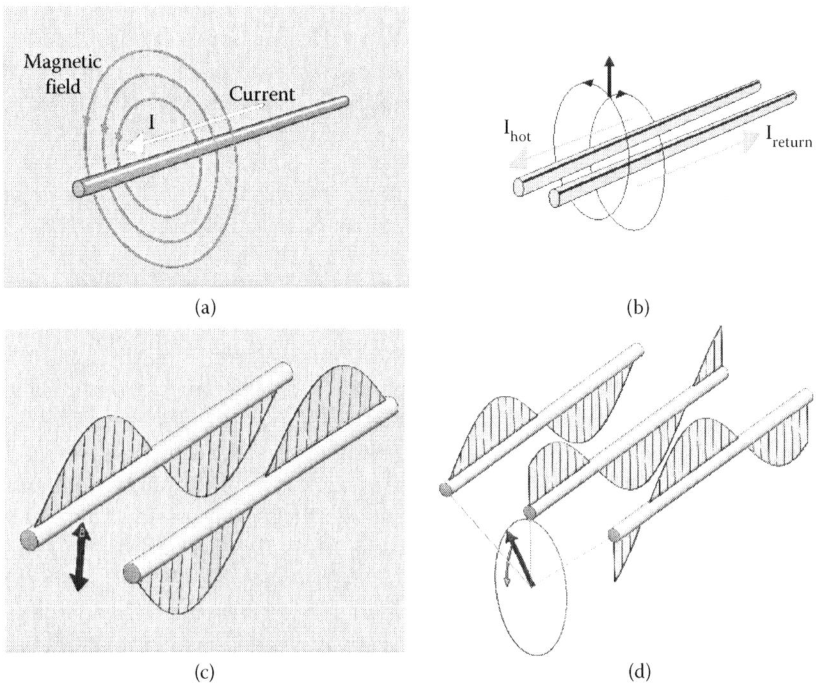

FIGURE 7.1
Net MFs from electric lines. (a) Single wire with DC electricity. (b) DC circuit. (c) Single-phase AC circuit. (d) Three-phase AC power line. (Original graphic by Joseph DeCapite [NIOSH]).

the same phase produce *linearly polarized* MFs (Figure 7.1c). *Elliptically polarized* fields are generated by the three-phase currents used in electric transmission and distribution lines (Figure 7.1d) as well as powerful electric motors and other industrial sources. Elliptical polarization can also arise from the sum of the fields from multiple single-phase sources with different *power factors* (the phase difference between the current and voltage). Similar relationships apply to the EF's polarization and the phases of its source voltages (Deno and Silva 1998).

This EMF cancellation between multiple electric wires also makes their net vector magnitude fall off more rapidly with the distance r from the line (Kaune and Zaffanella 1992). With a single wire (Figure 7.1a), the MF magnitude is proportional to $1/r$, a distance dependence that also applies to a multiwire power line with a non-zero *net current* (sum of the currents from all lines factoring in their phases) (Kaune and Zaffanella 1992). When the net current of a power line is zero (Figure 7.1c and 7.1d), the MF falls off as $1/r^2$. With the coils of wire found in electric motors and *solenoids*, the distance dependence is $1/r^3$.

These distance relationships between ELF-MFs and their source continue within our bodies and most other matter, except for ferromagnetic metals such as iron and nickel. Iron and steel strongly perturb MF in ways that either decrease or increase exposures, depending on the geometry.

In contrast, all matter interacts strongly with EFs, and in the everyday environment, usually decreases exposures. An extreme example is the *Faraday cage*, a conductive metal cage that has zero EF inside. Because the ionized water within our body makes it somewhat conductive, the same physical principles reduce the EF inside a human body to roughly 10^{-6} (one millionth) of its exterior magnitude (Kaune et al. 1990). This interaction with matter greatly

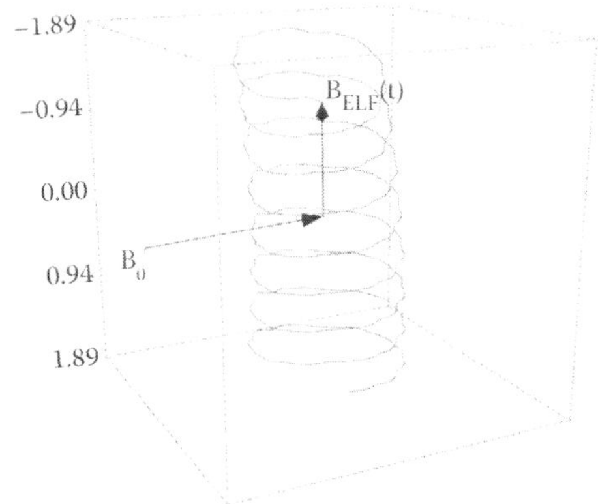

Magnetic Field Magnitudes (μT):	Freq. (Hz)	Mag. (μT)	Axis Ratio (%)
Static (B_0): 33.73	60	0.819	77
ELF+Gradient: 1.279	180	0.050	36
Fullband (ave.): 33.746	540	0.020	11

FIGURE 7.2
Trace of an MF vector $\mathbf{B}_0 + \mathbf{B}_{ELF}(t)$ measured on an electric line worker working on a distribution line from a bucket truck. (Bowman J. D. et al., Electric utility worker exposures to biologically based metrics for ELF and static magnetic fields measured by the Multiwave System III. In press.) In this graph, $\mathbf{B}_{ELF}(t)$ includes the MG, power-frequency, and harmonic components in Equation 7.1. Note: The $\mathbf{B}_0$ vector is actually 33.73 μT, but is truncated at the edge of the ±1.89 μT cube to get a better view of the trace. To improve the perspective, $\mathbf{B}_0$ is also rotated from its actual 67° downward inclination. (Computer generated plot by Joseph Bowman [NIOSH]).

complicates an EF's distance relationship to its source. The EF between two metal plates is unchanged between the high-voltage plate and the grounded plate where $V = 0$ volts. The field's strength is $E = V/d$, where d is the distance (meters) between the plates. Applying this relationship to the EF underneath a power line, the field strength decreases inversely with the height of the line but is approximately constant between the line and the ground.

More complicated are the MFs encountered in everyday environments due to the many sources in the environment. Figure 7.2 shows the trace of the MF vector measured with a 0–3000 Hz probe worn by a California electric line worker working on a distribution line from a bucket truck (Bowman et al. 2014). This complicated trace results from four sources in the worker's environment. First, a static MF with magnitude $B_0 = 33.7$ μT came from the earth's geomagnetic field perturbed by steel in the environment. Second is a 60 Hz MF component from the three-phase AC electricity of the line that gives the trace its periodic elliptical shape. Third are the 180- and 540 Hz *harmonic* frequencies (multiples of 60 Hz) that come from rectifiers and other nonlinear electronics plugged into the distribution system and that create the periodic wiggles in the trace. Finally, the periodic spiral is due to the worker's movement through gradients in the earth's MF near steel objects. With this typical example of ELF-MFs in everyday environments, the instantaneous vector can be represented by the sum of its frequency components:

$$\boldsymbol{B}(t) = \boldsymbol{B}_0 + \boldsymbol{B}_{motion\,gradient}(t) + \boldsymbol{B}_{f_{power}}(t) + \sum_{n>1} \boldsymbol{B}_{nf_{power}}(t) \tag{7.2}$$

where f_{power} is the power frequency (60 Hz in North America and Brazil, 50 Hz in the rest of the world; 400 Hz on airliners; and several other frequencies on trains). The harmonic index n is usually limited to 3 and 5 due to the dynamics of complex AC circuits, but it can have a spectrum with other values near nonlinear electric equipment, a common occurrence in manufacturing plants (Bowman and Methner 2001).

Summary of EMF Physical Characteristics

- An ELF-EMF is a vector that varies with time, space, and (generally) frequency.
- The degree of cancellation among ELF-MFs from electric circuits determines how rapidly the field decreases with distance, ranging from $1/r$ for an unbalanced current to $1/r^3$ for wire coils.
- MFs are largely unchanged by our bodies and most matter, whereas EFs interact strongly with matter, thereby shielding people from exposures in most cases.
- The time-varying components of ELF-MFs are generally the power frequency, its harmonics, and the MG field.

EMF Exposure Metrics

An exposure metric is typically a single number that summarizes a person's exposure to EFs, MFs, or both. Exposure metrics are essential components of occupational and environmental health research and practice, both for epidemiologic analyses and setting exposure limits. Examples for other agents are the effective irradiance in W/m^2 for UV radiation, the A-weighted sound pressure level in dBA (A-weighted decibels) for noise, or the respirable fraction in mg/m^3 for dusts that cause pneumoconiosis (ACGIH 2001).

To convert a person's exposures to time-dependent EMF vectors as in Figure 7.2 into a single number, an exposure metric identifies and combines frequency and spatial and temporal characteristics of the field (Bowman et al. 1998). Over short time scales (<1 s), a common metric is the root mean square (RMS):

$$\mathrm{RMS}\left[S(t)\right] = \sqrt{\frac{1}{T}\int_t^{t+T} S(t')^2\, dt'} \equiv x_t \tag{7.3}$$

where T is an averaging time, t is the measurement's start time, and $S(t)$ is a scalar function of an EMF vector. The simplest scalar function for an EMF vector is its magnitude:

$$\left|\mathbf{B}(t)\right| = \sqrt{B_x(t)^2 + B_y(t)^2 + B_z(t)^2} \tag{7.4}$$

where measurement probe(s) are aligned along the x, y, and z axes. Combining Equations 7.2 and 7.3 gives the *rms vector magnitude*:

$$B_t = \mathrm{RMS}\left[\left|\mathbf{B}(t)\right|\right] = \sqrt{B_{xt}^2 + B_{yt}^2 + B_{zt}^2} \tag{7.5}$$

A useful alternative metric is the peak vector magnitude $B_{pk,t} = \mathrm{Max}\ |\mathbf{B}(t)|$. For sinusoidal fields such as in Figure 7.1c and 7.1d, the peak and rms fields are related simply by $B_{pk,t} = \sqrt{2}B_t$ (Blume 2007, p. 233) although the relationship is more complicated for most environmental fields (Figure 7.2). In addition, an EMF meter usually applies a frequency *filter* to the total field (Equation 7.2) and gives metrics such as the static field magnitude $B_{0,t} = \sqrt{B_{0x,t}^2 + B_{0y,t}^2 + B_{0z,t}^2}$ or the peak ELF magnitude over one period $B_{pk,\mathrm{ELF},t} = \mathrm{Max}[|\mathbf{B}_{\mathrm{ELF}}(t')|$ where t' ranges over a single cycle.

An important variant of these exposure metrics is the *resultant*, which was developed to test the Wertheimer–Leeper (WL) hypothesis that residential MF from AC power lines is associated with increased risks of childhood cancers (Wertheimer and Leeper 1979). In the first two studies to test the WL hypothesis with measurements (Savitz et al. 1988; London et al. 1991), ELF-MF exposures in residences were assessed by spot measurements with a single-axis induction coil probe. To assess the rms vector magnitude (Equation 7.5) at a location within the home, three measurements were taken with the probe aligned in the x, y, and z directions, and the rms components were then combined with the resultant formula:

$$B_{r,t} \equiv \sqrt{B_{x,t}^2 + B_{x,t+\delta}^2 + B_{z,t+2\delta}^2} \tag{7.6}$$

where δ is the time between component measurements. Note that the resultant measures the three orthogonal components sequentially, whereas the rms vector magnitude (Equation 7.4) measures them simultaneously. These spot measurement protocols placed the induction coil probe on a stand that both aligned the probe in the three orthogonal directions and kept it stationary. This stationary probe holder eliminated the MG field (Equation 7.2) from the measurements, so these studies only assessed MF exposures from AC electricity, according to the WL hypothesis.

Because spot measurements are severely limited in assessing exposures to the highly variable MFs found in most homes and workplaces, the Electric Power Research Institute (EPRI) undertook an instrumentation development program to develop a personal meter with data-logging capability so that ELF-MF exposures could be measured for at least 24 hr in epidemiologic studies of the WL hypothesis (Bracken et al. 1993). This effort ultimately resulted in the EMDEX family of ELF-MF personal and spot measurement meters (Enertech Consultants, Campbell, CA) that were used in the overwhelming majority of epidemiologic studies described in Chapters 9–11.

The EMDEX meters measure the MF resultant with a three-axis probe at sampling rates ranging from 1.5 to 10 s over periods up to 24 hr. Like the spot measurement method described above, EMDEXs measure the rms components from the three probes sequentially, where the time δ between measurements increases with the meter's *sampling interval* (between 1.5 s and ~6 min, depending on the model and the meter's programming [Enertech Consultants 1993]). To eliminate both the MG-MF and higher frequencies, the EMDEXs have a *band-pass filter* with a lower bound at 40 Hz and an upper bound that depends on the model (800 Hz for the EMDEX II and 1000 Hz for the EMDEX Lite). The EMDEX's primary output is labeled the *broadband resultant* to distinguish it from measurements with the *harmonic mode*, whose filter has a 100 Hz lower bound (Enertech Consultants 1993).

Because ELF-MFs are not perturbed by the body, measurements of the ELF resultant with an EMDEX meter worn on the subject's body (typically the waist) give a reasonable assessment of personal MF exposures to the rms vector magnitude from AC electricity. The greatest inaccuracy is an MF whose vector trace varies in a nonperiodic manner over the sampling interval 2δ. (The trace in Figure 7.2 is nonperiodic due to the MG, but similar variations can result from rapid changes in the source current or the person's distance from the source.) When the MF trace diverges rapidly from periodicity, the resultant's errors relative to the rms vector magnitude have been as much as 3000% (Bowman and Methner 2001; McDevitt et al. 2002).

Exposure metrics for EFs are more difficult to define and measure because they are strongly perturbed by all objects including our bodies and, therefore, vary with body location and posture (Chartier et al. 1985). For health studies, most surveys of personal EF exposures have used the Positron monitor that only measures the rms EF component perpendicular to the meter's 8 × 14.3 cm case, an imperfect surrogate metric (Heroux 1991). The resulting uncertainty about the EF's impact on the body means that both area and personal measurements should be treated as semiquantitative exposure indicators.

In this chapter, the principal short-term exposure metrics are the rms component perpendicular to the body for the EFs and the ELF resultant for MFs because so much of these data were collected with the convenient Positron and EMDEX monitors in surveillance and epidemiologic studies.

In chronic disease studies, the EMDEX and Positron monitors collect thousands of measurements in their datalogger, so *long-term metrics* for time scales greater than 1 sec are needed to summarize exposure for epidemiologic analysis or exposure guidelines. Although a score of long-term metrics have been used (Morgan and Nair 1992; Zaffanella 1998; Yost 1999; Bowman et al. 2010), the metric used most often in studies of cancers and other chronic diseases has been the time-weighted average (TWA):

$$\begin{aligned} \mathrm{TWA}[S_t] &= \sum_{i=1}^{N} S_{t_i}\Delta t_i \Big/ \sum_{i=1}^{N} \Delta t_i \\ &= \frac{1}{N}\sum_{i=1}^{N} S_{t_i} \quad \text{if } \Delta t_i \text{ is constant.} \end{aligned} \tag{7.7}$$

where Δt_i is the time between two samples and S_t is a short-term scalar metric measured at time t.

A mechanistic rationale for using the TWA[S_t] in health studies is that its integral over the response time (i.e., the cumulative exposure) is proportional to the damage to the target organ if all intermediate biologic processes are linear functions of the metric S (Rappaport and Kupper 2008). For acute adverse effects such as electrostimulation with a millisecond response time, a useful long-term metric is the maximum of the short-term metric over the sampling period Max[S_t].

Epidemiologists, however, should be aware that scores of other short-term EMF exposure metrics have been defined (Bowman et al. 1998, 2010), such as the fullband magnitude (the rms vector magnitude of Equation 7.1) and the axial ratio (a measure of polarization) displayed in Figure 7.2. For EMF health studies, metrics derived from biophysical mechanisms should be better predictors of disease risks, as they are for other physical and chemical agents (Smith and Kriebel 2010). For example, the electrostimulation of voltage-gated ion channels in neurons by magnetic induction has been shown to cause adverse neurologic effects (Reilly 1998). Based on magnetic induction models (Equation 7.1), one of the safety guidelines for ELF-MF (IEEE 2002) recommends two short-term metrics as improvements on the rms vector magnitude: the peak vector magnitude of the MF's time derivative, $|\partial \boldsymbol{B}(t)/\partial t|_{pk}$, and the peak internal EF magnitude induced at the target organ, $|E|_{in,\,pk}$. Similarly, a study of magnetic navigation in birds and other animals (Vanderstraeten and Vorst 2005) suggests that the rms angle of oscillation of $\mathbf{B}(t)$ (Equation 7.2) relative to the static MF vector $\boldsymbol{B}_0$ as a short-term metric sensitive to the free radical and magnetosome mechanisms that are part of animal magnetoreception (Mouritsen and Ritz 2005).

Combining all the long-term and short-term metrics that have been proposed in the ELF literature results in a daunting number of possibilities—294 combinations in one compilation (Bowman et al. 2010). Ideally, health studies can sharply reduce the number of metrics to be evaluated by focusing on those derived from biologic mechanisms hypothesized as causes of the disease under investigation. Chapter 17 discusses some approaches for selecting EMF metrics (or "effect functions") for epidemiologic studies from the consideration of laboratory findings. The identification of the best exposure metric for health effects such as cancer and neurodegenerative diseases remains a major challenge for ELF-EMF research.

Summary of EMF Exposure Metrics

- An exposure metric is a single number that summarizes a person's exposure to an agent for either an epidemiologic analysis or an exposure limit.
- An exposure metric for ELF-EMF selects and combines the field's frequency, spatial, short-term (<1 s), and long-term characteristics into a single number.
- A convenient short-term metric for ELF-MF is the rms vector magnitude that is commonly measured by EMDEX instruments with an approximation called the "broadband resultant."
- All ELF exposure metrics derived from viable biophysical disease mechanisms involve other field characteristics besides the rms vector magnitude of the ELF-MF, yet most epidemiologic studies have measured exposures with the broadband resultant.

Population Statistics

Finally, EMF magnitudes are summarized in the rest of this chapter by tables of *population statistics* for the ELF resultant from sources and environments in everyday environments. As discussed in the survey paper (Kheifets et al. 2013), this choice of the ELF resultant for MF and the perpendicular component of the EF is motivated by the immense amount of such data available from ELF epidemiologic and surveillance studies, despite its questionable biologic significance. The tables use the geometric mean (GM) as its statistic for the central tendency of exposures in everyday environments because ELF-EMF exposures follow other occupational and environmental agents in belonging to log-normal distributions across a population (Rappaport and Kupper 2008). For between-person variability, the geometric standard deviation (GSD) is often used in exposure assessments (Rappaport and Kupper 2008), but this chapter reports instead the 95th percentile (P95) because it has the same units as the GM, facilitating comparisons. By assuming a log-normal distribution, the two variability statistics are related by $P95 = GM\ GSD^{z_{0.95}}$ where the standard normal quantile $z_{0.95} = 1.64$.

Note that the GM and P95 are occasionally used as long-term EMF metrics (e.g., Zaffanella 1998), but such statistically based exposure metrics should not be confused with population statistics. Likewise, the long-term metric TWA is sometimes called the arithmetic mean (AM), so to avoid confusion, this chapter uses the term AM only for the population statistics.

Summary of Population Statistics

- Because ELF-EMFs in most environments are distributed log-normally, the central tendency of a population's exposure is given as the GM and its exposure variability by the GSD or the 95th percentile (P95).
- To avoid confusion, the term "arithmetic mean" (AM) should only be used for population statistics, whereas TWA should be used for the arithmetic mean over time, a long-term exposure metric.

Electric Power Systems

Having covered the basic principles needed for exposure assessment, the remainder of this chapter surveys ELF-EMF sources and exposures encountered in the different environments of everyday environments: the electric power system, residences, schools, offices, manufacturing, other workplaces, and the MG fields. The section for each environment first discusses the main sources and then summarizes data on personal exposures. Due to the uncertain relevance to health effects of the ELF-MF resultant and the EF metric measured across these environments, this survey emphasizes determinants of exposure rather than numbers. Nonetheless, Table 7.1 gives a few benchmarks on ELF-EMF health effects to help put the exposures from these environments into perspective.

Modern society is powered by an intricate system of electric generation, transmission, and distribution lines ending with the electrical wiring in homes and workplaces (Leeper 2001; Blume 2007). The high voltages and currents in the electric power system create widespread exposures to ELF electric and MFs for electric utility workers and the general public. First, we discuss the main components of the electric power system that lead to ELF-EMF exposures (Blume 2007).

Generating Stations. Power from coal, oil, gas, hydro, nuclear fission, solar, and other renewable sources run electric generators in these plants. High MF exposures occur around generators, transmission lines, substations, and especially *bus bars*, the wide copper bars that carry thousands of amperes between the generators and the transmission substation. High EFs are found in the transmission substations and under the high-voltage lines that transmit electric power to distant customers. Although EMF exposures at generating stations can be high, most of the structures are occupied by the power sources such as steam, hydroelectric, and solar where EMF exposures are low or moderate (Table 7.2).

Transmission Lines. *High-voltage transmission lines* (HVTLs) transport electricity at voltages from 250 to 740 kV over long distances from a power plant to a local electric grid. HVTLs are generally overhead lines, but underground HVTLs are sometimes used in larger cities.

TABLE 7.1

Selected Benchmarks of ELF-EMF Exposures with Their Most Prominent Adverse Health Effects

Health Risk	Specifics	Metric	MF Limit (μT)	EF Limit (kV/m)
Neurologic effects	Threshold L (ACGIH 2001)	Maximum rms vector magnitude	1000[a]	92[a]
	High action level (EU 2013)		6000[b]	20[b]
Electromagnetic interference with pacemakers and other implants	Threshold limit value (ACGIH 2001)	Maximum rms vector magnitude	100	1
Cancer	Childhood leukemia (IARC 2002)	TWA ELF resultant	0.30[c]	Inadequate associations
	Occupational cancers (Bowman et al. 2013)		0.28[c]	

[a] Frequency-dependent limit at 60 Hz.
[b] Frequency-dependent limit at 50 Hz.
[c] Lower bound on exposure category with significant associations.

TABLE 7.2
Population Statistics for Selected ELF-EF and MF Exposures in Electric Utilities

	ELF-MF Resultant (μT)		ELF-EF (V/m)
Source	**GM/Values**	**P95/Max**	**Range**
Generator bus bar	8.82	37.23	
Boiler house	0.09	0.36	
High-voltage transmission lines			10,000–32,000
—Bare-hands work		2200	
—De-energized lines	12–942		
—Ground work	1.80		
Substations	4.68	156	15,000–47,000
Underground vaults	9.00	7700	
Administration building	0.04	1.18	
Occupation—Environment		**GM of TWAs**	
Mechanic—generating station	1.03		8.10
Plant operator—generating station	0.85		7.58
Electrician—substation	1.82		—
Lineman—HVTL (live line)	1.44		174.64
Lineman—distribution lines	1.39		10.95
White collar jobs—office	0.19		11.09

Note: A selection of measurements from the literature (Bracken et al. 1997, 2001, 2005; Guenell et al. 1996; Kelsh et al. 2000; Korpinen et al. 2009, 2011; Renew et al. 2003) summarized by the GM and P95 over a study population. Where population statistics are not available, the range, maximum, or single values are reported.

Substations. Distribution substations receive power from an HVTL, pass it through *step-down transformers*, and distribute it at lower voltages through several subtransmission lines or primary distribution lines with voltages from 2 to 35 kV. Substations are usually above ground, but they are also placed in underground vaults. The highest EFs occur in open substations where high-voltage lines and bus bars are uninsulated, as opposed to substations in buildings and underground vaults where insulation is often needed for safety.

Distribution Lines. A network of primary lines carries the power from substations to customers with their own transformers (industries, large commercial stores, and apartment buildings) and to neighborhoods with residences and smaller commercial customers. Along primary lines, small transformers are periodically mounted on a pole for overhead lines or in a cabinet for underground lines to step down the voltage to the levels used by appliances and other electrical equipment (120–600 V in different countries). Secondary distribution lines carry this low-voltage electricity to individual residences or commercial customers. The lower the voltage and the more urban the environment, the more likely distribution and transmission lines will be underground.

With work on overhead power lines, the main determinants of EF exposures are the line voltage and the worker's proximity to energized lines. Distances to energized lines are generally longer in Europe where the safety rules require power lines to be de-energized (dead lines) before maintenance work is performed (CENELEC 2004). In contrast, electric utilities in North America perform maintenance on live lines by the use of insulated gloves for lower voltages and "hot sticks" (insulated tools mounted on fiber glass poles) for higher voltages. The most extreme EMF exposures in electric utilities come from "bare-hands" work on live HVTLs (Bracken et al. 1997). To avoid exceeding the 20-kV/m safety limits, bare-hands workers wear metallic suits that shield them from the EFs.

MF exposures from electric line work are also determined by the distance from energized lines (either the line being repaired or neighboring lines) as well as the line current. Because bare-hands work on HVTLs requires proximity to live lines with currents >1000 A, this work practice again produces the highest MF exposures with a maximum of 2200 μT (Bracken et al. 1997).

Personal Exposures of Electric Utility Workers

Electric utility workers have the greatest EMF exposures from the electric power system because they are often (1) closer to the wires, (2) unshielded from EFs, and (3) exposed to electric shocks (ranging from nuisance to injurious) and *contact currents*, currents that flow between their hands and feet when touching surfaces with different electrical potentials. Note that shocks and contact currents inject ELF-EFs into the body, just as external EMF exposures do by induction (Bracken et al. 2009).

Table 7.2 gives a selection of ELF-EMF exposures by source or environment compared with personal TWA exposures. HVTLs and substations (both above and underground) are the greatest sources of exposures, followed by generating stations and distribution line work. The full-shift TWAs are substantially less than the exposures in a comparable environment, especially for power line work where the discrepancies are an order of magnitude or more. The discrepancy in EFs is due in part to the field's distortion by the worker's body during personal monitoring, whereas source measurements are made without the worker.

This large difference between MF exposures near sources and the TWAs is probably due to the long periods of time that line workers spend assembling supplies, traveling to the work site, and working underneath the power line (ground work). When the background MFs are far less than the exposures from working close to a strong source, the 8-hr TWA ELF-MF is approximately

$$TWA[B_{ELF}] \cong r\bar{B}_{hi}\,\Delta T/8\,hr \tag{7.8}$$

where $\bar{B}_{hi}$ is the average exposure in the high field environment near the source, ΔT is the task's duration, and r is the rate of repetition for the task. This is a useful rule-of-thumb for modeling TWA exposures from spot measurements near strong sources.

Summary of Electric Power Systems

- Important components of the electric power system are generating station, bus bars, substations, transformers, HVTLs, distribution lines, and grounding systems.
- Determinants of ELF-MF exposures from electric lines are the current, degree of cancellation, distance from the line, and duration of work near the line.
- Determinants of ELF-EF exposures from electric lines are the voltage, distortion, and/or shielding by metal objects exposure duration; and distance from energized lines.
- Electric shocks, contact currents, and ELF-EMF all induce EFs inside the body.

Residences

Sources

The principal residential sources of ELF-MF are appliances and the electrical distribution system. Nearby transmission lines, distribution lines (primary and secondary), and transformers in larger apartment buildings all emit ELF-MFs. Several of these power lines are adjacent to a typical home, averaging 3.1 lines per residence in Los Angeles County as an example (Bowman et al. 1999). A line's contribution to the home's ELF-MF distribution depends on its distance, location (overhead or underground), current, and the degree of MF cancellation between the wires, which is optimal if the line has no neutral wire.

Another source of ELF-MF is currents from water pipe grounds. Grounding systems are deployed throughout the electric power system to minimize danger and damage from equipment failures, lighting strikes, and accidental shocks to workers and the public (Blume 2007). In addition to large surges from lightning and major faults, ELF-MF is emitted by the small ground currents that constantly pass between electrical components to keep all their neutral wires at zero potential. In residences and commercial buildings, ground currents can be a significant MF source if the *system ground* is provided by bonding the building's neutral wires to a metal water pipe—an arrangement found in 36% of U.S. residences (Zaffanella 1993). Water pipe grounds create circuits whose MF emissions are not cancelled by MF vectors from adjacent conductors (Leeper 2001).

Consequently, residential ELF-MF vary widely from minimal exposures (<0.01 μT) to fields above 0.3 μT throughout the home or only in "hot spots" close to interior ground currents or exterior power lines with high currents. The ELF-MF exposures from selected power system sources are summarized in Table 7.3. This massive interresidence variability in ground current MFs (GM = 0.01 μT & P95 = 0.70 μT → GSD = 13.2) is due to factors such as metal versus plastic water pipes, the distance from living areas to water pipes carrying ground currents, and residential electric wiring practices that determine the proportion of home's neutral currents diverted to ground currents (Zaffanella 1993).

Residential ELF-MFs also display a diurnal cycle as currents in power lines fluctuate with their customer's daily activity cycle, air conditioner use, or both (Kaune et al. 1987; Zaffanella 1993). The same factors can create exposure variability over the seasons and between weekdays and the weekend. Thus, exposure assessment errors in epidemiologic studies can be reduced by adjusting residential measurements for the time of day, day of the week, and season.

Table 7.3 also summarizes average ELF-MFs in homes where exposures are lower in Europe than in North America. This can be explained by differences in the electrical distribution system where overhead distribution lines are more common in North America (77% in the United States vs. 15% in the United Kingdom). The minimal residential MF in Norway is attributed to secondary lines with no neutrals that create balanced currents by conservation of energy and therefore maximal cancellation of the MF from the phases (Mild et al. 1996).

Another component of residential MF is *high-frequency transients* (HFTs). The abrupt surge of electricity from flipping a switch produces HFTs in electrical circuits. Transients with frequencies from 60 Hz to 50 MHz and magnitudes up to 3 μT have been measured in residential MF (Guttman et al. 2001). A cruder approach is to measure a room's HFT with a Microsurge meter plugged into an AC wall socket (Havas and Stetzer 2004). This "dirty electricity" meter measures the voltage's rate of change (in units of volts per second) in the 4–150 kHz band.

TABLE 7.3

Selected Residential Exposures to ELF-MFs by Power System Sources, Country, and Area versus Personal Monitoring

	24-hr TWA of ELF-MF (μT)			
Power System Source[a]	**GM**[b]	**P95**[c]		
Apartment building transformers	0.59	1.30		
Transmission lines	0.09	0.49		
Overhead primary lines (no neutrals)	0.02	0.60		
Overhead secondary lines	0.04	1.58		
Underground distribution lines	0.03	0.50		
Ground currents	0.01	0.70		
	Area		**Personal**	
Country	**(Range of GMs)**		**GM**	**P95**
United States	0.06–0.07		0.089	0.389
Canada	0.05–0.11		0.081	0.360
United Kingdom	0.036–0.039	Europe	0.037	0.311
Germany	0.029–0.047	Europe	0.037	0.311
Norway	0.011–0.015	Europe	0.037	0.311

[a] Random survey of U.S. residences. (Zaffanella L. E., *Survey of Residential Magnetic Field Sources*, Volume 1:Goals, Results, Conclusions, EPRI Report No. TR-102759, Electric Power Research Institute, Palo Alto, CA, 1993.)
[b] GM ≈ Median over 24 hr and all rooms in the residence.
[c] P95 in the 5% of rooms with the highest MF.

Although dirty electricity measurements have not been correlated with MF exposures, they have been used in epidemiologic studies (Havas 2008; Milham and Morgan 2008).

As for ELF-EFs, HVTLs adjacent to homes are potential sources, but their EFs are shielded effectively by all materials, including the walls and insulation of household wiring. U.S. epidemiologic studies measured low ELF-EFs inside homes with GM = 4.4 V/m that were not correlated with the characteristics of nearby power lines (Kaune et al. 1987; Barnes et al. 1989; London et al. 1991).

Appliances and other electric equipment in homes also emit ELF-MFs either as a by-product of their currents (e.g., light, heating, and most electronics) or because electromagnetism is an essential part of their design (e.g., electric motors and the *cathode ray tubes* [CRTs] used in older televisions and computer monitors). The MF magnitude from appliances also decreases with distance from the source; but unlike power lines, the decline is proportional to $1/r^3$ (Preece et al. 1997).

Table 7.4 shows a sample of the MF emissions from residential appliance at distances where the head and trunk are likely to receive greater cumulative exposures (Behrens et al. 2004). Generally, appliances whose functioning depends on MFs have stronger emissions than others with comparable electric power. The reason is that electric motors and the other electromagnetic appliances have coils with multiple turns of wire whose number is directly proportional to the emitted MF from a given current. With CRTs, the MF systematically sweeps the cathode ray (an electron beam) across the screen horizontally at 15.75 kHz and vertically at 60 Hz (Kaune et al. 2000), so older TVs, computer monitors, and video games emitted both intermediate frequency and ELF-MFs. CRT monitors are now

TABLE 7.4

Selected ELF-MF from Home Appliances at Distances Typical of Personal Exposures

	Close (5 cm)		Manual Work Distance (0.5 m)		Far (≥1 m)	
Type	**Median**	**P95**	**Median**	**P95**	**Median**	**P95**
Light			Fluorescent lamp			
			0.10	0.34	0.02	0.08
Heat	Hair dryer		Electric range		Baseboard heater	
	13.01	45.82	0.07	0.22	0.04	0.09
			Electric oven			
			0.82	1.62		
Electronics	Cell phone[a]		Clock radio		Stereo	
	6.00	10.76	0.01	0.06	0.01	0.07
			Microwave			
			0.67	1.15		
Electric motors	Electric razor		Electric can opener		Heat pump	
	164.75	—	1.67	2.15	0.07	0.28
MF-based electronics			Computer w/CRT[b] monitor		TV with CRT	
			0.13	0.27	0.02	0.06
			Induction range[c]			
			1.00	1.72		

[a] 217 Hz pulses converted to rms.

[b] CRT, cathode ray tube.

[c] ELF modulation of 20–50 kHz carrier waves (total MF = 0.4–2 μT at this distance).

being replaced by plasma and liquid crystal displays with much lower ELF-MF emissions in more developed countries.

Table 7.4 includes cell phones and induction cooking stoves, whose ELF-MF emissions supplement to their better-known high-frequency EMF. The RF signals from 3G and 4G cell phones are emitted as 217 Hz pulses. The current from the phone's battery therefore pulses at the same rate and produces pulsed DC MFs that are predominantly in the ELF frequency range but with some higher harmonics (Tuor et al. 2005). Induction ranges cook food in metal pots with strong MF that consist of 20–50 kHz *carrier waves* with ELF *modulation* (Stuchly and Lecuyer 1987), similar to amplitude-modulated (AM) radio signals.

Personal Residential Exposures

Personal TWA MF exposures at home are determined by the duration of appliance use as well as the fields from the power distribution system and appliance emissions (discussed above). One-third of an adult's cumulative ELF-MF exposures are reportedly from home appliances, with higher proportions for children (Swanson and Kaune 1999; Behrens et al. 2004). This trend can be seen in Table 7.3 that compares the GMs of the 24-hr TWAs of the ELF-MF resultant from area and personal monitoring.

Summary of Residences

- The main determinants of residential ELF-MF exposures are the electric power system (including water pipe grounds as well as power lines) and appliance use.

- Appliance sources of ELF-MFs are lighting, heating, electronics, electric motors, and the direct use of MFs in applications such as induction cooking ranges.
- EFs in residences are low and not affected by outside electric lines.

Schools and Offices

Sources

Like residences, the primary sources of ELF-EMFs in schools and offices are the electrical distribution system (both outside and inside the building) and the electrical equipment that, in many cases, are similar to residential appliances. For example, the leading sources of ELF-MFs in California public schools (Enertech Consultants 1999) had 24-hr TWA exposures (Table 7.5) that mostly overlapped the comparable fields at homes (Table 7.3). The California school exposures are consistent with those measured in a Canadian school system (Sun et al. 1995) but higher than in Spanish classrooms (Tardon et al. 2002).

Additional MF sources in schools are fluorescent lighting, office equipment, and electrical panels (or switch boxes) that are all found in offices. An electric panel adjacent to an occupied room is often the greatest MF source in a school or office building. For example, the ELF-MFs near an electric panel attached to an outside power line were 13–70 μT in a U.S office building (Moss and Ragab 1995) and 0.88 μT in a Spanish school (Tardon et al. 2002).

Electrical equipment in schools and offices has ELF emissions that are, in many cases, similar to those of home appliances (Table 7.4). For example, ELF-MF emissions at working distances from computer monitors, photocopiers, scanners, FAX machines, and video projectors range from 0.05 to 0.14 μT. Most electrical equipment in offices are similar to those

TABLE 7.5

Selected ELF-MF Exposures in Schools

	GM (μT)	P95 (μT)
Leading Source	**24-hr TWA[a]**	
Ground currents	0.033	0.352
Distribution lines	0.024	0.063
Fluorescent lights	0.024	0.067
Electrical panel	0.023	0.117
Office equipment	0.023	0.073
Country	**Area Measurements**	
United States	0.040	0.210
Canada	0.033	0.281
Spain	0.014	0.037
Personal Exposure	**Full-Shift TWA**	
Students	0.078	0.197
Teaching occupations	0.114	0.399

[a] GM measured over 24 hr and all rooms in school. P95 measured in the school room with the highest MF.

in schools, but they often result in higher TWA exposures because they are used more frequently and are closer together than in schools.

Higher ELF-MF emissions have been introduced into offices, schools, and stores by computer servers, metal detectors, and electric article surveillance (theft detectors). For example, a room with banks of computers that service networks of personal computers for an office building had ELF-MFs ranging from 0.04 to 0.66 μT (Tepper et al. 1992). Metal detectors emit pulsed MFs with intermediate frequencies (1–100 kHz) that are perturbed by metal objects. Fixed metal detectors can have much higher MFs (up to 100 μT) than hand-held detectors (~5 μT).

Electronic article surveillance (EAS) devices detect electronic tags on books, clothing, and other articles by emitting EMFs with complicated waveforms that range from ELF to microwave frequencies. EAS systems also use EMF to deactivate the tags when the article leaves the facility and to reactivate it upon return. With EAS systems that use continuous wave ELF-MFs, measurements at the detector panels placed by doorways have GM = 340.5 μT with GSD = 2.07 and maximum = 843.9 μT (Moss and Roegner 1998; Harris et al. 2000; Cooper 2002; Trulsson et al. 2007; Joseph et al. 2012). The MF emissions from both EAS and metal detectors have very short ranges, so exposures are limited to people who use them regularly.

Personal Exposures in Schools and Offices

In a random survey of the U.S. population (Zaffanella 1998), the TWA MFs at schools (Table 7.5) are less than exposures at home (Table 7.3), work, and traveling. As in residences, the personal TWAs for students and teachers is higher than the area measurements in classrooms (Table 7.6), again showing that using electrical equipment is an important determinant of personal exposures. Personal ELF-MF exposures of workers in schools, offices, and a selection of other occupations are given in Table 7.6. These data were selected from a job–exposure matrix (JEM) of full-shift TWA measurements grouped by the 1988 International Classification of Occupations (ISCO) (Bowman et al. 2014).

The JEM is an essential tool for assessing exposures in retrospective chronic disease studies from contemporary measurements, historical data, expert judgments, or a combination. With the JEM for ELF-MF in Table 7.6, means and variances were calculated by pooling TWA measurements from multiple studies whose job titles were coded into the numerical ISCO system (ILO 1998). When the subjects' jobs also have ISCO codes, their exposures are then assessed by linking them to the mean exposures from the JEM. Variations of the JEM method are used by occupational epidemiologic studies described in Chapters 10 and 11.

Office occupations from the JEM (Table 7.6) provide a comparison of their TWA MF exposures with teaching professionals (from preschools to universities) and students (Table 7.5). A clear gradient in the MF GMs is apparent from students up to teachers and then to office jobs, probably due to how often electric equipment is used.

Note also that the students' P95 = 0.197 μT (Table 7.5) is above the GM = 0.15 μT for accounting, bookkeeping, and finance clerks (Table 7.6), showing that assessing exposures by the JEM means can create misclassifications in an individual's exposures. These *Berkson exposure assessment errors* result from assigning the JEM's mean exposure for an occupation to all subjects who held that job, and they can lead to both positive and negative errors in risk estimates (Carroll et al. 2006).

Despite these Berkson errors created by the JEMs, their ability to make unbiased quantitative assessments of exposures to an entire study population make them invaluable tools for occupational epidemiology.

TABLE 7.6

Personal Measurements of Full-Shift TWA ELF-MF Grouped by Selected Occupational Categories

Occupation	GM (μT)	P95 (μT)
Teaching professionals	0.11	0.40
Office Occupations		
Library and filing clerks	0.45	0.59
Accounting, bookkeeping, and finance clerks	0.15	0.87
Secretaries and keyboard-operating clerks	0.10	0.51
Manufacturing Occupations		
Ore and metal furnace operators	0.95	9.08
Sewing machine operators	0.83	1.88
Welders and flamecutters	0.80	8.93
Metal moulders and coremakers	0.52	6.08
Electrical and electronic equipment mechanics and fitters	0.23	2.30
Machinery mechanics and fitters	0.20	1.18
Food processing and related trades workers	0.14	0.85
Rubber and plastic products machine operators	0.11	0.39
Transportation Occupations		
Locomotive engine-drivers and related workers	13.26	65.65
Aircraft pilots	0.97	1.87
Ships' engineers	0.55	3.21
Ships' deck officers and pilots	0.22	1.06
Motor vehicle drivers	0.12	0.51
Transport laborers and freight handlers	0.10	0.37
Homemaker	0.06	0.08

Summary of Schools and Offices

- In schools and offices, sources of ELF-MF are generally similar to residential sources, but TWA exposures are usually higher because of the time using electric equipment.
- Students, on average, have less TWA exposures to ELF-MFs than teachers, who have less than office workers. However, exposure distributions overlap between these three groups.
- Although metal detectors and EAS (theft detectors) have introduced elevated MFs into schools, offices, and stores, their emissions have a short range, so exposures are limited to people who are near them regularly.

Transportation

Sources

Transportation has been a pervasive source of ELF-MF exposures since the beginning of the twentieth century, either from electric motors, the electrical components of internal combustion engines, or lighting and other electrical equipment inside the passenger compartment. Instantaneous MF exposures follow the same principles as other electrical equipment, and they vary inversely with the person's distance from the source (actually $1/r^3$ in most cases) and directly with its current, which is a function of the engine's power in electric transportation. The frequencies of transportation MFs vary widely within the 0–3000 Hz range due to the diversity of vehicle designs and with some designs, the frequency's dependence on the vehicle's speed.

Electric motors transport people and freight in many types of vehicles: intercity trains, urban rail, trolleys, subways, monorails, and battery-powered vehicles (e.g., hybrid cars, forklifts, milk delivery carts). Electric rail systems supply power to a vehicle either by a *third rail* or an *overhead line* (or *catenary*) whose electricity is carried to the train by a *pantograph*. The traction motors that power a train can be located either in locomotives or in individual passenger cars. After the current powers the traction motors, the electric circuit is completed through the train's metal wheels touching the regular rails. (Rubber-tired electric buses and trolleys close their traction circuits through a double pantograph contacting hot and neutral overhead lines.) Even a diesel locomotive can have substantial MF emissions because they have a *hybrid power system* with the diesel engine connected to a generator and traction motors at each axel that turns (or brakes) the wheels. These high currents on the rails and overhead catenary lines also expose railroad maintenance workers (Wenzl 1997; Yost 1999) and people living along the line (Brix et al. 2001) to EMF at the rail line's power frequency, which can be 0, 16.66, 25, 50, or 60 Hz in different countries and rail systems.

The proximity of electric rail passengers and crew to the traction motors and power circuit is a major determinant of their MF exposures. Trains with overhead lines tend to have higher MF exposures to the head and trunk, whereas those with third rails concentrate the exposures at the feet. Passengers have higher exposures near the traction motors in self-propelled rail cars than in trains with locomotives.

The traction motor's power is another determinant of MF exposures on electric vehicles. This relationship with locomotive power can be seen indirectly in a Swiss study where the electric locomotive engineers had higher exposures on alpine routes versus lowland routes and from hectic driving versus calm driving (Röösli et al. 2005). These higher exposure conditions both require greater locomotive acceleration and therefore greater traction motor power. Consequently, a logical hypothesis is that ELF-MF exposures from electric vehicles and all electric motors should generally increase with the motor's peak power that can usually be obtained from the equipment's specification information in manuals, name plates, or the Internet. However, this hypothesis has never been tested.

Finally, ELF exposures from electric transportation depend on whether the input electric power and traction motors are AC or DC—all four combinations are in use (e.g., AC power with AC motors, DC power with AC motors). With trains, the electrical and electronic circuit components that control the speed and convert current between AC and DC affect both the MF magnitude and frequency spectra (Chadwick and Lowes 1998; Dietrich and Jacobs 1999) and have changed historically as solid-state electronics have been developed for large currents and motors (Muc 2001). Note that an electric motor's rotating *armature* creates an oscillating MF even if it runs on DC currents. For example, battery-powered vehicles such as forklifts and milk delivery carts usually have static MF emissions near the battery and ELF-MF emissions near the motor.

Cars, buses, trucks, motorcycles, and other vehicles with internal combustion engines also emit ELF-MFs with a bewildering array of frequencies from their ignition system, the rotation of magnetized metal parts in the drive chain (including steel-belted radial tires), electric motors (e.g., fans), electronic components, and the MG-MFs from driving by steel structures (see section "Personal Transportation Exposures"). With motorcycles, MF emissions were traced to the engine's spark ignition system, located very close to the rider's groin (Chipkar 2007).

Personal Transportation Exposures

Table 7.7 summarizes the TWA ELF-MF exposures for the drivers and passengers in transportation vehicles, as well as maintenance workers on electric rail tracks. The JEM (Table 7.6) also has TWAs for selected transportation occupations.

The highest exposures were measured with intercity electric locomotive engineers, due to their proximity to the locomotive's high-powered traction motors, the catenary lines overhead, the neutral-return rail beneath, and the power circuits carrying high currents through the locomotive. Among railroad engineers, the Swiss exposure were the highest because of the magnetic induction forces needed to pull a train up an alpine pass and brake it on the way down. Remarkably, the "regenerative braking mode" that slows the locomotive by generating electric power emitted an 80% greater ELF-MF on average than climbing an alpine pass with the same slope (Röösli et al. 2005). In nonalpine countries, the passengers on intercity electric railroads had a GM exposure half that of the engineers because they also sat between the catenary lines above their car and the neutral return-rails below, even though the traction motors were far away in the locomotive.

Urban electric transit had MF exposures roughly an order of magnitude less than its intercity counterparts, probably due to the lower power needed to move a light rail train through a city. The one exception is the P95 = 11.31 μT for urban electric transit passengers that presumably occurred when they sat over their car's traction motor.

In addition, maximum exposures over 10 μT were measured on a maintenance crew for an intercity electric rail line, the operator of a battery-powered forklift, and a motorcycle rider. These high exposures can be explained for the first two sources because of the heavy loads moved by intercity trains and forklifts. However, the maximums above 20 μT

TABLE 7.7

Selected ELF-MF Exposures from Common Forms of Transportation

	TWA ELF-MF (μT)	
	GM	P95/Max
Swiss electric passenger train engineers	27.00	200.00
Other intercity electric train engineers	5.31	22.64
Electric intercity train passengers	3.08	6.25
Intercity electric rail maintenance	1.18	17.80
Urban Electric Transit (including Subways)		
—Engineers	0.29	1.40
—Passengers	0.81	11.31
Urban electric rail maintenance	0.19	0.83
Internal combustion vehicles (cars, buses, vans, pickup trucks)	0.19	0.80
—Hybrid cars	0.46	8.43
Motorcycles[a]	6.50	>20
Battery-powered forklift operator	1.17	125.00
All transportation (U.S. survey)	0.096	0.27

[a] Spot measurements with a single-axis MF meter.

measured on large "cruiser" motorcycles (Chipkar [2007] 2011) are much greater than the exposures from other internal combustion engines (Table 7.7) and can best be explained by the rider's proximity to the engine.

The transportation statistics in Table 7.7 were largely derived from a small number of measurements and are thus poor representatives of population exposures, especially given the complex and diverse transportation technologies reviewed above. The one exception is the "1000 Person" random survey of the U.S. population (Zaffanella 1998) where the TWA ELF-MF during travel had GM = 0.096 μT (P95 = 0.273 μT). This is more than home exposure in the U.S. survey (GM = 0.08 μT; P95 = 0.389 μT) but less than work exposures (GM = 0.103 μT; P95 = 0.500 μT). In the transportation occupations selected from the JEM (Table 7.6), the locomotive engine drivers, airline pilots, ships' engineers, and motor vehicle drivers had GMs greater than 85% of all other occupational categories (Bowman et al. 2014). (In making comparisons between Tables 7.6 and 7.7, note that the power frequency in transportation equipment can vary from 16.66 Hz in some electric railroad systems to 400 Hz in airplanes (Dietrich and Jacobs 1999), so these data might not accurately indicate biologic effects due to frequency dependence of magnetic induction. (Equation 7.1))

Summary of Transportation

- Important determinants of ELF-MF exposures from rail transportation are the power of the electric traction motors, the time spent close to the motor, the power circuit (the circuit within the vehicle from the electrical source to the traction motor and back to ground), or a combination of the above.
- The frequencies in transportation ELF-MF vary widely, due to both the electrical system's design (e.g., AC, DC) and the oscillating MF emitted by an electric motor's armature.
- The ignition system of an internal combustion engine can create elevated ELF-MF exposures with people who spend large amount of time close to the engine, such as motorcycle riders and chain saw operators (see section "Manufacturing and Other Occupations").

Manufacturing and Other Occupations

Sources

In manufacturing plants and other economic sectors, most electrical equipment use the five physical mechanisms outlined for home appliances (Table 7.4): light, heat, electronics, electric motors, and the direct use of MFs. However, manufacturing technologies can use electric currents that are orders of magnitude greater than home appliance currents, with similar increases in their MF emissions. For example, electric resistance furnaces for refining steel operate on the same principle as the electric ovens in kitchens, but their maximum reported emission is 8000 μT (Table 7.8) versus P95 = 1.62 μT for kitchen ovens (Table 7.4). Similarly, induction furnaces for metal heat treating emit a maximum of 8367 μT, whereas induction cooking ranges have P95 emissions = 1.7 μT.

Some MF sources in manufacturing use technologies not found in residences. Iron's ferromagnetic properties lead to some high ELF-MF exposures in many kinds of metal

TABLE 7.8

Selected TWA MF Exposures from Working with ELF Manufacturing Sources

Source	GM (μT)	Maximum (μT)
Metal Welding		
Spot resistance welding	967.50	11,436.8
Manual metal arc welding	141.42	
MIG (metal inert gas) arc welding	79.37	
TIG (tungsten inert gas) arc welding	12.68	141.4
Metal Heating		
Electrical resistance furnaces	567.27	8000.0
Induction heaters/furnaces[a]	9.85	8366.6
Electrochemical Processes[b]		
Rectifier room	69.62	781.2
Chlorine electrolysis cells	13.34	126.8
Metal electroplating	2.49	
Machining, Fabrication, etc.		
Nondestructive testing	14.66	5636.8
Sewing machine	0.83	11.05
Semiconductor fabrication	0.67	26.7
Machining operations (lathes, etc.)	0.45	
Battery charger	0.38	
Plastics extruder	0.23	
Chemical mixing machine	0.16	

[a] Intermediate frequency (3 kHz–10 MHz) induction heaters and furnaces.
[b] ELF component of static MF with an AC "ripple."

machining and fabricating. Because grinding magnetizes steel parts, degaussers remove this undesirable property by passing strong AC currents through the parts (Wenzl et al. 1997). Degaussers are also used to demagnetize steel parts after fault detection with magnetic fluorescent particles. Another form of nondestructive testing passes strong AC MFs through a steel part, exposing the operator to high MFs (Table 7.8).

Electrochemistry uses DC electricity to change salts into elements. Two examples are the *electrolysis* of NaCl (table salt) into chlorine gas and the chrome plating of metal parts. For industrial electrochemical processes, the DC electricity is generated by passing three-phase AC electricity through *rectifiers* that leaves an AC "ripple" on top of the DC current. Because the AC ripple degrades the electrochemical process, as in aluminum refining, the current is passed through many banks of rectifiers, leaving essentially pure static MF (Moss and Booher 1994). This ripple creates high exposures in the rectifier room and along the cells or tanks where the electrolysis takes place. As shown in Table 7.8, electroplating typically has much lower MF exposures than electrochemical plants that can use as much electricity as a medium-sized city (~60,000 A).

The list of ELF-MF sources in manufacturing (Table 7.8) is topped by metal welding, not the familiar *arc welding* but *spot resistance welding* used to bond steel plates in automobile bodies and other metal parts. Over the many types of welding, the applied current and the welder's distance from the current are major determinants of MF exposures. The frequency spectra of the welding currents varies immensely over welding technologies, involving DC, 50/60 Hz, its harmonics, and even RF.

Metal furnaces and heat treating equipment come next on the list, and also vary widely in current, frequency spectrum, and distance to the worker. *Resistance furnaces* in electrosteel plants use thousands of amperes, but their high temperatures keep workers at a distance, thereby moderating TWA exposures. Heat treating by magnetic induction localizes the high temperature to the metal part inside the solenoid, so TWA exposures are often elevated for operators of *induction furnaces* and *induction heaters.*

The lower exposures in Table 7.8 come from electric motors in, for example, sewing machines, lathes, and mixing machines. With motorized processes, the usual determinants of ELF-MF exposure (motor power, distance to the worker, and frequency spectrum) are moderated by the motor speed, which increases MF emissions, and the process's noise, which generally decreases the duration of exposure and thus the TWA. Among electric motors, the higher emissions from sewing machines with their high-speed motors (Table 7.8) are noteworthy because operators are predominantly women.

Personal Exposures in Manufacturing and Other Occupations

Workplaces have the highest personal ELF-MF exposures (GM = 0.103 μT; P95 = 0.500 μT) in the random survey of the U.S. population (Zaffanella 1998). The economic sectors with the highest occupational MF exposures are electric utilities (Table 7.2), transportation (Table 7.7), and manufacturing (Tables 7.6 and 7.8). Among the most exposed occupations in the ELF-MF JEM (Bowman et al. 2014), the only occupation from another sector is forestry workers and loggers (GM = 0.76 μT; P95 = 9.69 μT), presumably from their proximity to internal combustion engines of portable chain saws. Among the least exposed occupations are homemakers in the United States (GM = 0.06 μT; P95 = 0.08 μT); producers of market-oriented crops and animals (GM = 0.03 μT; P95 = 0.25 μT); childcare workers in the United Kingdom (GM = 0.03 μT; P95 = 0.17 μT); and operators of bleaching, dyeing, and cleaning machines (GM = 0.03 μT; P95 = 0.05 μT).

Summary of Manufacturing and Other Occupations

- The highest ELF-MF exposures in manufacturing are from spot resistance welding, resistance furnaces, induction heaters and furnaces, electrolysis, and magnetic nondestructive testing (NDT) of metal parts.
- The physical principles used in manufacturing ELF-MF sources are similar to home appliances (except for arc welding and NDT), but the currents and therefore the MF emissions can be orders of magnitude greater in manufacturing.
- The noise from large electric motors and the heat from electric furnaces reduce the TWA ELF-MF exposures because they tend to keep workers away at a distance.

Motion Gradient Magnetic Fields

Sources

The trace of the MF vector in Figure 7.2 is approximately helical, due to the lineman's motion through an MF gradient from perturbation of the geomagnetic field by steel objects. Although a pervasive part of the low-frequency MF environment, these MG-MFs are seldom considered in health studies because EMDEX-type instruments were designed to measure fields from power lines and therefore filter out frequencies <40 Hz. However, MG-MF magnitudes can be larger than other ELF-MF exposures where there are large steel structures, rapid motion, or strong DC MFs.

The ultimate example of an MG-MF comes from head motion in the spatial gradients of the DC MF around magnetic resonance imaging (MRI) scanners and has been shown to cause loss of balance, cognitive disturbances, and other neurologic disturbances (de Vocht et al. 2003; Glover et al. 2007; van Nierop et al. 2012a,b). These neurologic studies were conducted in MF gradients of 1–2.5 T/m close to 1.5 and 7 T MRIs (de Vocht et al. 2003). (Note: These spatial gradients in the static MF are different than the MRI's *gradient fields* that are pulsed MFs whose amplitudes gradually change along the length of the scanner's bore.)

The far more common source of MG-MF is the static geomagnetic field, whose main fields range from 30 μT near the equator to 60 μT near the poles (Campbell 1997). The geomagnetic field magnetizes steel structures that then perturb the static MF in their vicinity. Although MG-MFs in workplaces have not been studied systematically, their presence is indicated by a factory survey where the static MF at work locations (medians = 24.2–46.2 μT) were well below the local geomagnetic field of 55.0 μT (Bowman and Methner 2001). Therefore, steel in factories must be producing these large MF spatial gradients, exposing workers to MG-MF as they move through the plant. MG-MFs were also detected in cars, trucks, and buses as they drove by steel structures on the side of the road (Dietrich and Jacobs 1999).

Personal Exposures to MG-MFs

The most compelling explanation for loss of balance reported in the MG-MFs around MRIs (van Nierop et al. 2012b) is their induction of currents in the ionic fluids of the semicircular canals, the body's balance sensors (Glover et al. 2007; Roberts et al. 2011). According to the induced current mechanism (Equation 7.1), these ionic currents in the balance organs should be proportional to MF-MG's peak time derivative, dB_{MG}/dt. An alternative metric for MG-MF is its rms vector magnitude (Equation 7.5) that in combination with $\mathbf{B}_0$ and B_{ELF} also appears relevant to magnetoreception mechanisms by which animals navigate through MF gradients (Vanderstraeten and Gillis 2010).

Exposures to these two MG-MF metrics are given in Table 7.9. The most comprehensive data comes from exposure measurements with three-axis MF probes in a helmet on health care workers performing standard tasks in MRI scanner rooms (Groebner et al. 2011). In this study, the maximum Peak[*dB/dt*] was 1,400,000 μT/s. For perspective, a single-frequency sinusoidal MF has

$$\text{Peak}\left[dB/dt\right] = 2\pi f^2 \text{RMS}\left[B_{ELF}\right] \tag{7.9}$$

TABLE 7.9

Exposures to MG-MFs, Measured as TWAs of the RMS Vector Magnitudes for B_{MG} and Its Peak Time Derivative dB_{MG}/dt

	B_{MG} (μT)		dB_{MG}/dt (μT/s)	
Source	**GM**	**P95**	**GM**	**P95**
Working with MRIs	—	—	310,000	1,310,000
Riding a bucket truck next to a distribution line (see Figure 7.2)	0.98[a]	—	26.6[a]	—
Driving on city streets, rural roads, and expressways	0.50	1.87	—	

[a] Single measurement.

so this maximum MG-MF exposure for the MRI workers is the same in terms of magnetic induction as a 2600 μT_{rms} ELF-MF at 60 Hz—an exposure between the maximums ELF magnitudes from rectifier rooms and non-destructive testing in Table 7.8.

On the low end of the scale, MG-MFs from geomagnetic fields have not been studied systematically with the exception of a few surveys with the extinct Multiwave instruments. The vector trace in Figure 7.2 was measured with personal Multiwave III monitor and analyzed with Fourier transform techniques (Bowman et al. 2010). As shown in Table 7.9, the rms vector magnitudes of the MG-MF in this sample are 0.98 μT for B_{MG} and 26.6 μT/s for dB_{MG}/dt. For the power–frequency and harmonic components in this line worker measurement, the rms vector magnitude = 0.81 μT, a value that is slightly less than the MG-MF, but dB/dt = 337.2 μT/s, a value that is an order of magnitude greater than the MG-MF. In other words, magnetic induction effects that depend on the dB/dt metric will be little affected by the MG-MF in this case; but to assess mechanisms affected by the rms vector magnitudes, measurements of both the MG-MF and power–frequency components should be taken.

A second source of geomagnetic gradient field exposures was found by Multiwave MF monitoring of internal combustion vehicles (Dietrich and Jacobs 1999). Although the ELF-MF exposures reported in the *Transportation* section (Table 7.7) were attributed to the sources inside the vehicles, their survey data for the five vehicles consistently recorded frequency spectra that were attributed to MG-MFs from passing steel objects along the road. The TWA MG-MF from these test drives had a GM that is similar to the line worker's exposure in the bucket truck (Table 7.9). The MG-MF's maximum rms vector magnitude (11.16 μT) was measured when the vehicles drove on toll-free Interstate expressways, where they were likely due to passing under steel-reinforced concrete bridges at speeds around the U.S. limit (55 mi/hr = 88.5 km/hr). Obviously, the person's velocity is a determinant of MG-MF exposures. However, the strength of the MF gradient near the steel object is also a factor, and this depends on the steel's mass.

Comparing the MG-MF exposures in Table 7.9 with the ELF-MF exposures in residences (Table 7.3) and occupations (Table 7.6) suggests that MG-MF should not be neglected in assessing ELF-MF exposures in the environment, especially if biophysical mechanisms other than magnetic induction are important. The impact of this exposure assessment error on epidemiologic findings has never been studied.

Summary of MG-MFs

- MG-MFs are ELF fields caused by a person's motion through spatial gradients in the earth's MF near steel objects or from strong DC MF sources, such as MRIs.
- Determinants of MF-MF exposures are the person's velocity and the steepness of the MF gradient around the steel object or the DC MF source.
- Even though MF-MG exposures are very common and near MRIs have neurologic effects such as the MFs from AC electricity, MG-MFs are filtered out by the EMDEX meters used in ELF-MF epidemiologic studies. Whether this exposure assessment error affects epidemiologic results has never been studied.

Concluding Reflections

At the conclusion of this chapter, the broad diversity of ELF-EMF sources and personal exposures is clear. Table 7.10 gives an impression of that diversity by contrasting high and

TABLE 7.10

Overview of ELF-MF Exposures by Environment, Contrasting Dynamic Ranges within and between Sources, and TWA Personal Exposures

Environment		Sources			TWA from Personal Exposures		
		GM [μT]	Max [μT]	Hi:Lo GM Ratio	GM [μT]	GSD	Hi:Lo GM Ratio
Electric utilities	Hi	Underground vaults			Mechanics in substations		
		9.00	7700		3.8	—[a]	
	Lo	Reactor buildings			Clerical occupations in generating stations		
		0.02	0.06	450	0.14	—[a]	27
Residences	Hi	Hair dryers			US survey of home exposures (not in bed)		
		13.01	45.82		0.09	2.12	
	Lo	Ground currents (household median over space and time)			European 24-hr surveys		
		0.01	0.12	1301	0.04	2.96	2
Schools and offices	Hi	Electronic article surveillance			Library and filing clerks		
		76.2	843.9		0.45	1.16	
	Lo	Spanish classrooms			Teaching professionals		
		0.01	0.18	5443	0.11	1.89	4
Transportation	Hi	Swiss electric locomotives			Locomotive engine-drivers		
		13.26	195		13.26	2.26	
	Lo	Pickup truck with gasoline motor			Motor vehicle drivers		
		0.06	1.08	221	0.12	2.09	111
Manufacturing	Hi	Induction heaters and furnaces			Ore and metal furnace operators		
		141.84	8366		0.95	3.17	
	Lo	Chemical mixing machine			Rubber and plastic products machine operators		
		0.16	—	887	0.11	1.93	9
Motion gradient magnetic fields	Hi	Open MRIs			MRI technicians and physicists		
		$B_0 < 8$ T			0.31 T/s	2.09	
	Lo	Metal in geomagnetic fields			Gasoline vehicle passengers		
		$B_0 = 20 - 60$ μT		200,000	8.59 μT/s	1.96	36,100

Note: Within an environment, the dynamic ranges are given as the ratio of the geometric means between the high- and low-exposure situations. For personal exposures, the dynamic range within an occupation or region is given as the GSD.

[a] GSDs are impossible to estimate from Kelsh et al. (2000) because SEs are given without sample sizes.

low MF exposures in the five environments around which this chapter is organized (plus the MG-MF). Among other insights, these comparisons show that the high:low (Hi:Lo) ratios of ELF-MFs for sources within an environment are greater by at least an order of magnitude than the ratios for personal TWA exposures. This again suggests that people are not generally near high MF sources for long periods of time. Of the five environments surveyed, residences, schools, and offices have lower MF exposures, whereas high exposures are more likely to occur in electric utilities, transportation, and manufacturing. The transportation sector has the greatest variability with the Hi:Lo ratio for personal exposures = 111.

The comparisons in Table 7.10 also give some perspective on the statistical power of ELF-MF epidemiologic studies in these five environments and the populations with high and low exposures. Because power increases with the difference between high and low exposures if the sample size and other design factors are constant (Kelsey et al. 1996), the GSD is an indicator of the power of a study that assesses exposures for each subject in its study population. For example, studies of residential MF should have lower power in the United States (GSD = 2.12) than in Europe (GSD = 2.96) (although the higher GM in U.S. homes might be a countervailing factor). For occupational studies that assess exposures by a JEM within an environment, the ratio of the high GM to low GM likewise indicates a study's power. Therefore, a study in an electric utility where the Hi:Lo ratio = 27 should have greater power than occupational studies in manufacturing (Hi:Lo = 9) and office environments (Hi:Lo = 4), especially if exposure is assessed by primary work environment as well as occupation.

This broad survey of ELF-EMF exposures in everyday environments has been possible only by focusing on the GM and P95 of the TWA ELF resultant for MFs and the RMS component perpendicular to the body for EFs. As described in the section "Basic ELF-EMF Concepts," the choice of these metrics is dictated by the exposure assessment methods used by the best epidemiologic studies of ELF-EMF and chronic diseases, such as cancer, cardiovascular disease, and neurodegenerative diseases. However, these choices neglect many other EMF characteristics such as the frequency spectrum, polarization, intermittency, the static MF, MG-MF, and electric shocks that have been linked to health effects by physical, biologic, and epidemiologic research (Kheifets et al. 2009).

Surveying the ELF-EMF environment with only these two metrics is therefore like surveying aquatic life from the ocean's surface with only episodic submarine expeditions into deeper waters. Likewise, instruments such as the Multiwave III that measure ELF-EMFs in their full complexity and models of EMF's impacts on the body have been deployed in only a fraction of the environments in this survey, and never with the thoroughness of the ELF-MF resultant surveys by Zaffanella (1993, 1998). Consequently, our knowledge of the ELF-EMF environment as it affects the human body is somewhat superficial.

A major challenge for the EMF research community is therefore identifying biologically based exposure metrics that are possibly related to chronic diseases and developing instruments that can measure them reliably in large epidemiologic studies. Until better exposure assessment techniques are developed, exposure assessment errors and their biases on risk estimates will probably be present in all ELF-EMF epidemiologic studies, creating uncertainty about their implications for human health.

This chapter's analysis of the present state of ELF-EMF exposure assessment suggests that substantial resources will be required to develop measurement techniques adequate to resolve the questions of cancer and other chronic diseases through epidemiology. However, such investments in better exposure assessments would appear to be justified by the widespread exposures to occupational ELF-MF at levels associated with significant cancer risks (Bowman et al. 2013). As Kheifets et al. (2009) concluded in a recent review, research on improved exposure assessment is a high priority for strengthening ELF-EMF epidemiology so that answers can be found to the stubborn questions about cancer and other chronic diseases.

Acknowledgments

Thoughtful reviews by Chuck Kardous (National Institute for Occupational Safety and Health) and Danielle Vienneau (Swiss Tropical and Public Health Institute) are gratefully acknowledged.

Disclaimer

The findings and conclusions in this chapter are those of the author and do not necessarily represent the views of the National Institute for Occupational Safety and Health.

References

ACGIH. 2001. *Documentation of the Threshold Limit Values and Biological Exposure Indices*. 7th ed. Cincinnati, OH: American Conference of Governmental Industrial Hygienists.

Barnes, F., Wachtel, H., Savitz, D. and Fuller, J. 1989. Use of wiring configuration and wiring codes for estimating externally generated electric and magnetic fields. *Bioelectromagnetics* 10: 13–21.

Barr, R., Llanwyn Jones, D. and Rodger, C. J. 2000. ELF and VLF radio waves. *J Atmos Sol Terr Phys* 62: 1689–1718.

Behrens, T., Terschuren, C., Kaune, W. T. and Hoffmann, W. 2004. Quantification of lifetime accumulated ELF-EMF exposure from household appliances in the context of a retrospective epidemiological case-control study. *J Expo Anal Environ Epidemiol* 14(2): 144–153.

Blume, S. W. 2007. *Electric Power System Basics for the Nonelectrical Professional*. Edited by El-Hawary, M. E. *IEEE Press Series on Power Engineering*. Hoboken, NJ: Wiley-Interscience.

Bowman, J. D., Y. Guo, J. Figuerola, J. Fisker-Anderson, M.G. Yost, and the INTEROCC Study group. In press. A population-based job exposure matrix for power frequency magnetic fields: Upgrade for the INTEROCC study.

Bowman, J. D., Kelsh, M. A. and Kaune, W. T. 1998. *Manual for Measuring Occupational Electric and Magnetic Field Exposures*. DHHS(NIOSH) Publication No. 98-154. Cincinnati, OH: DHHS, CDC, National Institute for Occupational Safety and Health.

Bowman, J. D. and Methner, M. M. 2001. Hazard surveillance for industrial magnetic fields: II. Field characteristics from waveform measurements. *Ann Occup Hyg* 44: 615–633.

Bowman, J. D., Miller, C. K., Krieg, E. F. and Song, R. 2010. Analyzing digital vector waveforms of 0-3000 Hz magnetic fields for health studies. *Bioelectromagnetics* 31(5): 391–405.

Bowman, J. D., Ray, T. K. and Park, R. M. 2013. Possible health benefits from reducing occupational magnetic fields. *Am J Ind Med* 56(7): 791–805.

Bowman, J. D., Sahl, J. D., Krieg, E. F. and Kavet, R. 2014. Electric utility worker exposures to biologically-based metrics for ELF and static magnetic fields measured by the Multiwave System III. *In preparation*.

Bowman, J. D., Thomas, D. C., Jiang, L. Z., Jiang, F. and Peters, J. M. 1999. Residential magnetic fields predicted from wiring configurations: I. Exposure model. *Bioelectromagnetics* 20: 399–413.

Bracken, T. D., Kavet, R., Patterson, R. M. and Fordyce, T. A. 2009. An integrated job exposure matrix for electrical exposures of utility workers. *J Occup Environ Hyg* 6(8): 499–509.

Bracken, T., Senior, R., and Dudman, J. 2005. 60-Hertz electric-field exposures in transmission line towers. *J Occup Environ Hyg* 2(9): 444–455.

Bracken, T. D., Rankin, R. F., Senior, R. S., Kavet, R. I., and Geissinger, L. G. 2001. Magnetic-field exposures of cable splicers in electrical network distribution vaults. *Applied Occupational and Environmental Hygiene* 16(3): 369–379.

Bracken, T. D., Kheifets, L. I. and Sussman, S. S. 1993. Exposure assessment for power frequency electric and magnetic fields (EMF) and its application to epidemiologic studies. *J Expo Anal Environ Epidemiol* 3(1): 1–22.

Bracken, T. D., Senior, R. S., Rankin, R. F., Bailey, W. H. and Kavet, R. 1997. Magnetic field exposures in the electric utility industry relevant to occupational guideline levels. *Appl Occup Environ Hyg* 12(11): 756–768.

Brix, J., Wettemann, H., Scheel, O., Feiner, F. and Matthes, R. 2001. Measurement of the individual exposure to 50 and 16 2/3 Hz magnetic fields within the bavarian population. *Bioelectromagnetics* 22: 323–332.

Campbell, W. C. 1997. *Introduction to Geomagnetic Fields*. Cambridge, UK: Cambridge University Press.

Carroll, R. J., Ruppert, D., Stefanski, L. A. and Crainiceanu, C. 2006. *Measurement Error in Nonlinear Models: A Modern Perspective*. 2nd ed. Boca Raton, FL: CRC Press.

CENELEC. 2004. *Operation of Electrical Installations*. European Standard EN 50110. Brussels: European Committee for Electrotechnical Standardization.

Chadwick, P. and Lowes, F. 1998. Magnetic fields on British trains. *Ann Occup Hyg* 42(5): 331–335.

Chartier, V. L., Bracken, T. D. and Capon, A. S. 1985. BPA study of occupational exposure to 60-Hz electric-fields. *IEEE Trans Power Apparatus Syst* 104(3): 733–744.

Chipkar, R. J. (2007) 2011. *Motorcycle Cancer? Are Motorcycles Killing Us with Cancers of the Prostate, Colon, Kidney, Bone, Etc?* Chipkar Publishing. Available from: http://www.motorcyclecancer.com

Cooper, T. G. 2002. *Occupational Exposure to Electric and Magnetic Fields in the Context of the* ICNIRP *Guidelines*. Report NRPB-W24. Oxon, UK: National Radiation Protection Board.

Deno, D. E. and Silva, J. M. 1998. Method for evaluating human exposure to 60 Hz electric fields. *IEEE Trans Power Apparatus Syst* PAS-103: 1699–1706.

de Vocht, F., Vanwendeldejoode, B., Engels, H. and Kromhout, H. 2003. Neurobehavioral effects among subjects exposed to high static and gradient magnetic fields from a 1.5 tesla magnetic resonance imaging system—a case-crossover pilot study. *Magn Reson Med* 50: 670–674.

Dietrich, F. M. and Jacobs, W. L. 1999. *Survey and Assessment of Electric and Magnetic Field (EMF) Public Exposure in the Transportation Environment*. Report to the Department of Transportation. Report PB-99-130908/XAB. Springfield, VA: National Technical Information Service.

Enertech Consultants. 1993. *EMDEX II and EMCALC Users Manuals*. Campbell, CA: Enertech Consultants.

Enertech Consultants. 1999. *Electric and Magnetic Field Exposure Assessment of Powerline and Non-Powerline Sources for California Public School Environments*. Available from: http://www.ehib.org/emf/research4.html#SEA.

EU. 2013. Directive 2013/35/EU of the European Parliament and of the Council of 26 June 2013 on the minimum health and safety requirements regarding the exposure of workers to the risks arising from physical agents (electromagnetic fields). *Offic J Eur Union*. Available from: http://eur-lex.europa.eu/LexUriServ/LexUriServ.do?uri = OJ:L:2013:179:0001:0021:EN:PDF

Glover, P. M., Cavin, I., Qian, W., Bowtell, R. and Gowland, P. A. 2007. Magnetic-field-induced vertigo: a theoretical and experimental investigation. *Bioelectromagnetics* 28(5): 349–361.

Griffiths, D.J. 2013. *Introduction to Electrodynamics*. 4th ed. New York: Pearson.

Groebner, J., Umathum, R., Bock, M., Krafft, A. J., Semmler, W. and Rauschenberg, J. 2011. MR safety: simultaneous B0, dΦ/dt, and dB/dt measurements on MR-workers up to 7 T. *MAGMA* 24(6): 315–322.

Guenel, P., Nicolau, J., Imbernon, E., Chevalier, A., and Goldberg, M. 1996. Exposure to 50-Hz electric field and incidence of leukemia, brain tumors, and other cancers among French electric utility workers. *Am J Epidemiol* 144(12): 1107–1121.

Guttman, J. L., Kavet, R. and Johnson, G. B. 2001. Transient magnetic fields and currents in residences. *IEEE Int Symp* 1: 227–232.

Harris, C., Boivin, W., Boyd, S., Coletta, J., Kerr, L., Kempa, K., et al. 2000. Electromagnetic field strength levels surrounding electronic article surveillance (EAS) systems. *Health Phy* 78(1): 21–27.

Havas, M. 2008. Dirty electricity elevates blood sugar among electrically sensitive diabetics and may explain brittle diabetes. *Electromagn Biol Med* 27(2): 135–146.

Havas, M. and Stetzer, D. 2004. Dirty electricity and electrical hypersensitivity: five case studies. Wireless Protection website. http://www.wireless-protection.org/documents/Havas_Stetzer_WHO04.pdf (accessed February 3, 2014).

Heroux, P. 1991. A dosimeter for assessment of exposures to ELF fields. *Bioelectromagnetics* 12: 241–257.

IARC. 2002. *Non-Ionizing Radiation, Part 1: Static and Extremely Low-Frequency (ELF) Electric and Magnetic Fields, IARC Monographs on the Evaluaton of Carcinogenic Risks to Humans*. Lyon, France: International Agency for Research on Cancer.

IEEE. 2002. IEEE Standard for Safety Levels with Respect to Human Exposure to Electromagnetic Fields, 0-3 kHz. *IEEE Standards by the IEEE International Committee on Electromagnetic Safety*. New York, NY: Institute of Electrical and Electronic Engineers.

ILO. 1988. *International Standard Classification of Occupations (ISCO)*. Geneva, Switzerland: International Labor Organization.

ITU (International Telecommunications Union). 2008. *Frequency and Wavelength Bands*. Available from: http://life.itu.int/radioclub/rr/art02.htm

Jackson, J. D. 1999. *Classical Electrodynamics*. 3rd ed. New York: John Wiley.

Joseph, W., Vermeeren, G., Verloock, L. and Goeminne, F. 2012. In situ magnetic field exposure and ICNIRP-based safety distances for electronic article surveillance systems. *Radiat Prot Dosimetry* 148(4): 420–427.

Kaune, W. T., and Anderson, L. E. 1990. Physical aspects of ELF electric and magnetic fields: measurements and dosimetry. In Wilson, B. W., Stevens R. G. and Anderson L. E., *Extremely Low Frequency Electromagnetic Fields: The Question of Cancer*. Columbus, OH: Battelle Press, pp. 17–45.

Kelsey, J. L., Whittemore, A. S., Evans, A. S., and Thompson, W. D. 1996. Power calculations. In *Methods in Observational Epidemiology*. New York: Oxford University Press, pp. 327–332.

Kaune, W. T., Miller, M. C., Linet, M. S., Hatch, E. E., Kleinerman, R. A., Wacholder, S., et al. 2000. Children's exposure to magnetic fields produced by US television sets used for viewing programs and playing video games. *Bioelectromagnetics* 21: 214–227.

Kaune, W. T., Stevens, R. G., Callahan, N. J., Severson, R. K. and Thomas, D. B. 1987. Residential magnetic and electric fields. *Bioelectromagnetics* 8: 315–335.

Kaune, W. T. and Zaffanella, L. E. 1992. Analysis of magnetic fields produced far from electric power lines. *IEEE Trans Power Deliv* 7(4): 2082–2091.

Kelsh, M. A., Kheifets, L., and Smith, R. 2000. The impact of work environment, utility, and sampling design on occupational magnetic field exposure summaries. *Am Ind Hyg Assoc J* 61: 174–182.

Kheifets, L., Bowman, J. D., Checkoway, H., Feychting, M., Harrington, J. M., Kavet, R., et al. 2009. Future needs of occupational epidemiology of extremely low frequency electric and magnetic fields: review and recommendations. *Occup Environ Med* 66(2): 72–80.

Korpinen, L., Kuisti, H., Paakkonen, R., Vanhala, P., and Elovaara, J. 2011. Occupational exposure to electric and magnetic fields while working at switching and transforming stations of 110 kV. *Ann Occup Hyg* 55(5): 526–536.

Korpinen, L. H., Elovaara, J. A., and Kuisti, H. A. 2009. Evaluation of current densities and total contact currents in occupational exposure at 400 kV substations and power lines. *Bioelectromagnetics* 30(3): 231–240.

Leeper, E. A. 2001. *Silencing the Fields: A Practical Guide to Reducing AC Magnetic Fields*. Boulder, CO: Symmetry Books.

London, S. J., Thomas, D. C., Bowman, J. D., Sobel, E., Cheng, T. C. and Peters, J. M. 1991. Exposure to residential electric and magnetic fields and risk of childhood leukemia. *Am J Epidemiol* 134: 923–937.

McDevitt, J. J., Breysse, P. N., Bowman, J. D. and Sassone, D. M. 2002. Comparison of extremely low frequency (ELF) magnetic field personal exposure monitors. *J Exp Anal Environ Epidemiol* 12: 1–8.

Mild, K. H., Sandstrom, M. and Johnsson, A. 1996. Measured 50 Hz electric and magnetic fields in Swedish and Norwegian residential buildings. *IEEE Trans Instrum Meas* 45(3): 710–714.

Milham, S. and Morgan, L. L. 2008. A new electromagnetic exposure metric: high frequency voltage transients associated with increased cancer incidence in teachers in a California school. *Am J Ind Med* 51(8): 579–586.

Morgan, M. G. and Nair, I. 1992. Alternative functional relationships between ELF field exposure and possible health effects—report on an expert workshop. *Bioelectromagnetics* 13: 335–350.

Moss, C. E. and Booher, D. 1994. *NIOSH Health Hazard Evaluation Report No. 91-0337-2466, Aluminum Company of America, Badin, NC*. Cincinnati, OH: National Institute for Occupational Safety and Health.

Moss, C. E. and Ragab, M. 1995. *NIOSH Health Hazard Evaluation Report: County Board of Taxation, Cape May Court House, NJ*. Cincinnati, OH: National Institute of Occupational Safety and Health.

Moss, C. E. and Roegner, K. 1998. *NIOSH Health Hazard Evaluation Report: Library of Congress, Washington, D.C.* Cincinnati, OH: National Institute for Occupational Safety and Health.

Mouritsen, H. and Ritz, T. 2005. Magnetoreception and its use in bird navigation. *Curr Opin Neurobiol* 15(4): 406–414.

Muc, A. 2001. Electromagnetic fields associated with transportation systems. *Health Can* 5(29): 1–52.

Preece, A. W., Kaune, W. T., Grainger, P., Preece, S. and Golding, J. 1997. Magnetic fields from domestic appliances in the UK. *Phys Med Biol* 42: 67–76.

Rappaport, S. M. and Kupper, L. L. 2008. Exposure, dose, and damage. In: *Quantitative Exposure Assessment*. 127–142, El Cerito, CA: Stephen M. Rappaport, 127–142.

Reilly, J. P. 1998. *Applied Bioelectricity: From Electrical Stimulation to Electropathology*. New York: Springer-Verlag.

Renew, D. C., Cook, R. F., and Ball, M. C. 2003. A method for assessing occupational exposure to power-frequency magnetic fields for electricity generation and transmission workers. *J Radiol Prot* 23(3): 279–303.

Roberts, D. C., Marcelli, V., Gillen, J. S., Carey, J. P., Della Santina, C. C. and Zee, D. S. 2011. MRI magnetic field stimulates rotational sensors of the brain. *Curr Biol* 21(19): 1635–1640.

Röösli, M., Lortscher, M., Pluger, D. and Schrier, N. 2005. ELF (16 2/3 Hz) magnetic field exposure assessment in Swiss railway engineers. In: *Bioelectromagnetics 2005 Conference*. Dublin, Ireland.

Savitz, D. A., Wachtel, H., Barnes, F. A., John, E. M. and Tvrdik, J. G. 1988. Case-control study of childhood cancer and exposure to 60 Hz magnetic fields. *Am J Epidemiol* 128: 21–38.

SCENIHR. 1997. *Possible Effects of Electromagnetic Fields (EMF) on Human Health*. Edited by Scientific Committee on Emerging and Newly Identified Health Risks, Brussels: European Commission.

Smith, T. J. and Kriebel, D. 2010. *A Biologic Approach to Environmental Assessment and Epidemiology*. New York: Oxford University Press.

Stuchly, M. A. and Lecuyer, D. W. 1987. Electromagnetic fields around induction heating stoves. *J Microwave Power* 22(2): 63–69.

Sun, W. Q., Heroux, P., Clifford, T., Sadilek, V. and Hamade, F. 1995. Characterization of the 60-Hz magnetic-fields in schools of the Carleton Board-of-education. *Am Indust Hyg Assoc J* 56(12): 1215–1224.

Swanson, J. and Kaune, W. T. 1999. Comparison of residential power-frequency magnetic fields away from appliances in different countries. *Bioelectromagnetics* 20: 244–254.

Tardon, A., Velarde, H., Rodriguez, P., Moreno, S., and Raton, M. 2002. Exposure to extremely low frequency magnetic fields among primary school children in Spain. *J Epidemiol Community Health* 56(6): 432–433.

Tepper, A., Kiefer, M. and Moss, C. E. 1992. *NIOSH Health Hazard Evaluation Report No. 92-156-2256, Ford House Office Building, Washington, D.C. (Available Through NTIS).* Cincinnati, OH: U.S. Department of Health and Human Services, Public Health Service, Centers for Disease Control and Prevention, National Institute for Occupational Safety and Health.

Trulsson, J., Anger, G. and Estenberg, U. 2007. Assessment of magnetic fields surrounding electronic article surveillance systems in Sweden. *Bioelectromagnetics* 28(8): 664–666.

Tuor, M., Ebert. S., Schuderer, J. and Kuster, N. 2005. *Assessment of ELF Exposure from GSM Handsets and Development of an Optimized RF/ELF Exposure Setup for Studies of Human Volunteers*. Zurich, Switzerland: IT'IS Foundation.

Vanderstraeten, J. and Gillis, P. 2010. Theoretical evaluation of magnetoreception of power-frequency fields. *Bioelectromagnetics* 31(5): 371–379.

Vanderstraeten, J. and Vorst, A. V. 2005. Theoretical evaluation of dielectric absorption of microwave energy at the scale of nucleic acids—response. *Bioelectromagnetics* 26: 80.

van Nierop, L. E., Slottje, P., Kingma, H. and Kromhout, H. 2012a. MRI-related static magnetic stray fields and postural body sway: a double-blind randomized crossover study. *Magn Reson Med*. 70(1): 232–240.

van Nierop, L. E., Slottje, P., van Zandvoort, M. J., de Vocht, F. and Kromhout, H. 2012b. Effects of magnetic stray fields from a 7 Tesla MRI scanner on neurocognition: a double-blind randomised crossover study. *Occup Environ Med* 69(10): 759–766.

Wenzl, T. B. 1997. Estimating magnetic field exposures of rail maintenance workers. *Am Ind Hyg Assoc J* 58: 667–671.

Wertheimer, N., and Leeper, E. 1979. Electrical wiring configurations and childhood cancer. *American Am J Epidemiol* 109: 273–284.

WHO. 2007. *Environmental Health Criteria 238: Extremely Low Frequency Fields*. Geneva, Switzerland: World Health Organization.

Yost, M. G. 1999. Alternative magnetic field exposure metrics: occupational measurements in trolley workers. *Radiat Prot Dosimetry* 83(1–2): 99–106.

Zaffanella, L. E. 1993. *Survey of Residential Magnetic Field Sources.* Volume 1: Goals, Results, Conclusions. EPRI Report No. TR-102759. Palo Alto, CA: Electric Power Research Institute.

Zaffanella, L. E. 1998. *Survey of Personal Magnetic Field Exposure, Phase II: 1000-Person Survey*. Available from: http://www.emf-data.org/rapid6-report.html

8

Exposure to Radiofrequency Electromagnetic Fields in Our Everyday Environment

Patrizia Frei and Martin Röösli

CONTENTS

Radiofrequency Electromagnetic Fields: Characteristics and Typical Sources

Radiofrequency (RF) electromagnetic waves originate from charged particles and propagate through space in a wave-like behavior. The electrical and the magnetic components are orthogonal to each other with a fixed ratio of intensity. The propagation pattern depends on the type of source and the type of dispersion. For a point source with nondirectional dispersion, the electromagnetic field (EMF) level decreases with the square of distance from the source ($\sim 1/r^2$). For a point source with directional dispersion, the EMF level decreases inversely with the distance from the source ($\sim 1/r$). In practice, however, the propagation pattern may be more complex. Fixed site transmitters such as broadcasting or mobile phone antennae may be characterized by a mixture of directional and nondirectional propagation producing a decrease that is between $1/r$ and $1/r^2$. The EMF at a given location (i.e., receptor) can thus be calculated from the antennae emission characteristics and the geometry between the source and the receptor point. Physical characteristics of the environment and potential reflections, however, can make this calculation quite

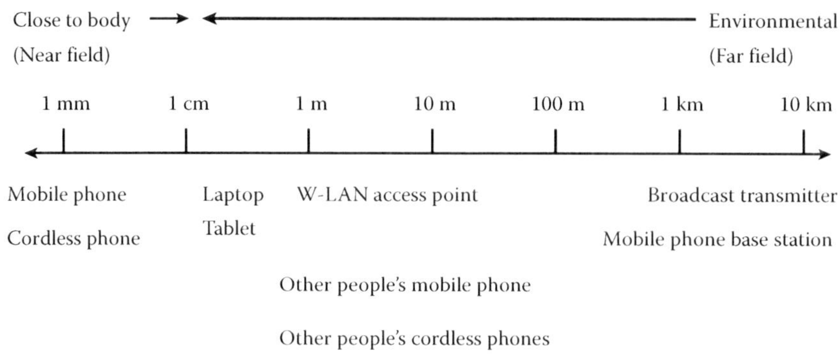

FIGURE 8.1
Illustration of near- and far-field RF sources.

complex, in particular if the receptor is not in the line of sight of the source (e.g., shielded by topography or buildings).

Physical characteristics differ between near- and far-field conditions, characterized by the distance from the source. Typically, far-field conditions occur at a distance above one wavelength, but it may differ for specific antennal characteristics (Figure 8.1). In the near-field range, physical conditions are complex, and interactions between absorbers and transmitters occur. In the far-field, the energy of the wave is conserved and decreases with distance according to an inverse or inverse-square law.

Far-field RF waves are typically measured as electrical field strength (V/m); thus, the term radiofrequency electromagnetic field (RF-EMF) is most common in this research area. RF-EMF values are, however, not additive and root-mean-square calculations are needed to sum up RF-EMF values from various exposure sources. Another measure is power flux density (W/m^2) that is additive. Under far-field conditions, we can convert between V/m and W/m^2 using the formula

$$S = \frac{E^2}{Z_0} \quad \text{respectively} \quad E = \sqrt{S \times Z_0}$$

where E represents the electrical field strength in V/m and S the power flux density in W/m^2. Z_0 is the free space impedance of 377 Ω. Conversion into the magnetic field (H) is also possible for far-field conditions using the formula $H = E/Z_0$.

In the near-field range, the specific absorption rate (SAR) is the most common measure of intensity. With the SAR, energy absorption of tissue is measured in watts per kilogram (W/kg). It depends on the field strength, the frequency of an exposure source, and the physiological characteristics of the absorbing tissue. Generally, the lower the frequency of a RF-EMF the further it can penetrate into the body.

RF-EMF range from 100 kilohertz (kHz) to 300 gigahertz (GHz) and are mainly used for heating processes (e.g., induction heating [300 kHz–1 MHz], microwave cooking [2.45 GHz]), wireless communication transmission (e.g., mobile phone, 800 MHz–2.4 GHz), or radar air traffic control (1–10 GHz). To be able to transmit information (e.g., audiovisual information), amplitude (AM) or frequency (FM) modulation is applied, that is, a property of the RF wave is systematically changing.

Mainly due to the rapid expansion of the mobile phone communication network over the past 20 years, exposure to RF-EMF in our environment has considerably increased (Frei et al. 2009b). The first generations (1G) of wireless telephone technology were analog telecommunications standards that were introduced in the 1980s. In the early 1990s, the 2G wireless telephone technology (global system for mobile communications standard, GSM), was launched. 2G networks are still used widely all over the world, although newer technologies have since been introduced. The 3G standard, including universal mobile telecommunications system (UMTS), was first offered in the early 2000s. The 4G standard technologies (e.g., long-term evolution [LTE]) were introduced a few years later. Newer telecommunications generations can, in general, be characterized by increased capacity and speed (see Chapter 19). Other common RF-EMF communication technologies include FM radio broadcasting, digital audio broadcasting (DAB), television broadcasting (DVB-T), digital enhanced cordless telecommunications (DECT) (i.e., cordless phones), and wireless local area networking (W-LAN, Wi-Fi). Typical RF-EMF emitting sources and their frequencies are shown in Table 8.1. In the past, various mobile phone standards, namely Nordic mobile telephone (NMT; no longer in use), GSM, and UMTS used specific allocated frequency bands. With the introduction of LTE, however, the frequency allocations in the mobile telecommunication network have, in general, become more flexible, and frequency bands are used for all types of mobile phone standards.

There are different medium access methods, such as time division multiple access (TDMA), that allows several users to share the same frequency. This scheme is, for example, used by DECT cordless phones and GSM mobile phones and mobile phone base stations. For GSM, one of eight time slots is occupied by one user. The other option is code division multiple access (CDMA) implemented in the UMTS and LTE standard. In this case, a code is assigned to the signal of each user so that they can use the same frequency band (see Chapter 19).

TABLE 8.1

Typical RF-EMF Sources in the Everyday Environment and Their Associated Frequencies in Europe

Source	Frequency (MHz)
FM radio broadcast	88–108
Digital video (TV) and digital audio broadcast (DAB)	174–230
Television broadcast (DVB-T)	470–790
Mobile phone handset (GSM, UMTS, LTE)	832–862
	880–915
	1710–1785
	1920–1980
	2500–2570
Mobile phone base station (GSM, UMTS, LTE)	791–821
	925–960
	1805–1880
	2110–2170
	2620–2690
DECT cordless phone	1880–1900
W-LAN	2400–2500
	5150–5350
	5470–5795
	5815–5875

Reference Values for RF-EMF

The only scientifically accepted effect of RF-EMF exposure on humans is the increase in body temperature caused by high-intensity RF-EMF radiation (see Chapter 17). Below this thermal threshold, no biological mechanism has yet been established (ICNIRP 2009; SCENIHR 2009). The International Commission on Non-Ionizing Radiation Protection (ICNIRP) has published guidelines that limit the exposure of the public to prevent heating effects due to RF radiation (ICNIRP 1998). These guidelines are described below.

The ICNIRP reference values are based on the amount of energy absorbed by the human body (SAR). The reference values are based on the criterion that the absorbed radiation must never increase the human body temperature by more than 1°C because this can cause interference with various body functions. Greater increases can lead to internal burns or death due to heat stroke. To prevent such heating effects from short-term RF-EMF exposure in the frequency range from 100 kHz to 10 GHz, the ICNIRP recommends both whole-body and localized SAR limits (head and trunk). Applying a safety factor of 10 for occupational exposure, these values are 0.4 and 10 W/kg for a whole-body and localized SAR, respectively. For public exposure, a safety factor of 50 is applied resulting in a basic restriction of 0.08 W/kg for whole-body exposure and 2 W/kg for localized exposures (ICNIRP 1998). The ICNIRP levels are obtained by averaging the SAR over 10 g of tissue. Because measuring SAR in living persons is impossible, the field strength (V/m) or the power density (W/m^2) measured outside of the human body is instead used for regulating far-field RF-EMF. Above 2 GHz, the ICNIRP recommends a field strength of 61 V/m or a power flux density of 10 W/m^2. Below 2 GHz, the values are decreasing with decreasing frequency, although constant between 10 and 400 MHz (28 V/m).

The ICNIRP reference values have been adopted by more than 30 countries, mainly in Europe (Valberg et al. 2007; Grandolfo 2009). North America and some Asian countries apply a maximum SAR value of 1.6 W/kg averaged over 1 g of tissue for exposure from mobile phones as proposed by the Institute of Electrical and Electronics Engineers (IEEE) (http://standards.ieee.org/). For far-field exposure, some countries, such as Switzerland, Italy, and Belgium, have instituted additional reference values that are substantially below the ICNIRP values.

Near-Field and Far-Field RF-EMF Sources

As mentioned, depending on the distance from the source, near- and far-field conditions to RF-EMF occur in everyday life. Near-field sources are, for example, mobile and cordless phones, whereas far-field sources include mobile phone base stations, broadcast transmitters, or base stations of cordless phones (Figure 8.1). Far-field sources can also be called environmental far-field sources. Near-field sources are generally responsible for highly localized exposure, for example, in the head area, and exposure is in general limited to short time periods. Exposure from environmental far-field sources results in a more homogeneous whole-body exposure that is lower than the maximum exposure due to an operating mobile phone on the head, but it occurs usually over prolonged time periods. Exposure from mobile and cordless phones can be considered both near-field and environmental far-field sources; whereas the personal mobile and cordless phones are used in proximity of the body, mobile or cordless phones used by people nearby are generally distant enough to cause a far-field exposure.

Exposure from mobile phones has changed over time and is characterized by decreasing power output in newer generation phones compared with the older generation phones. A feature used by GSM networks aimed at reducing the power output of mobile phones is adaptive power control (APC). APC starts with the maximum output power of the mobile phone handset (250 mW for GSM 900 or 125 mW for GSM 1800) and reduces the power over time to the lowest level compatible with a good signal quality (Lonn et al. 2004). When changing an area covered by one mobile phone base station or a cluster of a few mobile phone base stations, the power gets back to the maximum and is downregulated again. UMTS networks, the third generation, use a more efficient power control technology that starts with minimum power output and adapts faster to the optimal transmission level.

Mobile phones in standby mode emit occasionally, even when not being actively used. This type of transmission is called organizational communication or location update. Location updates are necessary to maintain constant connectivity with the network. In particular, when moving in a car or train, a mobile device periodically sends information about its position while changing location. Modern wireless devices such as smart phones and tablet computers may provide constant access to the Internet and may have installed applications using push functions for regular updates. It can be expected that such modern devices give rise to more organizational communication than older generation mobile phones, although a systematic evaluation is not yet available (Urbinello and Röösli 2013).

The maximum output power of the headset of cordless telephones using the DECT standard is 10 mW. Because the base station for cordless phones is usually relatively close to the handset, this is considerably lower than for mobile phones that are typically more distant from their base station and require more power for radio communications. Compared with mobile phones, emission power of almost all current cordless phone models is constant. Many DECT base stations are emitting constantly, even when not being used for making calls.

Assessment of Exposure to RF-EMF Sources for Epidemiological Purposes

In an epidemiological study, the most important prerequisite of an exposure assessment method is to be able to reliably differentiate between high- and low-exposed individuals or to produce a reliable exposure ranking between individuals (see Chapter 4). Therefore, for hazard identification it is not essential that the exact exposure level of an individual be perfectly determined (Heid et al. 2004) (see Chapter 4). For methods that require active participation from potential study participants, selection bias can be of concern, especially if a large effort for study participants is involved (see Chapter 5). Another issue is information bias that can be introduced if an exposure assessment method relies on subjective information provided by the study participants (see Chapter 4). Last but not least, when choosing an exposure assessment method, the cost and feasibility of the method are also important criteria that have to be taken into account. In the following, an overview about state-of-the-art RF-EMF exposure assessment methods and results is given. A more general overview on epidemiological exposure assessments can be found in Chapter 4.

Far-Field RF-EMF Sources

Distance to the Closest Fixed Transmitter

Using the distance of the geo-coded residency of the study participants to the closest fixed site transmitter as exposure proxy is often appealing because of the low cost in deriving such a measure. In addition, it is particularly attractive for assigning exposures to a large study population because exposure can be assessed without contact with study participants, which minimizes information and selection bias. This exposure proxy has been used in the past in some epidemiological studies (Navarro et al. 2003; Santini et al. 2003). However, more extensive studies using measurements in the bedroom of study participants or 24-hr personal measurements have shown that distance to mobile phone base station is a poor exposure predictor (Radon et al. 2006; Bornkessel et al. 2007; Berg-Beckhoff et al. 2009; Frei et al. 2010), whereas somewhat better performance was found for broadcast transmitters (Merzenich et al. 2008).

Spot Measurements in Bedroom (Stationary Devices)

Spot measurements are point-in-time measures, and they do not consider exposure at other places a person spends time nor the temporal variation of the field. Depending on the type of measurement device, they enable the separate capture of different exposure sources, including exposures from indoor sources. Spot measurements that allow capturing the most relevant exposure sources in the everyday environment have been shown to be a reasonable proxy for personal exposure when compared to 24-hr personal measurements (Frei et al. 2010). In particular, they are most accurate for estimating residential or bedroom exposure and might be used in the context of studying acute effects of RF-EMF exposure on sleep outcomes. Several studies to date have been conducted using spot measurements in the bedrooms of study participants (Hutter et al. 2006; Preece et al. 2007; Berg-Beckhoff et al. 2009; Tomitsch et al. 2010). One downside of spot measurements is the potential for selection bias, because this approach requires active participation (see Chapter 5). Not only might this influence recruitment but also participants may, for example, manipulate measurements by turning on or rearrange around indoor devices. When using spot measurements in a study dealing with subjectively reported outcomes, information bias might also play a role because it is rarely possible to conceal the purpose of a study from the participants. This is expected to be, especially, the case in studies dealing with idiopathic environmental intolerance attributed to EMF (IEI-EMF) (see Chapter 16).

Personal Measurements with Exposimeters

Exposimeters were introduced in the early 2000s and are portable devices for the measurement of individual exposure. There are several types of exposimeters on the market, with the devices from SATIMO (Courtaboeuf, France; http://www.satimo.fr/) being the most widely used in epidemiological research. Unlike stationary devices, exposimeters can record large amounts of personal exposure measurements at fixed locations, for example, at home, but they also are small enough to be carried by the participants, thus measuring exposure during their daily life activities. Therefore, they can be used to investigate the spatial and temporal variability of RF-EMFs. The use of exposimeters is widely recommended to characterize the exposure distribution in a defined population (Neubauer et al. 2007; Ahlbom et al. 2008; Röösli et al. 2010), and personal exposure measurement studies have been conducted in various countries. In addition to the cost of measurement devices, exposimeter measurement studies require a large organizational effort and are, therefore, very expensive to run. In most studies, the participants not only have to carry around the exposimeter but also fill in an activity diary that is a demanding and time-consuming task. Keeping a diary would likely deter some individuals from participating and therefore may introduce selection bias. Measurements can also be manipulated if a study participant places the exposimeter where high RF-EMF exposures are expected, thus yielding nonrepresentative results. For these reasons, exposimeter measurements are not often feasible in large-scale epidemiological studies or for assessing long-term exposures.

The exposimeters have to be relatively small to be conveniently carried by study participants. The antenna size determines the minimum dimensions of an RF meter, and measurement accuracy is a challenge. Measurement accuracy of the EME Spy 120 exposimeter from SATIMO was thoroughly tested in an anechoic chamber, that is, a room shielded from external RF-EMF by radiation-absorbent material (Lauer et al. 2012). Isotropy (whether the measurements depend on the orientation of the incoming wave), multiple signal detection in the same frequency band, and cross-talk between adjacent frequency bands are common issues for all personal meters. To solve the problem of cross-talk in the RF-EMF frequency range, where some bands are directly adjacent to each other (e.g., DECT and GSM 1800) is a particular challenge from a physical point of view for a small antenna device. Thus, a combination of signal analysis and physical filtering, as implemented in the EME Spy 120, was promising for the past. However, it is unclear whether this approach is still suitable for future technologies, when the same frequency bands are used by different types of technology (e.g., UMTS, LTE). For example, tests by Lauer et al. (2012) revealed substantial measurement uncertainties depending on the slot configurations of the GSM bands that affect the accuracy of the signal analysis for elimination of cross-talk.

A further challenge is the detection limit for band selective measurements because typical RF-EMF values in the environment are small. Measurements below the detection limit have to be adequately considered in the data analysis. Replacing the values below the detection limit to the value of the detection limit considerably overestimates exposure contributions from minor RF-EMF sources and would lead to an underestimation of the exposure range in the population. Other simple approaches are often not effective and produce biased summary statistics (Helsel 2006). It was found that the robust regression on order statistics (ROS) method produces more reliable summary statistics for such data (Röösli et al. 2008). Fortunately, the detection limit has been considerably lowered in newer exposimeter devices, and the problem may be of minor relevance in the future.

Exposimeter readings can be influenced by the body of the person wearing the measurement device because the human body interacts with RF-EMF (Radon et al. 2006; Blas et al. 2007; Knafl et al. 2008). This leads to an underestimation of the actual field strength; field strength depends on the body mass index (BMI) of a person. The higher the BMI, the higher the underestimation of an exposimeter reading (Neubauer et al. 2010).

Finally, although exposimeters measure exposure from near-field sources, measurements taken during personal mobile and cordless phone use are not expected to represent exposure of an individual using the phone (Inyang et al. 2008). The reason for this is that measurements during personal phone calls significantly depend on the distance between the emitting device and the exposimeter. Thus, it is impossible to differentiate between exposure from other people's mobile phones and the personal mobile phone when being used for calling or data transmission or from organizational communication.

In principle, two types of personal exposure data acquisition may be used both with merits and limitations. Some studies have (randomly) selected volunteers who carry the meter and fill in an activity diary that is referred to as a population survey (Röösli et al. 2010).

In this case, the unit of observation is the person. Exposure levels in various microenvironments, such as residential areas, schools, or trains, may be determined based on the recorded diary information. The advantage of a population survey is the direct estimation of the exposure distribution in the population if participants are randomly selected. Other studies collected data in various microenvironments by means of hired people or the researchers themselves. In this case, the compliance with the study protocol is expected to be better. In particular, recording of the microenvironments in the diary may be more accurate and reliable than in a random population survey. Furthermore, the personal mobile phone can be turned off to measure environmental RF-EMF only. This is not possible in a survey with volunteers.

Geospatial Propagation Models

Exposure from fixed site transmitters (e.g., broadcasting, mobile phone base stations) can be modeled using geospatial propagation models. Such models have been developed for the use in epidemiological studies in Germany (Neitzke et al. 2007), Switzerland (Bürgi et al. 2010), South Korea (Ha et al. 2007), and the United Kingdom (Briggs et al. 2012). Usually, such models are validated against spot measurements. In Switzerland, the values obtained by the geospatial propagation model at the residencies of study participants in the urban and suburban area of Basel, Switzerland, were found to be moderately correlated with personal measurements over 1 week (Frei et al. 2010). Restricting personal measurements to records only taken when at home yielded a substantially higher Spearman rank correlation in the range of 0.5 to 0.6. It was shown that the quality of input parameters of the model, typically including input such as topography, building geography, and specific data for the fixed site transmitters (e.g., radiation pattern, frequency, maximum power output), are crucial to the performance of the model.

The advantage of such a model is that it can be applied to large study populations, because exposure can be assigned on the basis of residential location, and as such does not introduce information and selection bias. A geospatial propagation model also has the advantage of enabling assignment of long-term exposures because the input data can be continuously updated. These models, however, do not provide information on lifestyle-related factors such as indoor sources at home or other behavioral aspects that are relevant to the total exposure. Thus, modeling is convenient to assess RF-EMF exposure, such as to base stations, that is not lifestyle related. It may be argued that assessing exposure that is related to lifestyle is problematic for health research, in particular with respect to quality-of-life outcomes, because it may be the consequence and not the cause of health-related quality-of-life status (reverse causality).

Exposure Prediction Models

In Switzerland, a prediction model for RF-EMF exposure to all relevant sources was developed by combining data from a geospatial model for residential exposure with behavioral characteristics collected using questionnaires data (Frei et al. 2009a). It could be shown that this approach is feasible. In this particular model, it was found that the modeled RF-EMF at the participants' homes derived from the geospatial propagation model was a relevant exposure predictor. In addition, housing characteristics, that is, the type of house wall (concrete vs. wood/brick) and lifestyle characteristics, such as mobile phone and W-LAN ownership or time spent in public transport, turned out to be important exposure predictors. The modeled values showed acceptable correlations with the personal measurements over a week (Frei et al. 2010), and are expected to be of use in improving exposure assessment in large collective with moderate effort, assuming a robust propagation model exists. A disadvantage may be potential information bias and selection bias, although this can be expected to be of less concern compared with personal measurements.

Near-Field RF-EMF Sources

Most studies rely on self-reported mobile phone use as a proxy for exposure to mobile phone radiation. Some studies also use records from the network operators because they are considered to be more objective than self-reported use. This assumption has been tested in many studies comparing self-reported with the operator-recorded mobile phone use among adults and children. Studies among adults show, rather consistently, that recall of past mobile phone use is afflicted with large random errors, even if the recall period is as short as 6 months (Parslow et al. 2003; Vrijheid et al. 2006, 2009a; Inyang et al. 2009; Aydin et al. 2011a, 2011b). Systematic recall errors were found to be moderate. Almost all comparison studies found that participants underestimated their number of calls and overestimated the duration of calls. Only few studies addressed differential exposure misclassification or recall bias, that is, whether agreement between self-reported mobile phone use and operator-recorded phone use differs according to health status (see Chapter 4). In the framework of the brain tumor study INTERPHONE, little differential exposure misclassification between cases and controls was found, on average, in a subsample of the participants from Australia, Canada, and Italy (Vrijheid et al. 2009a). However, for the highest category of cumulative number of calls, overestimation of exposure was more pronounced in cases than in controls. Furthermore, the ratio of self-reported phone use divided by recorded phone use increased with increasing time before the interview, in cases but not in controls. Cases also had a higher variance of these ratios than controls. In a brain tumor study of children and adolescents, little indication for recall bias was found (Aydin et al. 2011b).

Despite the many advantages, operator-recorded mobile phone traffic data also have some disadvantages, in particular if collected retrospectively. More effort is needed to obtain data and study participants have to remember their previous phone numbers to facilitate linkage of the operator-recorded data to the participant. Furthermore, subscriptions may be made for someone else (e.g., children or partners); and for occupational mobile phones, legal issues may complicate acquisition of operator data. It is also impossible to verify whether individual calls were made or taken by the subscription holder or someone else, given that a mobile can easily be lent to others. Operator data also do not help researchers determine how a mobile phone has been used, for example, if hands-free devices were used or on which side of the head the phone is typically held during calls, although recently apps have been developed to tackle this problem. This helps to minimize recall bias for reporting of side of use as has been observed previously (Schüz 2009). Reliance on operator data may also be limited by the fact that network operators are often obliged by law to delete data after a certain time period, thereby precluding acquisition of long-term exposure data.

A very rough exposure proxy sometimes used in epidemiological research is the subscriber status (e.g., Frei et al. 2011). The start date of a subscription is most useful for differentiating between long- and short-term users, but it does not provide information on the exact amount of mobile phone use. A comparison between self-reported mobile phone use and subscriber status found only a fair agreement, with a kappa value of 0.30 (95% confidence interval [CI], 0.23–0.36), with a low sensitivity (30%) and a high specificity (94%) (Schüz and Johansen 2007). The implication of such a situation on the study results is discussed in Chapter 4.

Exposure Distribution in the Population

Far-Field RF-EMF Sources

Far-field RF-EMF exposure at outdoor sites is generally far below the ICNIRP reference levels, although there may be areas close to antennae where levels can be higher and even exceed compliance levels. Low average exposure levels were found in studies where stationary measurements of one or several frequency bands in the RF range were performed (Hutter et al. 2006; Keow and Radiman 2006; Alanko and Hietanen 2007; Bornkessel et al. 2007; Neitzke et al. 2007; Schmid et al. 2007a, 2007b; Breckenkamp et al. 2008; Berg-Beckhoff et al. 2009; Joseph et al. 2009; Tomitsch et al. 2010) or where exposure from fixed site transmitters (Bürgi et al. 2010), from mobile phone base stations (Neitzke et al. 2007), or broadcast transmitters (Ha et al. 2007) was modeled. A review compiling data from mobile phone base station measurements in 23 countries across five continents from 2000 onward reported an average value of approximately 0.5 V/m (Rowley and Joyner 2012), but it is not clear how representative the measurement sites were for the population exposure. Personal measurement studies using exposimeters collect data at the place where people usually spend time. Such studies, to date, have been conducted in Germany (Thomas et al. 2008a, 2008b, 2010; Kühnlein et al. 2009; Heinrich et al. 2010), France (Viel et al. 2009a, 2009b) Switzerland (Frei et al. 2009b), the Netherlands (Bolte and Eikelboom 2012), Belgium (Joseph et al. 2008), Hungary (Thuróczy et al. 2008), and Slovenia (Valic et al. 2009). In the random population surveys with adults from France, the Netherlands, and Switzerland, mean exposure levels over 24 hr or 1 week from all measured sources were between 0.20 and 0.26 V/m.

It is generally very difficult to directly compare exposure levels across studies due to different recruitment strategies and analysis methods. In a joint effort, researchers from five European countries (Belgium, Hungary, Slovenia, Switzerland, and the Netherlands) where exposimeters of the same type had been used have reanalyzed their data using defined microenvironments (outdoors, offices, urban homes, trains, and car/bus) and the exact same data analysis method (Joseph et al. 2010, 2012). It was found that in these specific microenvironments, levels of exposure and absorption were low, and in the same order of magnitude, across the considered countries (Figure 8.2). Exposure to mobile telecommunication was, in general, important and the most dominant source in all microenvironments. Exposure levels were highest in public transport where a lot of people are using mobile phones and reflections from the vehicle shell occur. Because people usually spend most of their time of the day at home, this home exposure is particularly relevant in terms of cumulative exposure (Frei et al. 2009b).

Near-Field RF-EMF Sources

There has been a steep rise in popularity of mobile phones since their introduction (Figure 8.3). So far, assessment of exposures from mobile phones focused on worst-case scenarios for determining maximum SAR values, although new (Cardis et al. 2011; Lauer et al. 2013) approaches have also been developed to estimate dose. As mentioned, duration of mobile phone use or time since first subscription is the most common exposure measure in epidemiological research. In the population-based INTERPHONE study, median duration of self-reported life-time mobile phone use among regular user was 115 hr (INTERPHONE Study Group 2010). Ten percent used a mobile phone for >1640 hr, and these figures are expected to continue to rise in future.

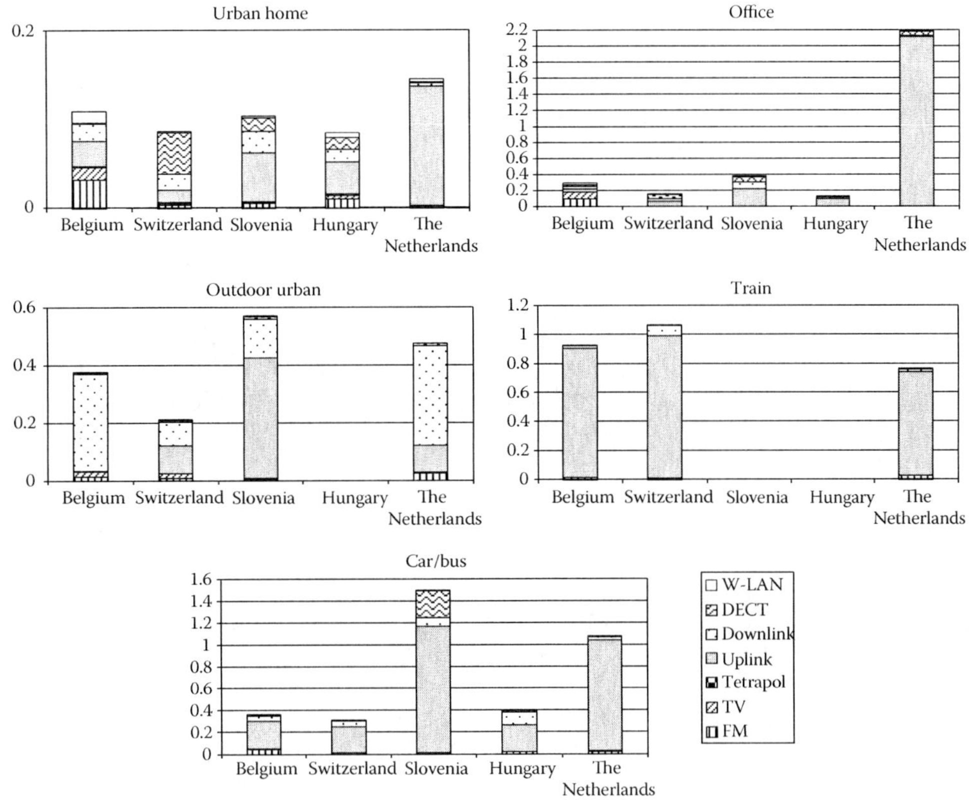

FIGURE 8.2
Mean total exposures (mW/m^2) in different microenvironments and contributions of the different sources for five European countries. (Adapted from Joseph et al. 2010.)

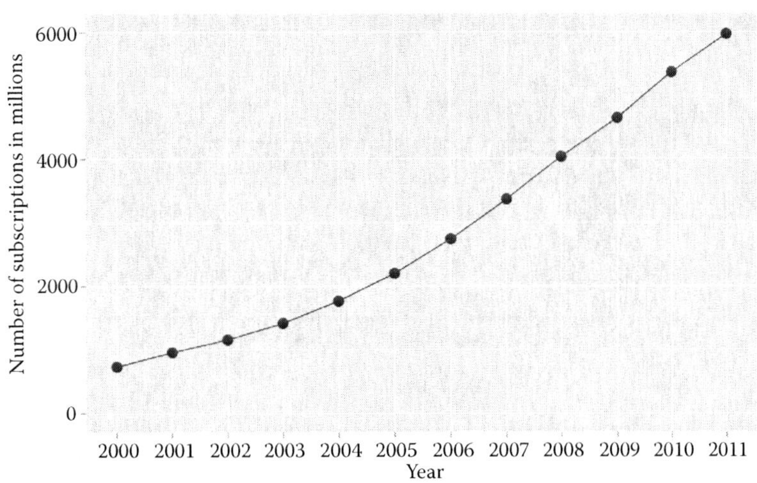

FIGURE 8.3
Number of mobile phone subscriptions worldwide. (Adapted from ITU World Telecommunication/ICT Indicators database, http://www.itu.int/ITU-D/ict/statistics/.)

In addition, consumer behavior is subject to rapid change. The functionality of modern smartphones and other wireless devices goes far beyond mere calling and texting, and exposure from data transmission will become more and more relevant. Thus, more sophisticated methods are needed for future assessment of exposure from near-field sources that take into account all aspects of exposure, including data transmission, organizational communication, and type of network. Promising avenues are apps installed on smart phones recording this type of information directly and ideally directly transmitting to a study server.

A European study using software-modified phones demonstrated that the power regulation of GSM phones is less effective than theoretically achievable in reducing the output power. The average output power of approximately 60,000 phone calls made by 500 volunteers in 12 countries was approximately 50% of the maximum (Vrijheid et al. 2009b). The values varied between operators and countries but were little related to use behavior such as duration of calls, making the calls indoors, or when moving. UMTS networks, the third generation, use an improved power control technology, and the average output power is typically <1% of the maximum (Gati et al. 2009; Kelsh et al. 2011; Persson et al. 2011). Thus, the average output power of UMTS phones is 100–500 times lower than that of typical GSM phones during average use, and the observed correlation between mobile phone use and output power (Erdreich et al. 2007) is expected to be substantially modified by the type of network.

DECT cordless phones have no power regulation and emit constantly at maximum level of 10 mW. Nevertheless, in most situations, the exposure from cordless phones is lower compared with GSM mobile phones. However, UMTS mobile phones are expected to emit less compared with cordless phones, unless the connection quality is very bad. For the exposure assessment, this means that use of cordless phones has become more relevant since the introduction of UMTS phones and continues to become even more important, given that many people are switching from GSM to UMTS phones (Lauer et al. 2013).

Combination of Near- and Far-Field RF-EMF Exposure

Few attempts to date have been undertaken to carry out an integrated whole-body exposure assessment for all types of RF-EMF exposure in the everyday environment, including both near- and far-field sources. In a study by Lauer et al. (2013), realistic data on far- and near-field exposures from Switzerland were used to assess the relative importance of these two types of exposure sources for the total absorbed radiation of the whole body. For the year 2007 in the Swiss population, it was estimated that if the GSM 900 standard is used for mobile phone calls, 8.1 min/week would be required to obtain the same whole-body exposure as from the far-field sources. If the GSM 1800 or DECT cordless phone standard is used, the calling time can be increased to 13.1 and 19.8 min/week, respectively. If, however, a UMTS standard is used, the call time can be increased to 20.9 hr/week. This demonstrates that exposure from environmental far-field sources has, in relative terms, become a more relevant exposure contribution because the output power of mobile phones has been decreased since their introduction. Whether this trend will continue in the future is not clear. Voice over Internet Protocol (VoIP) calls on the Internet is not subject to power regulation and is conducted with constant power output. At the moment, it is also unclear how effective the power control of the fourth generation LTE phones, currently mainly used for data transmission, will be.

The exposure contribution from location updates and push functions implemented in applications on smart phones has also neither yet been fully considered in exposure assessment. A recent study, however, suggests that this exposure contribution has become more relevant in smart phones compared with previous phones (Urbinello and Röösli 2013).

In summary, new approaches other than call duration are needed to quantify RF-EMF exposure from devices close to the body. More dosimetric studies should be done to help quantify the contribution of various consumer behaviors as well as location updates to the total RF-EMF exposure to clarify which exposure surrogate are most appropriate for epidemiological RF-EMF research.

References

Ahlbom, A., Bridges, J., de Seze, R., Hillert, L., Juutilainen, J., Mattsson, M. O., et al. 2008. Possible effects of electromagnetic fields (EMF) on human health—opinion of the scientific committee on emerging and newly identified health risks (SCENIHR). *Toxicology* 246(2–3): 248–50.

Alanko, T. and Hietanen, M. 2007. Occupational exposure to radiofrequency fields in antenna towers. *Radiat Prot Dosimetry* 123(4): 537–9.

Aydin, D., Feychting, M., Schüz, J., Andersen, T. V., Poulsen, A. H., Prochazka, M., et al. 2011a. Impact of random and systematic recall errors and selection bias in case–control studies on mobile phone use and brain tumors in adolescents (CEFALO study). *Bioelectromagnetics* 32(5): 396–407.

Aydin, D., Feychting, M., Schüz, J., Klaeboe, L. and Röösli, M. 2011b. Predictors of mobile phone use and overestimation of recalled mobile phone use among children and adolescents in a case-control study (CEFALO-study). Paper read at European Bioelectromagnetics Association, Rome, 22–24 February, 2011.

Berg-Beckhoff, G., Blettner, M., Kowall, B., Breckenkamp, J., Schlehofer, B., Schmiedel, S., et al. 2009. Mobile phone base stations and adverse health effects: phase 2 of a cross-sectional study with measured radio frequency electromagnetic fields. *Occup Environ Med* 66(2): 124–30.

Blas, J., Lago, F. A., Fernandez, P., Lorenzo, R. M. and Abril, E. J. 2007. Potential exposure assessment errors associated with body-worn RF dosimeters. *Bioelectromagnetics* 28(7): 573–6.

Bolte, J. F. and Eikelboom, T. 2012. Personal radiofrequency electromagnetic field measurements in the Netherlands: exposure level and variability for everyday activities, times of day and types of area. *Environ Int* 48: 133–42.

Bornkessel, C., Schubert, M., Wuschek, M. and Schmidt, P. 2007. Determination of the general public exposure around GSM and UMTS base stations. *Radiat Prot Dosimetry* 124(1): 40–7.

Breckenkamp, J., Neitzke, H. P., Bornkessel, C. and Berg-Beckhoff, G. 2008. Applicability of an exposure model for the determination of emissions from mobile phone base stations. *Radiat Prot Dosimetry* 131: 474–81.

Briggs, D., Beale, L., Bennett, J., Toledano, M. B. and de Hoogh, K. 2012. A geographical model of radio-frequency power density around mobile phone masts. *Sci Total Environ* 426: 233–43.

Bürgi, A., Frei, P., Theis, G., Mohler, E., Braun-Fahrländer, C., Fröhlich, J., et al. 2010. A model for radiofrequency electromagnetic field predictions at outdoor and indoor locations in the context of epidemiological research. *Bioelectromagnetics* 31(3): 226–36.

Cardis, E., Varsier, N., Bowman, J. D., Deltour, I., Figuerola, J., Mann, S., et al. 2011. Estimation of RF energy absorbed in the brain from mobile phones in the Interphone study. *Occup Environ Med* 68(9): 686–93.

Erdreich, L. S., Van Kerkhove, M. D., Scrafford, C. G., Barraj, L., McNeely, M., Shum, M., et al. 2007. Factors that influence the radiofrequency power output of GSM mobile phones. *Radiat Res* 168(2): 253–61.

Frei, P., Mohler, E., Bürgi, A., Fröhlich, J., Neubauer, G., Braun-Fahrländer, C., et al. 2009a. A prediction model for personal radio frequency electromagnetic field exposure. *Sci Total Environ* 408: 102–8.

Frei, P., Mohler, E., Bürgi, A., Fröhlich, J., Neubauer, G., Braun-Fahrländer, C., et al. 2010. Classification of personal exposure to radio frequency electromagnetic fields (RF-EMF) for epidemiological research: evaluation of different exposure assessment methods. *Environ Int* 36(7): 714–20.

Frei, P., Mohler, E., Neubauer, G., Theis, G., Bürgi, A., Fröhlich, J., et al. 2009b. Temporal and spatial variability of personal exposure to radio frequency electromagnetic fields. *Environ Res* 109(6): 779–85.

Frei, P., Poulsen, A. H., Johansen, C., Olsen, J. H., Steding-Jessen, M. and Schüz, J. 2011. Use of mobile phones and risk of brain tumours: update of Danish cohort study. *BMJ* 343: d6387.

Gati, A., Hadjem, A., Wong, M.-F. and Wiart, J. 2009. Exposure induced by WCDMA mobiles phones in operating networks. *IEEE Trans Wireless Comm* 8(12): 5723–7.

Grandolfo, M. 2009. Worldwide standards on exposure to electromagnetic fields: an overview. *Environmentalist* 29(2): 109–17.

Ha, M., Im, H., Lee, M., Kim, H. J., Kim, B. C., Gimm, Y. M., et al. 2007. Radio-frequency radiation exposure from AM radio transmitters and childhood leukemia and brain cancer. *Am J Epidemiol* 166(3): 270–9.

Heid, I. M., Kuchenhoff, H., Miles, J., Kreienbrock, L. and Wichmann, H. E. 2004. Two dimensions of measurement error: classical and Berkson error in residential radon exposure assessment. *J Expo Anal Environ Epidemiol* 14(5): 365–77.

Heinrich, S., Thomas, S., Heumann, C., von Kries, R. and Radon, K. 2010. Association between exposure to radiofrequency electromagnetic fields assessed by dosimetry and acute symptoms in children and adolescents: a population based cross-sectional study. *Environ Health* 9: 75.

Helsel, D. R. 2006. Fabricating data: how substituting values for nondetects can ruin results, and what can be done about it. *Chemosphere* 65(11): 2434–9.

Hutter, H. P., Moshammer, H., Wallner, P. and Kundi, M. 2006. Subjective symptoms, sleeping problems, and cognitive performance in subjects living near mobile phone base stations. *Occup Environ Med* 63(5): 307–13.

ICNIRP. 1998. Guidelines for limiting exposure to time-varying electric, magnetic, and electromagnetic fields (up to 300 GHz). International Commission on Non-Ionizing Radiation Protection. *Health Phys* 74(4): 494–522.

ICNIRP. 2009. Exposure to high frequency electromagnetic fields, biological effects and health consequences (100 kHz-300 GHz). International Commission on Non-Ionizing Radiation Protection. Available from: http://www.icnirp.org/documents/RFReview.pdf (accessed 15 July 2009).

INTERPHONE Study Group. 2010. Brain tumour risk in relation to mobile telephone use: results of the INTERPHONE international case-control study. *Int J Epidemiol* 39(3): 675–94.

Inyang, I., Benke, G., McKenzie, R. and Abramson, M. 2008. Comparison of measuring instruments for radiofrequency radiation from mobile telephones in epidemiological studies: implications for exposure assessment. *J Expo Sci Environ Epidemiol* 18(2): 134–41.

Inyang, I., Benke, G., Morrissey, J., McKenzie, R. and Abramson, M. 2009. How well do adolescents recall use of mobile telephones? Results of a validation study. *BMC Med Res Methodol* 9: 36.

Joseph, W., Vermeeren, G., Verloock, L., Heredia, M. M. and Martens, L. 2008. Characterization of personal RF electromagnetic field exposure and actual absorption for the general public. *Health Phys* 95(3): 317–30.

Joseph, W., Verloock, L., Tanghe, E. and Martens, L. 2009. In-situ measurement procedures for temporal RF electromagnetic field exposure of the general public. *Health Phys* 96(5): 529–42.

Joseph, W., Frei, P., Röösli, M., Thuróczy, G., Gajsek, P., Trcek, T., et al. 2010. Comparison of personal radio frequency electromagnetic field exposure in different urban areas across Europe. *Environmental Res* 110(7): 658–63.

Joseph, W., Frei, P., Röösli, M., Vermeeren, G., Bolte, J., Thuróczy, G., et al. 2012. Between-country comparison of whole-body SAR from personal exposure data in urban areas. *Bioelectromagnetics* 33(8): 682–94.

Kelsh, M. A., Shum, M., Sheppard, A. R., McNeely, M., Kuster, N., Lau, E., et al. 2011. Measured radiofrequency exposure during various mobile-phone use scenarios. *J Expo Sci Environ Epidemiol* 21(4): 343–54.

Keow, M. A. and Radiman, S. 2006. Assessment of radiofrequency/microwave radiation emitted by the antennas of rooftop-mounted mobile phone base stations. *Radiat Prot Dosimetry* 121(2): 122–7.

Knafl, U., Lehmann, H. and Riederer, M. 2008. Electromagnetic field measurements using personal exposimeters. *Bioelectromagnetics* 29(2): 160–2.

Kühnlein, A., Heumann, C., Thomas, S., Heinrich, S. and Radon, K. 2009. Personal exposure to mobile communication networks and well-being in children—a statistical analysis based on a functional approach. *Bioelectromagnetics* 30(4): 261–9.

Lauer, O., Frei, P., Gosselin, M. C., Joseph, W., Röösli, M. and Fröhlich, J. 2013. Combining near- and far-field exposure for an organ-specific and whole-body RF-EMF proxy for epidemiological research: a reference case. *Bioelectromagnetics* 34(5): 366–74.

Lauer, O., Neubauer, G., Röösli, M., Riederer, M., Frei, P., Mohler, E., et al. 2012. Measurement setup and protocol for characterizing and testing radio frequency personal exposure meters. *Bioelectromagnetics* 33(1): 75–85.

Lönn, S., Forssen, U., Vecchia, P., Ahlbom, A. and Feychting, M. 2004. Output power levels from mobile phones in different geographical areas; implications for exposure assessment. *Occup Environ Med* 61(9): 769–72.

Merzenich, H., Schmiedel, S., Bennack, S., Bruggemeyer, H., Philipp, J., Blettner, M., et al. 2008. Childhood leukemia in relation to radio frequency electromagnetic fields in the vicinity of TV and radio broadcast transmitters. *Am J Epidemiol* 168(10): 1169–78.

Navarro, E. A., Segura, J., Portoles, M. and Gomez-Perretta de Mateo, C. 2003. The microwave syndrome: a preliminary study in Spain. *Electromagn Biol Med* 22(2): 161.

Neitzke, H. P., Osterhoff, J., Peklo, K. and Voigt, H. 2007. Determination of exposure due to mobile phone base stations in an epidemiological study. *Radiat Prot Dosimetry* 124(1): 35–9.

Neubauer, G., Feychting, M., Hamnerius, Y., Kheifets, L., Kuster, N., Ruiz, I., et al. 2007. Feasibility of future epidemiological studies on possible health effects of mobile phone base stations. *Bioelectromagnetics* 28(3): 224–30.

Neubauer, G., Cecil, S., Giczi, W., Petric, B., Preiner, P, Fröhlich, J., et al. 2010. The association between exposure determined by radiofrequency personal exposimeters and human exposure: a simulation study. *Bioelectromagnetics* 31(7): 535–45.

Parslow, R. C., Hepworth, S. J. and McKinney, P. A. 2003. Recall of past use of mobile phone handsets. *Radiat Prot Dosimetry* 106(3): 233–40.

Persson, T., Tornevik, C., Larsson, L. E. and Loven, J. 2012. Output power distributions of terminals in a 3G mobile communication network. *Bioelectromagnetics*. 33(4): 320–5.

Preece, A. W., Georgiou, A. G., Dunn, E. J. and Farrow, S. C. 2007. Health response of two communities to military antennae in Cyprus. *Occup Environ Med* 64(6): 402–8.

Radon, K., Spegel, H., Meyer, N., Klein, J., Brix, J., Wiedenhofer, A., et al. 2006. Personal dosimetry of exposure to mobile telephone base stations? An epidemiologic feasibility study comparing the Maschek dosimeter prototype and the Antennessa SP-090 system. *Bioelectromagnetics* 27(1): 77–81.

Röösli, M., Frei, P., Bolte, J., Neubauer, G., Cardis, E., Feychting, M., et al. 2010. Conduct of a personal radiofrequency electromagnetic field measurement study: proposed study protocol. *Environ Health* 9: 23.

Röösli, M., Frei, P., Mohler, E., Braun-Fahrländer, C., Bürgi, A., Fröhlich, J., et al. 2008. Statistical analysis of personal radiofrequency electromagnetic field measurements with nondetects. *Bioelectromagnetics* 29(6): 471–8.

Rowley, J. T. and Joyner, K. H. 2012. Comparative international analysis of radiofrequency exposure surveys of mobile communication radio base stations. *J Expo Sci Environ Epidemiol* 22(3): 304–15.

Santini, R., Santini, P., Danze, J. M., Le Ruz, P. and Seigne, M. 2003. Survey study of people living in the vicinity of cellular phone base stations. *Electromagn Biol Med* 22(1): 41.

SCENIHR. 2009. *Health Effects of Exposure to EMF. Scientific Committee on Emerging and Newly Identified* Health Risks. Brussels: European Commission.

Schmid, G., Lager, D., Preiner, P., Überbacher, R. and Cecil, S. 2007a. Exposure caused by wireless technologies used for short-range indoor communication in homes and offices. *Radiat Prot Dosimetry* 124(1): 58–62.

Schmid, G., Preiner, P., Lager, D., Überbacher, R. and Georg, R. 2007b. Exposure of the general public due to wireless LAN applications in public places. *Radiat Prot Dosimetry* 124(1): 48–52.

Schüz, J. 2009. Lost in laterality: interpreting "preferred side of the head during mobile phone use and risk of brain tumour" associations. *Scand J Public Health* 37(6): 664–7.

Schüz, J. and Johansen, C. 2007. A comparison of self-reported cellular telephone use with subscriber data: agreement between the two methods and implications for risk estimation. *Bioelectromagnetics* 28(2): 130–6.

Thomas, S., Heinrich, S., von Kries, R. and Radon, K. 2010. Exposure to radio-frequency electromagnetic fields and behavioural problems in Bavarian children and adolescents. *Eur J Epidemiol* 25(2): 135–41.

Thomas, S., Kühnlein, A., Heinrich, S., Praml, G., Nowak, D., von Kries, R., et al. 2008a. Personal exposure to mobile phone frequencies and well-being in adults: a cross-sectional study based on dosimetry. *Bioelectromagnetics* 29(6): 463–70.

Thomas, S., Kühnlein, A., Heinrich, S., Praml, G., von Kries, R. and Radon, K. 2008b. Exposure to mobile telecommunication networks assessed using personal dosimetry and well-being in children and adolescents: the German MobilEe-study. *Environ Health* 7(1): 54.

Thuróczy, G., Molnár, F., Jánossy, G., Nagy, N., Kubinyi, G., Bakos, J., et al. 2008. Personal RF exposimetry in urban area. *Ann Telecomm* 63(1): 87–96.

Tomitsch, J., Dechant, E. and Frank, W. 2010. Survey of electromagnetic field exposure in bedrooms of residences in lower Austria. *Bioelectromagnetics* 31(3): 200–8.

Urbinello, D. and Röösli, M. 2013. Impact of one's own mobile phone in stand-by mode on personal radiofrequency electromagnetic field exposure. *J Expo Sci Environ Epidemiol* 23: 545–8.

Valberg, P. A., van Deventer, T. E. and Repacholi, M. H. 2007. Workgroup report: base stations and wireless networks-radiofrequency (RF) exposures and health consequences. *Environ Health Perspect* 115(3): 416–24.

Valic, B., Trcek, T. and Gajsek, P. 2009. Personal exposure to high frequency electromagnetic fields in Slovenia. *Joint Meeting of the Bioelectromagnetics Society and the European BioElectromagnetics Association*, 14–19 June 2009, Davos, Switzerland, pp. 262–263. Available from: http://bioem2009.org/uploads/Abstract%20Collections.pdf (accessed 4 January 2010).

Viel, J. F., Cardis, E., Moissonnier, M., de Seze, R. and Hours, M. 2009a. Radiofrequency exposure in the French general population: band, time, location and activity variability. *Environ Int* 35(8): 1150–4.

Viel, J. F., Clerc, S., Barrera, C., Rymzhanova, R., Moissonnier, M., Hours, M., et al. 2009b. Residential exposure to radiofrequency fields from mobile-phone base stations, and broadcast transmitters: a population-based survey with personal meter. *Occup Environ Med* 66(8): 550–6.

Vrijheid, M., Armstrong, B. K., Bedard, D., Brown, J., Deltour, I., Iavarone, I., et al. 2009a. Recall bias in the assessment of exposure to mobile phones. *J Expo Sci Environ Epidemiol* 19(4): 369–81.

Vrijheid, M., Cardis, E., Armstrong, B. K., Auvinen, A., Berg, G., Blaasaas, K. G., et al. 2006. Validation of short term recall of mobile phone use for the Interphone study. *Occup Environ Med* 63(4): 237–43.

Vrijheid, M., Mann, S., Vecchia, P., Wiart, J., Taki, M., Ardoino, L., et al. 2009b. Determinants of mobile phone output power in a multinational study: implications for exposure assessment. *Occup Environ Med* 66(10): 664–71.

9

Childhood Leukemia and Extremely Low-Frequency Magnetic Fields: Critical Evaluation of Epidemiologic Evidence Using Hill's Framework

Leeka Kheifets and John Swanson

CONTENTS

Introduction

The first study linking extremely low-frequency magnetic fields (ELF-MFs) to childhood cancer was published in 1979 and has been followed by many other studies. The International Agency for Research on Cancer (IARC) and the World Health Organization (WHO) have classified MFs as "possibly carcinogenic to humans," or as a Group 2B carcinogen; this classification was mostly based on consistent epidemiological evidence of an association between exposure to these fields and childhood leukemia (CL), and on laboratory studies in animals and cells, that were not supportive of exposure to ELF-MF causing cancer. Because CL is the outcome for which the scientific evidence is strongest, it can be regarded as the critical effect in risk assessment and risk evaluation and therefore attracts particular attention.

Leukemias are the most common cancers to affect children, accounting for 25%–35% of all childhood malignancies. The major morphologic types are acute lymphoblastic leukemia (ALL) and acute myeloblastic leukemia (AML). The rate of leukemia for children ≤15 years old has been estimated to be 4.5 per 100,000 per year in the developed world and 2.7 per 100,000 per year in the developing world (Ferlay et al. 2012). In developed countries, the incidence of leukemia rises rapidly after birth, peaking at 3 years of age before declining and then rising steadily again throughout life. Thus, unlike many cancers, it has a peak incidence early in life and a short latency. These characteristics have resulted in many etiologic hypotheses, most notably those involving exposure to infections. Ionizing radiation, definitely when given at large doses and by extrapolation also at small doses, is one of the few established risk factors for leukemia (Committee to Assess Health Risks from Exposure to Low Levels of Ionizing Radiation 2006).

Exposures acting before birth and early in life have long been thought to be important determinants of leukemia, but it is unfortunate that the evidence regarding the majority of suggested exposures is limited and often contradictory. Although there have been numerous calls to stop pursuing ELF-MF and to focus on other factors, as early as the 1990s (Campion 1997), a clear new risk factor that should be pursued instead of ELF-MF is yet to be identified (Linet et al. 1999). This makes pursuit of understanding a potential role of ELF-MF that much more important; of course, after 30 years of research without resolution, progress will require new approaches.

In this chapter, we review the literature, discuss possible interpretations, and present new promising designs that might provide ways to reduce existing uncertainty.

Overview of Epidemiologic Literature

Since that first seminal publication in 1979, most attention has focused on a potential association between residential MF exposure and CL. More than 30 studies on CL and exposure to ELF-MF have since been published. The earliest studies of ELF-MF and CL used wire codes, a categoric metric developed by Wertheimer and Leeper (1979) where exposure is estimated by combining the likely current load carried by electrical power lines outside homes estimated from physical characteristics (e.g., the thickness of the wires), distance to the wires, and the spatial arrangements of the wires (e.g., wiring configurations) (Wertheimer and Leeper 1979). After the development of ELF-MF measurement instruments, studies used spot measurements, 24- to 48-hr measurements in the child's bedroom as well as shorter measurements in other areas inside and outside the home, personal exposure measurements, and calculated fields (Swanson 1999; Kheifets and Oksuzyan 2008). For children, it is reasonable to approximate total exposure by background field in the home (Forssen et al. 2002). Accordingly, this is what most epidemiologic studies of children have measured. Some have also investigated exposure in schools, but without affecting the results much. A few studies have attempted to measure actual personal exposure with meters, but issues of the child's behavior changing with age, and particularly after diagnosis and treatment, mean these are usually regarded as prone to bias. Accordingly, epidemiologic evidence for children almost entirely concerns background fields in homes.

Some have separately assessed exposure from appliances, because the highest short-term exposure comes from proximity to appliances when in use. For the most part, appliance use by children and women during pregnancy was not associated with increased risks of CL, with only sporadic increased odds ratios (ORs) for use of TVs, electric blankets, hair dryers, and video arcades (Kheifets et al. 2010d). However, it should be noted that estimation of exposure from appliances is not well developed. Little is known about the influence of the type and age of the appliance on potential exposure. Previous efforts that evaluated exposure from one appliance at a time are clearly unsatisfactory, with nondifferential exposure misclassification of particular concern (see Chapter 4). Much work needs to be done to develop methodology for combining exposure from many appliances used by an individual to make evaluation of exposure from appliances worthwhile.

For largely practical reasons, most epidemiologic studies of children have investigated the home occupied at diagnosis, with only a few investigating previous homes or the home at birth. Table 9.1 summarizes these studies.

Almost all individual studies of MFs and CL have found increased risks associated with the highest exposure levels; most of them, however, have involved a small number of highly exposed cases and thus lacked precision. This has given rise to a variety of interpretations. Often, individual studies are regarded as "negative," including by their own authors, because they find no *statistically significant* association, when in fact they still supply evidence for an association. It is unfortunately common to confuse lack of significance with lack of evidence or lack of positive association, but as well documented in textbooks (Royall 1997; Rothman and Greenland 1998) it is a fallacy nonetheless, for the significance reflects numbers in categories as much as strength of the association. Others focus on selected positive findings within some studies.

Combining results in a "pooled analysis" overcomes some of these difficulties and represents the most powerful approach to provide a cohesive assessment of the epidemiologic data of ELF-MF and CL. Pooled analysis, considered the gold standard for synthesizing results from multiple studies, allows for comparison across different metrics and studies, free of artifacts introduced by analytic differences, and for derivation of statistically more stable results (Kheifets et al. 2006b). Pooled analysis uses raw data from the component studies and thus can apply identical analyses to all included studies. The choices of, for example, cut points, reference groups, and metrics in a pooled analysis may differ from the choices made in the original studies and may result in changes in the study-specific effect estimates. Despite strengths, results from pooled analyses are prone to the same biases operating in the original studies.

In the pooled analysis by Greenland et al. (2000), of the 16 studies, 12 were included in pooling of measured or calculated fields; included were a total of 2656 cases and 7084 controls. For this analysis, the metric of choice was the time-weighted average. The estimated OR for CL was 1.68 (95% CI, 1.23–2.31) for exposures >0.3 μT compared with exposures <0.1 μT, controlling for age, sex, and study.

Using more stringent inclusion criteria, Ahlbom et al. (2000) included nine studies using measured and calculated fields. There were a total of 3203 cases and 10,338 controls in the pooled sample. Using the geometric mean as the metric of choice for studies with measured fields, the estimated OR for CL was 2.00 (95% CI, 1.27–3.13) for exposures greater than or equal to 0.4 μT compared with exposures <0.1 μT, controlling for age, sex, socioeconomic status (SES) (in measurement studies only), and East/West (in German study only).

More recently, Kheifets et al. (2010a) identified 14 studies published since the two previous pooled analyses from 2000, of which seven met their inclusion criteria. A total of 10,865 cases and 12,853 controls were included in the pooled analysis that focused on 24-hr MF measurements or calculated fields in residences. The OR for exposures >0.3 μT compared with <0.1 μT was 1.44 (95% CI, 0.88–2.36). Without the most influential study, from Brazil, which is suspected to be particularly prone to bias, the ORs increased and became similar to previous pooled analysis. All three analyses, while focusing on overlapping but distinct sets of studies, come to similar conclusions. A fourth pooled analysis was designed principally to test the hypothesis that the CL association was with, specifically, exposure at nighttime. The findings did not support the hypothesis (Schuz et al. 2007). Figure 9.1 summarizes findings of the pooled analyses.

To examine whether results change with adjustments for potential confounders and to what extent results are limited to a particular subgroup, Kheifets et al. conducted sensitivity

TABLE 9.1

Epidemiologic Studies on the Association between ELF-MF and CL

Study	Measurement	Lowest and Highest Exposure Categories (μT)	Cases/Controls	OR (95% CI) Highest Exposure Category Compared with Lowest	Included in the Pooled Analysis
Wertheimer and Leeper 1979	Wiring configuration	Low current	84/107 52/29	2.3– birth address	Greenland
		High current	92/126 63/29	3.0– death address	
Fulton et al. 1980	Wiring configuration	Very low + low Very high	74/134 10/26	0.54 (0.21–1.41)	Greenland
Tomenius 1986	Spot[a]	<0.3 ≥0.3	239/202 4/10	0.3–	Greenland
Savitz et al. 1988	Spot[a]	<0.065 ≥0.25 *and*	75/134 10/12	1.5 (0.6–3.6) low power	Greenland
		<0.2 ≥0.2	31/NR[b] 5/NR	1.9 (0.7–5.6) low power	
Myers et al. 1990	Calculated fields	≤0.1 ≥0.1	358/567 1/4	0.4 (0.04–4.3)	
London et al. 1991	24-hr bedroom	<0.067 ≥0.27	85/69 20/11	1.5 (0.7–3.3)	Greenland
Feychting and Ahlbom 1993	Calculated fields	≤0.09 ≥0.3	27/475 7/32	3.8 (1.4–9.3)	Ahlbom Greenland
Olsen et al. 1993	Calculated fields	<0.1 ≥0.4	829/1658 3/1	6.0 (0.8–44)	Ahlbom Greenland
Verkasalo et al. 1993	Calculated fields	<0.01 ≥0.2	NR 3/NR	1.6 (0.32–4.5)	Ahlbom Greenland
Coghill et al. 1996	24-hr in home	0.012 0.448	NR NR	NR	Greenland
		$<5\ Vm^{-1}$ $20\ Vm^{-1}$	17/30 13/5	4.69 (1.17–27.78)	
Linet et al. 1997	24-hr bedroom, weighted by spot measurements	<0.065 ≥0.2	206/215 58/44	1.5 (0.91–2.6)	Ahlbom Greenland
Tynes and Haldorsen 1997	Calculated fields	<0.05 ≥0.14	139/546 1/14	0.3 (0.0–2.1)	Ahlbom Greenland
Fajardo-Gutierrez et al. 1997	Wiring configurations	Low Very high	13/20 82/65	2.05 (0.95–4.43)	Greenland
Michaelis et al. 1998	24-hr bedroom	<0.2 ≥0.2	NR 9/8	2.3 (0.8–6.7)	Ahlbom Greenland
Dockerty et al. 1999	24-hr bedroom	<0.1 ≥0.2	31/33 5/1	15.5 (1.1–225)	Ahlbom Greenland
Green et al. 1999[c]	48-hr personal	<0.03 ≥0.14	14/33 29/33	4.5 (1.3–15.9)	Greenland

TABLE 9.1 ***(Continued)***

Epidemiologic Studies on the Association between ELF-MF and CL

Study	Measurement	Lowest and Highest Exposure Categories (μT)	Cases/Controls	OR (95% CI) Highest Exposure Category Compared with Lowest	Included in the Pooled Analysis
McBride et al. 1999	48-hr personal	<0.08 0.27–1.61 *and* <0.2 ≥0.2	149/147 32/37 239/287 54/52	0.7 (0.4–1.3) 1.1 (0.7–1.8)	Ahlbom Greenland
UKCCS Investigators 1999	Two-phase[d]	<0.1 ≥0.4	995/977 5/3	1.7 (0.4–7.0)	Ahlbom
Bianchi et al. 2000	Calculated fields	≤0.001 >0.1	92/401 3/3	4.5 (0.9–23.2)	Kheifets
Schuz et al. 2001	24-hr bedroom	<0.1 ≥0.4	456/1188 3/3	5.9 (0.8–44.1)	Kheifets
Perez et al. 2005	Spot measurements	0.1 1.0	NR	45.15–	
Kabuto et al. 2006	1-wk bedroom	<0.1 ≥0.4	276/542 6/5	2.6 (0.8–8.9)	Kheifets
Mejia-Arangure et al. 2007[e]	Spot measurements	≤1.0 ≥6.00	14/43 10/13	3.7 (1.05–13.1)	
Lowenthal et al. 2007	None – distance only (meters)	>300 m 0–50 m	19/9 760/790	2.06 (0.87–4.91) ever 1.07 (0.99–1.17) per year of residence	Kheifets
Feizi and Arabi 2007	Calculated fields	<0.45 ≥0.45	45/54 15/5	3.60 (1.11–12.39)	
Yang et al. 2008	Transformers or power lines within 50 m		123 (case-only study)	4.39 (1.42–13.54) for interaction of XRCC1 Ex9þ16A gene and exposure	
Abdul Rahman et al. 2008	None – distance only (meters)	≤200 m >200 m	52/31 76/97	2.30 (1.18–4.49)	
Kroll et al. 2010	Calculated fields Distance	<0.1 ≥0.4 ≥600m 0–49m	9645/9647 2/1 9378/9447 5/3	2.00 (0.18–22.04) 1.65 (0.39–6.89)	Kheifets
Malagoli et al. 2010	Calculated fields	<0.1 ≥0.4	27/129 1/2	RR[f]: 2.1 (0.2–26.2)	Kheifets
Does et al. 2011	30-min indoor	Quart. 1: 0–0.01 Quart. 4: ≥0.05 a priori: ≤0.10 a priori: >0.30	55/61 66/68 215/245 3/6	1.18 (0.71–1.96) 0.57 (0.14–2.36)	

Continued

TABLE 9.1 (*Continued*)

Epidemiologic Studies on the Association between ELF-MF and CL

Study	Measurement	Lowest and Highest Exposure Categories (μT)	Cases/Controls	OR (95% CI) Highest Exposure Category Compared with Lowest	Included in the Pooled Analysis
Wunsch-Filho et al. 2011	24-hr bedroom and	<0.1	113/394	1.09 (0.33–3.61) 24 hr	Kheifets
	night exposure	≥0.3	11/34	1.52 (0.46–5.01) night	

[a] Tomenius 1986: Maximal uniaxial value outside front door; Savitz et al. 1988: Arithmetic mean of low-power measurement in three or more locations (child's bedroom, parent's bedroom, other room occupied by child >1 hr/day, front door).
[b] NR, not reported.
[c] Included only in wire code analysis.
[d] 48-hr home measurement used if shorter measurement or other indication showed high ELF-MF.
[e] Only included children with Down's syndrome.
[f] RR, relative risk.

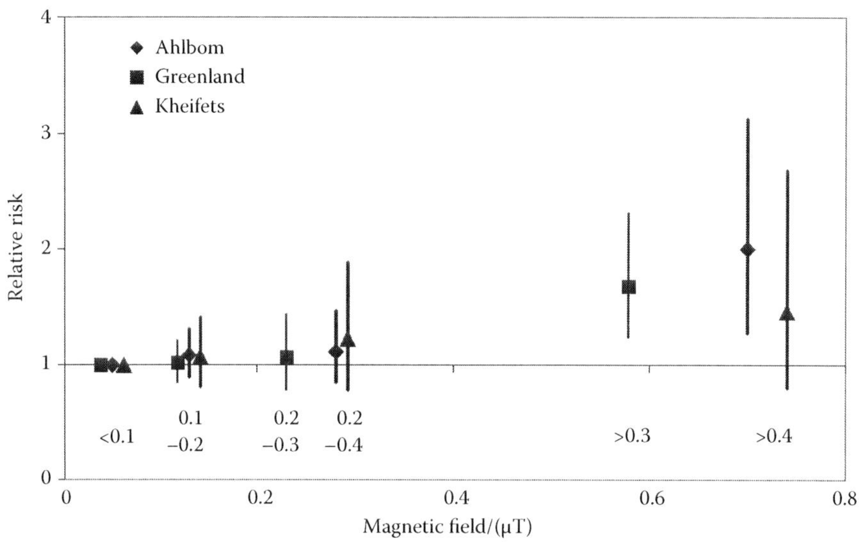

FIGURE 9.1
Results of the three pooled analyses.

and subgroup analyses on the studies in their pooled analysis. Not all potential confounders were available in all studies. Analyses adjusting for confounding were carried out on the subset of studies and subjects for which data on the confounder were available. Most adjustments did not make appreciable changes in the OR estimates. Risks were a little higher for ALL and for a younger age group and a little lower for addresses at birth, despite a suggestion from one study that exposure at birth might carry particular risks (Lowenthal et al. 2007). Both an adjustment for mobility and restriction to subjects who lived in a single residence before diagnosis did not change the risk estimates appreciably. All CIs included the null value.

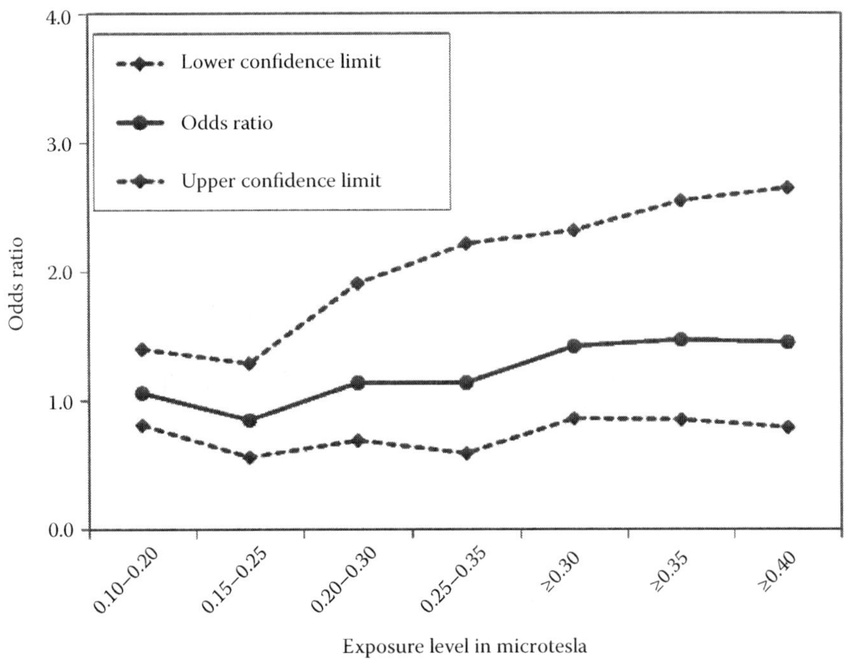

FIGURE 9.2
ORs (95% CI) for moving window of exposure levels, adjusted for age, sex, SES, and study. Reference level, <0.1 μT. (From Kheifets L et al., *Br J Cancer* 103, 1128–35, 2010a.)

The OR estimates using categorical cut points and involving relatively small numbers of subjects are vulnerable to unstable results. To address this concern, Kheifets et al. calculated ORs using a moving window of exposure levels (Figure 9.2).

These results, along with continuous analysis of Ahlbom et al. and floated case–control ratios based on a quadratic spline fit with a single knot (at 0.2 μT) in Greenland, suggest a possible trend of increasing risk with increasing exposure; however, the estimates are imprecise and wide CIs in all pooled analyses fit a variety of exposure–response relationships, including no increase in risk.

Bias Analysis

The conventional summary estimate assumes no bias is present. Monte-Carlo and Bayesian bias-modeling analyses were also performed using hierarchical prior distributions on parameters governing bias (Greenland 2003, 2005; Greenland and Kheifets 2006). In these analyses, three types of biases were considered: selection bias, confounding, and exposure misclassification. The distributions of parameters governing bias were allowed to vary with type of exposure assessment (measured vs. calculated), with prevalence, with voltage (high vs. low), and across studies. The analyses allowing for misclassification resulted in point estimates that were larger than the estimates from conventional analyses. This is because they assumed prior distributions for classification rates whose means corresponded to independent nondifferential misclassification. This misclassification tends to produce bias toward the null, resulting in upward adjustment away from the null. Nonetheless, the high degree of uncertainty about bias sources (misclassification, confounding, and selection) resulted in the bias analyses producing much wider interval estimates and larger p values, with lower 95% limits falling below the conventional lower limit. The net result is that the uncertainty about the size of the effect is much greater after bias allowances, although on the balance the association still appears to be positive (posterior probability of RR > 1 as high as 98% in some analyses). Even with some allowance for possible biases, it is hard to explain away the observed associations completely, although it should be noted that such allowances and the results they produce are highly sensitive to assumptions about bias sources.

Attributable Fraction

The observed associations between ELF-MF and CL are small and involve only the highest and most infrequent levels of exposure. Potential public health impacts of an effect, assuming there is a causal relationship, can be evaluated as a proportion of the case load of a disease that is attributable to exposure (see Chapter 18). Such calculations are based on an estimate of the effect of exposure on the disease incidence, the incidence rates, and an estimate of the exposure distribution within a given population.

Because both high exposure levels and CL are rare, worldwide, the number of cases possibly attributable to ELF-MF exposure ranges from 100 to 2400 (Greenland and Kheifets 2006; Kheifets et al. 2006a). There are considerable uncertainties in these estimates, particularly in the assumptions regarding exposure distribution. Most importantly, the assumption of causality, needed for these calculations is not presently satisfied. Nevertheless, estimates of the possible public health impact are needed to provide a potentially useful input into policy analysis under different scenarios. Given these caveats, attributable fraction calculations support the idea that the public health impact of residential fields, if any, is likely to be limited.

The preceding discussion has related to MFs. A smaller number of studies, seven, have looked at electric fields, with a variety of exposure assessment techniques, and almost uniformly null results (Kheifets et al. 2010d).

Critical Evaluation

The consistent association found between CL and average MF exposure above 0.3–0.4 μT could be due to chance, selection bias, misclassification, other factors that confound the association, or true causal relationship.

In 1965, Austin Bradford Hill detailed nine criteria to assist with assessing evidence of causation (Hill 1965). Importantly, Hill said, "None of my nine viewpoints can bring indisputable evidence for or against the cause-and-effect hypothesis and none can be required sine qua non." Consistent with Hill's intention, we critically evaluate epidemiologic evidence of CL and ELF-MF following a modified version of his criteria as a framework.

1. Strength of Association and Chance. The pooled analyses were based on large numbers including at high exposure levels, and hence resulted in RR estimates with tight CIs. When compared, they demonstrate consistency in the size of their effect estimates. It appears very unlikely that random variability (chance) alone could have produced these results (e.g., the conventional $p < .0001$).

   However, the magnitude of the association between CL and ELF-MF is not large and has possibly been weakened somewhat in a recent pooled analysis. Small risks are notoriously hard to evaluate, because small risks are more vulnerable to subtle confounding and biases that can go undetected. More generally, it should be borne in mind that no analytic approach can compensate for the fundamental limitation that all these studies, except one, are observational case–control designs (see Chapter 2) with retrospective exposure assessment (see Chapter 4) and potential for selection bias (see Chapter 5).

2. Consistency. Consistent epidemiologic evidence carries a good deal of weight when considering the potential health effects of ELF-MF. Consistency in results is key because epidemiologic studies of ELF-MF are difficult to design, conduct, and interpret for many methodologic reasons. The association is remarkably consistent across studies. It has been reported in many countries, albeit mainly countries with similar industrialized societies, because these are the countries with registries and health systems conducive to this type of research. It is also found in studies with two different designs: (1) studies that calculate the field by entering the home and that are thus subject to selection and participation bias but that capture all sources of residential exposure; and (2) studies that calculate the field in the home from high-voltage power lines outside the home, with no subject participation, and that are less (if at all) subject to bias, but the exposure assessment is more limited.
3. Specificity. Kheifets et al. (2010b) performed a pooled analysis, methodologically parallel to the pooled analyses for CL, on childhood brain tumors. The main finding was that compared with CL, there was little indication of an elevation in the highest exposure category and much less of a monotonic exposure–response relationship (with RRs < 1 in intermediate-exposure categories). Likewise, studies of various adult cancers and various noncancer endpoints have failed to identify associations as strong and consistent as for CL (World Health Organization 2007). This suggests the CL association is specific to just that disease.

   The orthodox view is that an association that is confined to CL alone, such as appears to be the case, increases the likelihood of it being causal. But it is at least arguable that, with a physical agent as simple as MFs, applied to every cell of the body equally, it would be more likely to be expected to cause multiple cancers or other diseases, if any at all. There is an analogy with ionizing radiation, another relatively simple physical agent, that causes multiple cancers. On this view, the failure to find associations with other cancers weakens the causal hypothesis for CL and makes it more likely that the CL results are perhaps confounding by a factor that is indeed specific to CL.
4. Temporal Relationship. Cause should precede effect. Leukemia, although the most common childhood malignancy, in absolute terms is quite rare, necessitating investigation by retrospective, case–control designs. Although temporality is likely not an issue for the studies of CL, the retrospective nature of exposure assessment is, as all of the difficulties with retrospective exposure assessment are likely to have led to substantial exposure misclassification that, in turn, is likely to interfere with detection of an association between exposure and disease.
5. Dose–Response (Biological Gradient). Interpretation of the pooled analyses has sometimes focused on the clearer evidence for an excess risk above a threshold of 0.3 or 0.4 μT than below these fields. However, a true threshold for risk at these fields, rather than a progressive increase with increasing exposure, would be biologically implausible. Almost certainly, measurement errors in both measured

and calculated fields are not only present in all studies but also vary considerably from study to study. Furthermore, it is likely more severe at low levels and thus may disguise or distort the true shape of any dose–response relationship. Target exposure, often described as the average exposure during the period before disease diagnosis, is not measured consistently among studies. Also, measured exposure probably does not reflect the biologically relevant exposure, which remains unknown.

It is generally assumed that misclassification in ELF-MF and leukemia studies is nondifferential; this means exposure misclassification does not differ by disease status. Nondifferential misclassification translates into a bias of the effect estimate toward the null in most situations, although misclassification in middle categories can lead to the distortion of the dose–response curve.

Nevertheless, the results of the latest pooling suggested a possible trend of increasing risk with increasing exposure (dose–response); however, the estimates were imprecise. Kheifets et al. (2011) tested various exposure–response relationships to see how well they fit the data. There are many approximations and assumptions involved in this approach, but their study suggests that a strict linear no-threshold response fits even more poorly than the threshold but that better fits are obtained by superlinear responses.

6. Plausibility (Biological Plausibility). There is no dispute that both electric and MFs interact with the various components of matter (e.g., atoms, molecules, electrons). However, merely having an interaction is not sufficient to produce an effect at the cell, organ, or organism level. A survey of nearly 20 biophysical mechanisms that have been proposed to explain effects at low levels reported in epidemiologic studies found that some of the mechanisms are impossible and that some require specific conditions for which there is limited or no evidence as to their existence in a way that would make them relevant to human exposure (Swanson and Kheifets 2006). Others are predicted to become plausible above some level of field. They conclude that at low fields, below say 5 μT, no mechanism has been identified despite extensive investigation. At fields of the order of 50 μT, although no mechanism is yet established in living systems, more than one possibility becomes theoretically plausible, and the fundamental problem of implausibility is removed (although a lesser problem will still remain until an actual mechanism that is operating at these levels and is relevant to the particular disease in question, for example, CL, is identified). Mechanisms cease to be an issue for fields ≥500 μT, for which there are known (or likely) effects and mechanisms (although health effects have not been demonstrated except at still higher fields).

   The absence of a plausible biophysical mechanism at lower fields cannot be taken as proof that health effects of environmental electric and MFs are impossible. As Hill stated, "What is biologically plausible depends upon the biological knowledge of the day." Nevertheless, we consider the implausibility argument as one of the two strongest arguments against the causality of MFs in the development of CL.

7. Coherence. A U.S. study reported poorer survival in children with ALL exposed to ELF-MF >0.3 mT but based on small numbers (Foliart et al. 2006). As a follow-up of this observation, a large pooled analysis of >3000 children diagnosed with ALL

in eight countries, including the original U.S. study, found no statistically significant associations between exposure to ELF-MF and event-free survival or overall survival of ALL. Reassuringly, these results provide no evidence that ELF-MF has a role in predicting outcome of childhood ALL, but they are not directly relevant to the development of CL (Schuz et al. 2012).

Toxicologic studies have failed to find carcinogenic effects of MFs. These include standard tests on mice and rats conducted according to established protocols of the National Toxicology Program (except with larger numbers) (Mandeville et al. 1997; Yasui et al. 1997; Boorman et al. 1999; McCormick et al. 1999). Negative findings in such studies would normally be regarded as fairly strong evidence against causation in humans. More generally, despite many reports of positive findings, no laboratory effect of MFs at environmental levels on any biological system whether *in vitro* or *in vivo* has proved able to be robustly replicable.

Negative results from those toxicologic studies included failures to observe increases in, specifically, adult leukemia and lymphoma. Furthermore, studies on mice genetically disposed to develop leukemia have also been negative (Bellossi 1991; Harris et al. 1998; McCormick et al. 1998; Sommer and Lerchl 2004). However, there is increasing understanding of how CL, and in particular ALL, in humans, is associated with specific genetic mutations. So, it is possible that the pathway through which MFs affect leukemia rates depends on one or a few specific genetic mutations, and because rodent models do not display these mutations, the negative results are irrelevant. Only recently more relevant CL models have been developed, but MFs are yet to be tested in these models.

8. Experiment (Reversibility). It is not realistic to remove exposures to ELF-MF from modern societies to conduct the experiment of testing whether CL rates decline. However, the historical introduction of ELF-MF into those societies can be regarded as an experiment in itself.

   Exposures to ELF-MF were effectively nonexistent before public electricity supplies were first introduced in the late nineteenth century and rose to their present values over the twentieth century. If MFs are a risk factor for CL, an increase in exposures would lead to an increase in leukemia incidence, although the increase might be undetectable if the effects were weak or uncommon and might be masked by other changes. Comparison of electrification or changes in electricity consumption (surrogates for exposure) to changes in CL rates, a type of ecologic correlation, have been used to argue both for (Milham and Ossiander 2001) and against (Jackson 1992) the association between MFs and CL. Although exposures and reported leukemia rates have both risen dramatically over the twentieth century, they have done so at different times, with the major increase in leukemia rates preceding exposure increase by 20–30 years. In addition, there are so many approximations and assumptions involved in connecting the two trends that we cannot regard the ecologic evidence as providing any meaningful evidence for or against a causal link (Kheifets et al. 2006c).

9. Consideration of Alternate Explanations (Analogy). *Bias.* Because practically all epidemiologic studies of ELF-MF and CL have been case–control studies, it has been proposed that control selection bias (see Chapter 5)—a common and potentially serious problem of all case–control studies—may be fully or, at least, partially responsible for the consistently described epidemiologic association

between ELF-MF and CL. In a case–control study, control selection bias occurs when the ratio of the selection probabilities of exposed and unexposed cases is different than the ratio of the selection probabilities of exposed and unexposed controls. The overall requirement of the controls is that they are representative of the source population of the cases.

Studies using measurements that thus require entry to the home generally have low participation rates and might have led to selection bias (Mezei and Kheifets 2006; Schuz and Ahlbom 2008). Studies estimating calculated fields do not require participation and thus are less vulnerable to selection bias, but they neglect sources of MFs other than high-voltage power lines and thus are likely to introduce exposure misclassification and loss of statistical power.

Exposure to MFs in homes is associated with SES: bigger, more expensive homes are set back further and are therefore further from the electricity cables along the street that are often the main source of exposure. Similarly, exposure is related to population density and is thought to be higher in apartment buildings compared with single-family homes. Studies that select controls from sources with a socioeconomic bias (random-digit telephone dialing or insurance lists) or that require subject participation (which almost always varies with SES) are therefore likely to introduce a bias. Most such studies have adjusted for a measure of SES without it affecting the result greatly if at all.

However, Mezei and Kheifets (2006) identified three instances where an ELF-MF study, potentially including bias, had been partially reanalyzed in a way that should have reduced bias. In all three cases, the RR reduced, but in no case was the risk eliminated (Gurney et al. 1995; Ebi et al. 2000; Hatch et al. 2000). One interpretation is that bias is therefore unlikely to explain the whole risk even in studies where it is present. An alternative is that if partial correction for bias reduces the RR, elimination of bias could eliminate the excess risk.

Studies that are based on comprehensive registers and registries and that do not require subject participation (because they assess exposure from information on sources, primarily high-voltage power lines, outside the home) should be free from this type of bias. Such studies have found similar risks to measurement-based studies (in the Ahlbom pooled analysis, RR of 2.13 compared with 1.87), arguing against bias as the explanation for the latter.

In addition, the absence of an equivalent association for brain tumors argues against bias as the main explanation for the observed CL association, because the studies have similar methodologies and bias should therefore operate similarly for both diseases. However, the small numbers, especially for brain tumors, mean there is considerable uncertainty in this comparison.

Confounding

Since the early days of ELF-MF research, investigators searched for possible other factors that would explain the observed association (see Chapter 6). The hypothesized confounders of the relation between ELF-MF and CL include SES, residential mobility, residence type, viral contacts, environmental tobacco smoke, dietary agents, and traffic density (World Health Organization 2007). None of these variables have been found to confound the association, although some have been identified as potential risk factors. For a factor to be a confounder it has to exert an effect considerably larger than the observed association and be strongly correlated with exposure. Owing to limited knowledge of the etiology of CL and an absence of strong risk factors, it is not surprising that substantial confounding has not been identified. The same observation, however, makes it difficult to exclude a possibility of some (yet to be identified) confounder or of the combination of several factors. The pooled analyses looked at the possible effect of several putative risk factors. However, for none of them did adjustment materially change risk estimates. Although it is impossible to discuss the effect of yet unsuspected risk factors, it seems to us that substantial confounding from factors that do not represent an aspect of the electric or MFs is less likely.

But exposure to MFs from high-voltage power lines is correlated with distance from those lines and therefore with any other factors that vary with this distance. This possibility was thrown into focus by a 2005 U.K. study (Draper et al. 2005) that found elevated rates of CL at distances out to at least 600 m, well beyond the distance explicable by the MF risk found in the pooled analyses (Kroll et al. 2010). If the whole finding is not chance, some other factor must operate in addition to or instead of MFs. Previous studies of power lines have analyzed distance but not generally to these larger distances.

This therefore prompts an avenue of research in identifying any other factors associated with distance from a power line that could produce elevated risks at distances beyond the range of the MFs. A 1999 suggestion was that the corona ions produced on the surface of overhead-line conductors when the electric field is high enough to ionize air are blown away; attach themselves to existing airborne pollutants, thereby charging them; and thus increase the probability of retention when inhaled (Fews et al. 1999). This has been questioned on physical grounds as the magnitude of the effect appears not to be large enough to be significant (Jeffers 2007; World Health Organization 2007). Epidemiologically, the key test would be whether elevated risks are found downwind, rather than upwind, of power lines, and a crude test of this did not find such an effect (Draper et al. 2005).

Other suggested physical factors associated with high-voltage power lines include the direct effect of corona ions or other chemicals produced by them; herbicides used to control vegetation growth under power lines; and chemicals derived from the supports for the lines (e.g., preservatives from wood pole lines, galvanizing from metal pylons). Few of these seem to have sufficient plausibility to merit further investigation. Alternatively, power lines may be co-located with other potential risk factors (e.g., motorways, railways). More promising still is the possibility that power lines, either because of the areas they are routed through when they are first built, or because of the effect they have on the surrounding areas after they are built, may be associated with demographic or socioeconomic factors. Great strides are being made in the epidemiology of CL in the area of response to, for example, infections and population mixing, and it certainly seems worth investigating whether power lines play any part in this picture.

Both high-voltage power lines and the grounding features of low-voltage distribution networks that make them produce MFs can result in small voltages impressed on plumbing within homes. A child can thus experience a small contact current by, for example, touching a faucet while in contact with ground potential. Those currents pass through the bone marrow of the arm, providing a possible mechanistic link to CL, and they are correlated with MFs. They thus meet some of the criteria for an alternative explanation of the epidemiology, but the magnitudes and degrees of correlation do not yet amount to an alternative mechanism (Kavet et al. 2000). Also, no association with contact current was found in one study that examined this hypothesis (Does et al. 2011).

Ways Forward

Overall, with more than three decades of epidemiologic investigation on the relationship of ELF-MF to CL, little can be gained from further repetition of investigations of risks at moderate and low exposure levels, unless such studies can be designed to test specific hypotheses, such as selection bias or aspects of exposure not previously captured. In addition, further study is warranted only if investigations are of high methodologic quality, of sufficient size, and with sufficient numbers of highly exposed subjects, and sophisticated exposure assessment.

Most previous studies of CL used case–control methodology and are susceptible to biases as discussed above. The cohort study is less likely to be subject to biases, particularly to selection and participation biases, but the low incidence of CL and the relatively low frequency of highly exposed residences would require an enormous study size, rendering the cohort study unfeasible in the general population. But, if a cohort could be identified in which the number of highly exposed is larger or the risk of CL were larger, the required cohort size would be correspondingly smaller and might conceivably become feasible. New approaches to advancing the epidemiology of MFs concentrate on fresh study designs that minimize bias and all are variations on a cohort design. The first approach maximizes the number of highly exposed; the other two approaches focus on populations with high risk of disease.

One initiative depends on the presence in some apartment buildings or blocks of flats of indoor substations, adjacent to living areas. In some circumstances, the apartment immediately above (or next to) the substation can receive an elevated exposure from it. Assembling a cohort of children who have lived in such buildings and comparing different apartments in the same building, which are expected to have similar socioeconomic characteristics, may be a way of avoiding socioeconomic bias, and assessing exposure without requiring subject participation. The study, known as "Transexpo," will be feasible only as an international collaboration, because of the low prevalence of such exposure situations in any one country (Kheifets et al. 2013). Pilot studies have been performed in Finland, Hungary, Israel, and Switzerland, confirming that the location of an apartment in relation to the built-in transformer is sufficient to identify highly exposed apartments (Ilonen et al. 2008; Thuroczy et al. 2008; Hareuveny et al. 2011; Röösli et al. 2011).

A second approach concerns children with Down's syndrome. Such children have 10- to 20-fold increased risks of ALL. A key question, however, is whether the genetic characteristics of the ALL contracted by children with Down's syndrome is sufficiently similar to that of the general population (Mezei et al. submitted).

A third approach is to study the possible joint effects (interactions) of MFs and genetic cofactors on CL by sampling from birth cohorts. Certain chromosomal anomalies, a common example being the TEL-AML1 translocation, are implicated in the causal pathway for ALL. But those anomalies are thought to be found in children in general 100 times more often than occurrence of overt leukemia, consistent with these markers being a "first hit" that has to be followed by a "second hit." A cohort of children born with one of the relevant genetic translocations would allow investigation of whether MFs were associated with the development of overt leukemia in these children. A related investigation would be whether MFs were associated, not with the development of overt leukemia, but with the presence at birth of these genetic translocations. Such approaches are being explored (Greenland and Kheifets 2009).

Conclusions

Leukemias are the most common cancers in children, but their etiology remains largely unknown. Residential exposure to MFs as a possible risk factor has been examined in more than 30 studies with most finding increased risks, based, however, on case–control studies with a small number of highly exposed cases. Based on the pooled analysis, systematic reviews have concluded that CL is the outcome for which the scientific evidence of association with MFs is strongest; thus, it is regarded as the critical effect in risk assessment. The consistent association found between CL and average MF exposure >0.3–0.4 μT could be due to chance, selection bias, misclassification, other factors that confound the association, or to a true causal relationship. Unfortunately, the possible explanations so far identified all seem unlikely, with arguments for and against each possibility summarized in the following table.

Assessment		Explanation	Arguments For	Arguments Against
Increasingly unlikely (↓)	?	Bias	Recognized as a problem for all case–control studies; demonstrated to be operating in some ELF-MF studies	Does not apply to calculated-field studies; not demonstrated whether magnitude sufficient to explain entire risk
	??	Confounding (by as yet unidentified risk factor)	Association of ELF-MF with various socioeconomic and other factors	Associations with actual risk factor would have to be strong; no good evidence in favor yet found
	??	Magnetic fields	Consistency across studies in different countries, as well as across measured-field and calculated-field studies	Lack of plausible mechanism; absence of robust laboratory evidence
	??	Other physical aspect of power lines	Avoid the lack of mechanism and lack of laboratory support with MFs	Lack of plausibility and evidence for suggestions so far advanced
	???	Misclassification	Large misclassification clearly present even in the best of studies	Misclassification is likely to be nondifferential and thus probably conceals effect if any
	????	Electric fields	Has been suggested by some authors	Very limited support both in human and laboratory studies
	????	Chance	Becomes more likely when the alternative explanations are also improbable	Consistency of an association across many studies

The epidemiologic evidence on MFs and CL presents a substantial but unresolved body of data. It seems likely that more than one of these explanations, or of other explanations not yet identified, will be operating. Even though it appears that the attributable fraction of leukemias would not be large even if MFs are causal, the public health case for resolving the issue, as well as both public pressure and scientific rigor, mandate continuing research. However, any investigations need to be of high methodologic quality; of sufficient size and with sufficient numbers of highly exposed subjects; have sophisticated exposure assessment; and most importantly, be designed to test specific hypotheses, such as examination of a susceptible subpopulation, selection bias, or aspects of exposure not previously captured.

Acknowledgment

We thank Dr. Madhuri Sudan for contributions to the development of this chapter. J.S. worked on this chapter with the permission of National Grid, but the views expressed are his alone.

References

Abdul Rahman, H. I., Shah, S. A., Alias, H. and Ibrahim, H. M. 2008. A case-control study on the association between environmental factors and the occurrence of acute leukemia among children in Klang Valley, Malaysia. *Asian Pac J Cancer Prev* 9(4): 649–52.

Ahlbom, A., Day, N., Feychting, M., Roman, E., Skinner, J., Dockerty, J., et al. 2000. A pooled analysis of magnetic fields and childhood leukaemia. *Br J Cancer* 83: 692–8.

Bellossi, A. 1991. Effect of pulsed magnetic fields on leukemia-prone AKR mice. No-effect on mortality through five generations. *Leuk Res* 15(10): 899–902.

Bianchi, N., Crosignani, P., Rovelli, A., Tittarelli, A., Carnelli, C. A., Rossitto, F., et al. 2000. Overhead electricity power lines and childhood leukemia: a registry-based, case-control study. *Tumori* 86(3): 195–8.

Boorman, G. A., McCormick, D. L., Findlay, J. C., Hailey, J. R., Gauger, J. R., Johnson, T. R., et al. 1999. Chronic toxicity/oncogenicity evaluation of 60 Hz (power frequency) magnetic fields in F344/N rats. *Toxicol Pathol* 27(3): 267–78.

Campion, E. W. 1997. Power lines, cancer, and fear. *N Engl J Med* 337(1): 44–6.

Coghill, R. W., Steward, J. and Philips, A. 1996. Extra low frequency electric and magnetic fields in the bedplace of children diagnosed with leukaemia: a case-control study. *Eur J Cancer Prev* 5(3): 153–8.

Committee to Assess Health Risks from Exposure to Low Levels of Ionizing Radiation. 2006. *Health Risks from Exposure to Low Levels of Ionizing Radiation, Beir VII Phase 2*. Washington, DC: The National Academies Press.

Dockerty, J. D., Elwood, J. M., Skegg, D. C. and Herbison, G. P. 1999. Electromagnetic field exposures and childhood leukaemia in New Zealand. *Lancet* 354(9194): 1967–8.

Does, M., Scelo, G., Metayer, C., Selvin, S., Kavet, R. and Buffler, P. 2011. Exposure to electrical contact currents and the risk of childhood leukemia. *Radiat Res* 175(3): 390–6.

Draper, G., Vincent, T., Kroll, M. E. and Swanson, J. 2005. Childhood cancer in relation to distance from high voltage power lines in England and Wales: a case-control study. *BMJ* 330(7503): 1290.

Ebi, K. L., Kheifets, L. I., Pearson, R. L. and Wachtel, H. 2000. Description of a new computer wire coding method and its application to evaluate potential control selection bias in the Savitz et al. childhood cancer study. *Bioelectromagnetics* 21(5): 346–53.

Fajardo-Gutierrez, A., Velasquez-Perez, L., Martinez-Mendez, J. and Martinez-Garcia, C. 1997. *Exposicion a campos electromagneticos y su asociacion con leucemia en ninos residentes de la ciudad de Mexico*. Mexico: Unidad de Investigacion Medica en Epidemiologia Clinica Hospital de Pediatria Centro Medico Nacional Siglo XXI.

Feizi, A. A. and Arabi, M. A. 2007. Acute childhood leukemias and exposure to magnetic fields generated by high voltage overhead power lines— a risk factor in Iran. *Asian Pac J Cancer Prev* 8(1): 69–72.

Ferlay, J., Shin, H. R., Bray, F., Forman, D., Mathers, C. and Parkin, D. M. 2012. *GLOBOCAN 2008 v1.2, Cancer Incidence and Mortality Worldwide: IARC CancerBase No. 10 [Internet]*. Lyon, France: International Agency for Research on Cancer.

Fews, A. P., Henshaw, D. L., Wilding, R. J. and Keitch, P. A. 1999. Corona ions from powerlines and increased exposure to pollutant aerosols. *Int J Radiat Biol* 75(12): 1523–31.

Feychting, M. and Ahlbom, A. 1993. Magnetic fields and cancer in children residing near Swedish high-voltage power lines. *Am J Epidemiol* 138(7): 467–81.

Foliart, D. E., Pollock, B. H., Mezei, G., Iriye, R., Silva, J. M., Ebi, K. L., et al. 2006. Magnetic field exposure and long-term survival among children with leukaemia. *Br J Cancer* 94(1): 161–4.

Forssen, U. M., Ahlbom, A. and Feychting, M. 2002. Relative contribution of residential and occupational magnetic field exposure over twenty-four hours among people living close to and far from a power line. *Bioelectromagnetics* 23(3): 239–44.

Fulton, J. P., Cobb, S., Preble, L., Leone, L. and Forman, E. 1980. Electrical wiring configurations and childhood leukemia in Rhode Island. *Am J Epidemiol* 111(3): 292–6.

Green, L. M., Miller, A. B., Villeneuve, P. J., Agnew, D. A., Greenberg, M. L., Li, J., et al. 1999. A case-control study of childhood leukemia in southern Ontario, Canada, and exposure to magnetic fields in residences. *Int J Cancer* 82(2): 161–70.

Greenland, S. 2003. The impact of prior distributions for uncontrolled confounding and response bias. *J Am Stat Assoc* 98(461): 47–54.

Greenland, S. 2005. Multiple-bias modelling for analysis of observational data. *J R Stat Soc Series A Stat Soc* 168(2): 267–306.

Greenland, S. and Kheifets, L. 2006. Leukemia attributable to residential magnetic fields: results from analyses allowing for study biases. *Risk Anal* 26(2): 471–82.

Greenland, S. and Kheifets, L. 2009. Designs and analyses for exploring the relationship of magnetic fields to childhood leukaemia: a pilot project for the Danish National Birth Cohort. *Scand J Public Health* 37(1): 83–92.

Greenland, S., Sheppard, A. R., Kaune, W. T., Poole, C. and Kelsh, M. A. 2000. A pooled analysis of magnetic fields, wire codes, and childhood leukemia. *Epidemiology* 11(6): 624–34.

Gurney, J. G., Davis, S., Schwartz, S. M., Mueller, B. A., Kaune, W. T. and Stevens, R. G. 1995. Childhood cancer occurrence in relation to power line configurations: a study of potential selection bias in case-control studies. *Epidemiology* 6(1): 31–5.

Hareuveny, R., Kandel, S., Yitzhak, N. M., Kheifets, L. and Mezei, G. 2011. Exposure to 50 Hz magnetic fields in apartment buildings with indoor transformer stations in Israel. *J Expo Sci Environ Epidemiol* 21(4): 365–71.

Harris, A. W., Basten, A., Gebski, V., Noonan, D., Finnie, J., Bath, M. L., et al. 1998. A test of lymphoma induction by long-term exposure of E mu-Pim1 transgenic mice to 50 Hz magnetic fields. *Radiat Res* 149(3): 300–7.

Hatch, E. E., Kleinerman, R. A., Linet, M. S., Tarone, R. E., Kaune, W. T., Auvinen, A., et al. 2000. Do confounding or selection factors of residential wiring codes and magnetic fields distort findings of electromagnetic fields studies? *Epidemiology* 11(2): 189–98.

Hill, A. B. 1965. The environment and disease: association or causation? *Proc R Soc Med* 58: 295–300.

Ilonen, K., Markkanen, A., Mezei, G. and Juutilainen, J. 2008. Indoor transformer stations as predictors of residential ELF magnetic field exposure. *Bioelectromagnetics* 29(3): 213–18.

Jackson, J. D. 1992. Are the stray 60-Hz electromagnetic fields associated with the distribution and use of electric power a significant cause of cancer? *Proc Natl Acad Sci U S A* 89(8): 3508–10.

Jeffers, D. 2007. Modelling and analyses do not support the hypothesis that charging by power-line corona increases lung deposition of airborne particles. *Radiat Prot Dosimetry* 123(2): 257–61.

Kabuto, M., Nitta, H., Yamamoto, S., Yamaguchi, N., Akiba, S., Honda, Y., et al. 2006. Childhood leukemia and magnetic fields in Japan: a case-control study of childhood leukemia and residential power-frequency magnetic fields in Japan. *Int J Cancer* 119(3): 643–50.

Kavet, R., Zaffanella, L. E., Daigle, J. P. and Ebi, K. L. 2000. The possible role of contact current in cancer risk associated with residential magnetic fields. *Bioelectromagnetics* 21(7): 538–53.

Kheifets, L., Afifi, A., Monroe, J. and Swanson, J. 2011. Exploring exposure—response for magnetic fields and childhood leukemia. *J Expo Sci Environ Epidemiol* 21(6): 625–33.

Kheifets, L., Afifi, A. and Shimkhada, R. 2006a. Public health impact of extremely low-frequency electromagnetic fields. *Environ Health Perspect* 114(10): 1532–7.

Kheifets, L., Ahlbom, A., Crespi, C. M., Draper, G., Hagihara, J., Lowenthal, R. M., et al. 2010a. Pooled analysis of recent studies on magnetic fields and childhood leukaemia. *Br J Cancer* 103(7): 1128–35.

Kheifets, L., Ahlbom, A., Crespi, C. M., Feychting, M., Johansen, C., Monroe, J., et al. 2010b. A pooled analysis of extremely low-frequency magnetic fields and childhood brain tumors. *Am J Epidemiol* 172(7): 752–61.

Kheifets, L., Breslow, N., Mezei, G. and TransExpo Study Group. 2013. *TransExpo: International Study of Childhood Leukemia and Residences Near Electrical Transformer Rooms Protocol*, Palo Alto, CA.

Kheifets, L., Mezei, G. and Greenland, S. 2006b. Comment concerning "childhood leukemia and residential magnetic fields: are pooled analyses more valid than the original studies?" (Bioelectromagnetics 27: 1–7 [2006]). *Bioelectromagnetics* 27(8): 674–5; discussion 675–6.

Kheifets, L. and Oksuzyan, S. 2008. Exposure assessment and other challenges in non-ionizing radiation studies of childhood leukaemia. *Radiat Prot Dosimetry* 132(2): 139–47.

Kheifets, L., Renew, D., Sias, G. and Swanson, J. 2010d. Extremely low frequency electric fields and cancer: assessing the evidence. *Bioelectromagnetics* 31(2): 89–101.

Kheifets, L., Swanson, J. and Greenland, S. 2006c. Childhood leukemia, electric and magnetic fields, and temporal trends. *Bioelectromagnetics* 27(7): 545–52.

Kroll, M. E., Swanson, J., Vincent, T. J. and Draper, G. J. 2010. Childhood cancer and magnetic fields from high-voltage power lines in England and Wales: a case-control study. *Br J Cancer* 103(7): 1122–7.

Linet, M. S., Hatch, E. E., Kleinerman, R. A., Robison, L. L., Kaune, W. T., Friedman, D. R., et al. 1997. Residential exposure to magnetic fields and acute lymphoblastic leukemia in children. *N Engl J Med* 337(1): 1–8.

Linet, M. S., Ries, L. A. G., Smith, M. A., Tarone, R. E. and Devesa, S. S. 1999. Cancer surveillance series: recent trends in childhood cancer incidence and mortality in the United States. *J Natl Cancer Inst* 91(12): 1051–8.

London, S. J., Thomas, D. C., Bowman, J. D., Sobel, E., Cheng, T. C. and Peters, J. M. 1991. Exposure to residential electric and magnetic fields and risk of childhood leukemia. *Am J Epidemiol* 134(9): 923–37.

Lowenthal, R. M., Tuck, D. M. and Bray, I. C. 2007. Residential exposure to electric power transmission lines and risk of lymphoproliferative and myeloproliferative disorders: a case-control study. *Intern Med J* 37(9): 614–19.

Malagoli, C., Fabbi, S., Teggi, S., Calzari, M., Poli, M., Ballotti, E., et al. 2010. Risk of hematological malignancies associated with magnetic fields exposure from power lines: a case-control study in two municipalities of northern Italy. *Environ Health* 9: 16.

Mandeville, R., Franco, E., Sidrac-Ghali, S., Paris-Nadon, L., Rocheleau, N., Mercier, G., et al. 1997. Evaluation of the potential carcinogenicity of 60 Hz linear sinusoidal continuous-wave magnetic fields in Fischer F344 rats. *FASEB J* 11(13): 1127–36.

McBride, M. L., Gallagher, R. P., Theriault, G., Armstrong, B. G., Tamaro, S., Spinelli, J. J., et al. 1999. Power-frequency electric and magnetic fields and risk of childhood leukemia in Canada. *Am J Epidemiol* 149(9): 831–42.

McCormick, D. L., Boorman, G. A., Findlay, J. C., Hailey, J. R., Johnson, T. R., Gauger, J. R., et al. 1999. Chronic toxicity/oncogenicity evaluation of 60 Hz (power frequency) magnetic fields in B6C3F1 mice. *Toxicol Pathol* 27(3): 279–85.

McCormick, D. L., Ryan, B. M., Findlay, J. C., Gauger, J. R., Johnson, T. R., Morrissey, R. L., et al. 1998. Exposure to 60 Hz magnetic fields and risk of lymphoma in PIM transgenic and TSG-p53 (p53 knockout) mice. *Carcinogenesis* 19(9): 1649–53.

Mejia-Arangure, J. M., Fajardo-Gutierrez, A., Perez-Saldivar, M. L., Gorodezky, C., Martinez-Avalos, A., Romero-Guzman, L., et al. 2007. Magnetic fields and acute leukemia in children with Down syndrome. *Epidemiology* 18(1): 158–61.

Mezei, G. and Kheifets, L. 2006. Selection bias and its implications for case-control studies: a case study of magnetic field exposure and childhood leukaemia. *Int J Epidemiol* 35(2): 397–406.

Mezei, G., Sudan, M., Izraeli, S. and Kheifets, L., submitted. Epidemiology of childhood leukemia in the presence and absence of Down syndrome.

Michaelis, J., Schuz, J., Meinert, R., Zemann, E., Grigat, J. P., Kaatsch, P., et al. 1998. Combined risk estimates for two German population-based case-control studies on residential magnetic fields and childhood acute leukemia. *Epidemiology* 9(1): 92–4.

Milham, S. and Ossiander, E. M. 2001. Historical evidence that residential electrification caused the emergence of the childhood leukemia peak. *Med Hypotheses* 56(3): 290–5.

Myers, A., Clayden, A. D., Cartwright, R. A. and Cartwright, S. C. 1990. Childhood cancer and overhead powerlines: a case-control study. *Br J Cancer* 62(6): 1008–14.

Olsen, J. H., Nielsen, A. and Schulgen, G. 1993. Residence near high voltage facilities and risk of cancer in children. *BMJ* 307(6909): 891–5.

Perez, C. B., Pineiro, R. G. and Diaz, N. T. 2005. Campos electromagneticos de baja frecuencia y leucemia infancil en Cuidad de La Habana. *Rev Cubana Hig Epidemiol* 43(3): 10.

Röösli, M., Jenni, D., Kheifets, L. and Mezei, G. 2011. Extremely low frequency magnetic field measurements in buildings with transformer stations in Switzerland. *Sci Total Environ* 409(18): 3364–9.

Rothman, K. J. and Greenland, S. 1998. *Modern Epidemiology*. 2nd ed. Philadelphia, PA: Lippincott Williams & Wilkins.

Royall, R. M. 1997. *Statistical Evidence: A Likelihood Paradigm. 1st ed. Monographs on Statistics and Applied Probability*. London: Chapman & Hall.

Savitz, D. A., Wachtel, H., Barnes, F. A., John, E. M. and Tvrdik, J. G. 1988. Case-control study of childhood cancer and exposure to 60-Hz magnetic fields. *Am J Epidemiol* 128(1): 21–38.

Schuz, J. and Ahlbom, A. 2008. Exposure to electromagnetic fields and the risk of childhood leukaemia: a review. *Radiat Prot Dosimetry* 132(2): 202–11.

Schuz, J., Grell, K., Kinsey, S., Linet, M. S., Link, M. P., Mezei, G., et al. 2012. Extremely low-frequency magnetic fields and survival from childhood acute lymphoblastic leukemia: an international follow-up study. *Blood Cancer J* 2: e98.

Schuz, J., Grigat, J. P., Brinkmann, K. and Michaelis, J. 2001. Residential magnetic fields as a risk factor for childhood acute leukaemia: results from a German population-based case-control study. *Int J Cancer* 91(5): 728–35.

Schuz, J., Svendsen, A. L., Linet, M. S., McBride, M. L., Roman, E., Feychting, M., et al. 2007. Nighttime exposure to electromagnetic fields and childhood leukemia: an extended pooled analysis. *Am J Epidemiol* 166(3): 263–9.

Sommer, A. M. and Lerchl, A. 2004. The risk of lymphoma in AKR/J mice does not rise with chronic exposure to 50 Hz magnetic fields (1 microT and 100 microT). *Radiat Res* 162(2): 194–200.

Swanson, J. 1999. Residential power-frequency electric and magnetic fields: sources and exposures. *Radiat Prot Dosimetry* 83(1–2): 9–14.

Swanson, J. and Kheifets, L. 2006. Biophysical mechanisms: a component in the weight of evidence for health effects of power-frequency electric and magnetic fields. *Radiat Res* 165(4): 470–8.

Thuroczy, G., Janossy, G., Nagy, N., Bakos, J., Szabo, J. and Mezei, G. 2008. Exposure to 50 Hz magnetic field in apartment buildings with built-in transformer stations in Hungary. *Radiat Prot Dosimetry* 131(4): 469–73.

Tomenius, L. 1986. 50-Hz electromagnetic environment and the incidence of childhood tumors in Stockholm County. *Bioelectromagnetics* 7(2): 191–207.

Tynes, T. and Haldorsen, T. 1997. Electromagnetic fields and cancer in children residing near Norwegian high-voltage power lines. *Am J Epidemiol* 145(3): 219–26.

UK Childhood Cancer Study Investigators. 1999. Exposure to power-frequency magnetic fields and the risk of childhood cancer. *Lancet* 354(9194): 1925–31.

Verkasalo, P. K., Pukkala, E., Hongisto, M. Y., Valjus, J. E., Jarvinen, P. J., Heikkila, K. V., et al. 1993. Risk of cancer in Finnish children living close to power lines. *BMJ* 307(6909): 895–9.

Wertheimer, N. and Leeper, E. 1979. Electrical wiring configurations and childhood cancer. *Am J Epidemiol* 109(3): 273–84.

World Health Organization. 2007. *Environmental Health Criteria 238: Extremely Low Frequency Fields*. Geneva, Switzerland: World Health Organization.

Wunsch-Filho, V., Pelissari, D. M., Barbieri, F. E., Sant'Anna, L., de Oliveira, C. T., de Mata, J. F., et al. 2011. Exposure to magnetic fields and childhood acute lymphocytic leukemia in Sao Paulo, Brazil. *Cancer Epidemiol* 35(6): 534–9.

Yang, Y., Jin, X., Yan, C., Tian, Y., Tang, J. and Shen, X. 2008. Case-only study of interactions between DNA repair genes (hMLH1, APEX1, MGMT, XRCC1 and XPD) and low-frequency electromagnetic fields in childhood acute leukemia. *Leuk Lymphoma* 49(12): 2344–50.

Yasui, M., Kikuchi, T., Ogawa, M., Otaka, Y., Tsuchitani, M. and Iwata, H. 1997. Carcinogenicity test of 50 Hz sinusoidal magnetic fields in rats. *Bioelectromagnetics* 18(8): 531–40.

10

Adult Cancer and Extremely Low-Frequency Magnetic Fields

Gabor Mezei and Ximena P. Vergara

CONTENTS

Introduction

In industrialized societies such as the United States, about half of adult men and a third of women develop cancer during their lifetime (Siegel et al. 2013). Cancer is typically characterized as uncontrolled or unregulated proliferation of cells, with a tendency to invade neighboring tissues and organs and to spread to more distant parts of the body through metastasis. Cancer types are classified based on the organ and cell type from which they originate, and by morphological characteristics of the cancer cells themselves. Cancer types may be further characterized and categorized using various molecular techniques or based on the presence of specific genetic changes in the cancer cells. Carcinogenesis, according to current thinking, is a multistep process that may include genetic (alteration in DNA composition and structure) and epigenetic (alteration in gene expression without DNA modification) changes. External stimuli, such as environmental and occupational exposures, may contribute to some of these changes and play a role in tumor causation. Extremely low-frequency magnetic fields (ELF-MF) are part of the nonionizing range of the electromagnetic spectrum. Due to its low energy level, ELF-MF is not able to break chemical bonds or directly affect the cells' DNA, resulting in mutations. ELF-MF is hypothesized to influence epigenetic changes, resulting in altered cell proliferation and differentiation, apoptosis, or modified adaptive responses (Vijayalaxmi and Prihoda 2009), thus potentially playing a role in cancer development.

Exposure to ELF-MF and its relationship to cancer have been investigated since the late 1970s, when Wertheimer and Leeper (1979) first reported an association between a surrogate of ELF-MF exposure, wire coding of power lines, and childhood cancer (see Chapter 9). Following this first report of increased cancer risk among children in association with ELF-MF exposure, researchers also began to investigate cancer risk in adult populations. Adult cancer represents significantly higher societal burden (with typically increasing incidence by age for most cancers), and adults are more likely to encounter higher than average exposures in various occupational settings and use high ELF-MF exposure sources in residential environments (e.g., electric blankets and appliances).

Many assessments and reviews have been published on the relationship between ELF-MF and adult cancers (IARC 2002; Johansen 2004; Schüz et al. 2009; SSM 2010; Repacholi 2012). Among the most comprehensive reports, we highlight the monograph of the International Agency for Research on Cancer (IARC) and the Environmental Health Criteria (EHC) of the World Health Organization (WHO 2007). The IARC report reviewed the relevant literature published before 2001, whereas the WHO EHC covered the literature published between 2001 and January 2005. We further assessed the epidemiologic literature published between 2005 and 2012.

In this chapter, we review the available epidemiologic literature on ELF-MF and adult cancer, and we discuss methodological considerations and potential directions for further research in this area.

Breast Cancer

Breast cancer is a heterogeneous disease with most common forms developing in the breast ducts and lobules. Occurring mainly in women, the number of incident cases was estimated at 1.4 million worldwide in 2008 (IARC 2013). In industrialized countries, it is the most common cancer among women. In the United States, for example, one in eight women is expected to be diagnosed with breast cancer in her lifetime (Siegel et al. 2013). Due to its public health impact, even weak risk factors for breast cancer are of potentially great importance. According to a hypothesis put forth by Stevens and colleagues, ELF-MF, similarly to light at night, may reduce the nocturnal production of the pineal gland hormone melatonin that has a tumor suppressor effect. The suppression in melatonin production, in turn, would result in increased risk of breast cancer (Stevens and Davis 1996; Davis et al. 2002). This hypothesis appeared to be supported by some sporadic evidence from animal studies; these studies, however, overall remained inconsistent (IARC 2002). Epidemiologic studies with subgroup analyses were suggestive of an increased risk among younger (premenopausal) women and for estrogen receptor-positive cancers; but, overall, the 2002 IARC review considered both animal and epidemiologic studies inadequate to draw firm conclusions. The IARC review of epidemiologic literature considered studies of residential exposures (proximity to electrical installations, such as power lines and measured ELF-MF fields), electric blanket use, and occupational exposure to ELF-MF.

In 2007, the WHO EHC summarized the literature published since 2001. In its evaluation of the epidemiologic evidence, the WHO EHC examined four residential exposure case–control studies (Davis et al. 2002; London et al. 2003; Schoenfeld et al. 2003; Kliukiene et al. 2004), four electric blanket use case–control studies (McElroy et al. 2001; Davis et al. 2002; Schoenfeld et al. 2003; Zhu et al. 2003), three occupational exposure cohort studies (Pollan et al. 2001; Hakansson et al. 2002; Kliukiene et al. 2003), and seven occupational exposure case–control studies (Band et al. 2000; Koc and Polat 2001; Van Wijngaarden et al. 2001a; Gardner et al. 2002; Labreche et al. 2003; Teitelbaum et al. 2003; Forssen et al. 2005). Of these, the largest occupational study, relying on a measurement-based job exposure matrix (JEM) (see Chapter 7) and including >20,000 female breast cancer cases, demonstrated no increased risk in any of the subgroup analyses for pre- and postmenopausal women and estrogen receptor-positive and -negative tumors. The WHO EHC concluded that the evidence for an association between ELF-MF and breast cancer was weakened with the addition of more recent studies (studies that tended to be larger and less susceptible to bias than previous studies), and the evidence overall was not in support of an association. Relatively few ELF-MF epidemiologic studies have been published on breast cancer since 2005, overall, providing no new evidence for altering the conclusion reached by the WHO EHC (Table 10.1).

Residential Studies, 2005–2012

To investigate whether residential ELF-MF levels increased breast cancer risk, Davis and Mirick (2007) examined incident breast cancer cases diagnosed among women with information on medication use. Controls, identified through random digit dialing, were frequency matched to cases by 5-year age group. Unlike a previous analysis of the same study population investigating residential ELF-MF levels and its relationship to breast cancer risk overall (Davis et al. 2002), the more recent analysis examined melatonin suppressing medication use among those with high nighttime ELF-MF levels. No evidence in support of an association was observed with high nighttime ELF-MF levels, regardless of medication use.

Chen et al. (2010) conducted a meta-analysis of 10 case–control studies of residential ELF-MF exposure and breast cancer published between 2000 and 2009. In five of these studies, exposure assessment was based on measurements, and the other five studies examined electric blanket use. The overall summary odds ratios (ORs) were not supportive of an association (OR = 1.02; 95% confidence interval [CI], 0.92–1.12). These results are consistent with previous reviews in which the availability of data does not support an association between residential ELF-MF and breast cancer.

These results were further confirmed in a recent population-based case–control study of several cancers, including 29,202 female breast cancer cases diagnosed between 1974 and 2008, and residential proximity to overhead power lines (Elliott et al. 2013) (Table 10.1).

Occupational Studies, 2005–2012

Within the meta-analysis previously discussed, Chen et al. (2010) also examined five occupational studies most of which incorporated ELF-MF measurements (Van Wijngaarden et al. 2001a; Kliukiene et al. 2003; Labreche et al. 2003; Forssen et al. 2005; McElroy et al. 2007). Overall, summary estimates of these studies did not support an association between occupational ELF-MF and breast cancer (OR = 0.93; 95% CI, 0.79–1.10). Chen et al. (2010) also combined residential and occupational studies. Overall, there was no association between

TABLE 10.1

Epidemiologic Studies of Residential and Occupational Exposure to Magnetic Fields and Adult Breast Cancer, Brain Tumors, and Leukemia Published Subsequent to WHO EHC 238

Author	Year	Location and Study Period	Study Base and Subject Identification	Study Type	Exposure Assessment Approach	JEM Used	Study Design	RR Measure	Risk Estimates and Confidence Intervals	Number of Cases/ Controls or Person/Years
Forssen	2006	Sweden: Dxed between 1987 and 1999	Cases from cancer registry; sex-, age-matched controls from population registers	O	Census-based occ hx, ELF-MF JEM	Floderus et al. Forssen et al.	Case–control (P)	OR	Acoustic neuroma, exposure at least 10 years before reference date:	All: 793/101,762
									< 0.12 μT: 1	196/21,241
									0.11–0.19 μT: 1.01 (0.83–1.23)	287/39,562
									0.20–0.30 μT: 1.02 (0.81–1.29)	128/16,459
									> 0.30 μT: 1.07 (0.78–1.48)	53/6,716
Mester	2006	Germany, six regions: 1999 –	Newly Dxed cases aged 18–80 years; sex-, age-, region-matched controls from population registers	O	Questionnaire	—	Case–control (P)	OR	Lymphomas, electrical fitters and related electrical and electronics workers:	All: 710/710
									1–10 years: 0.9 (0.5–1.6)	24/31
									> 10 years: 2.2 (1.0–5.1)	19/10
									Lymphomas, plumbers, welders, sheet metal and structural metal preparers, and erectors:	All: 710/710
									1–10 years: 1.9 (1.0–3.5)	31/19
									> 10 years: 0.5 (0.2–1.2)	8/20
Davis and Mirick	2007	U.S., Two counties in Washington State: Dxed between 1992 and 1955	Incident cases from surveillance; age-matched controls RDD from nearby locations	R	48 hr ELF-MF measurements	—	Case–control (P)	OR	Breast cancer per 0.1 μT, NSAID[a] duration of regular use:	All: 558/588
									Never: 0.95 (0.81–1.11)	392/412
									< 5 year: 0.80 (0.53–1.20)	40/42
									5–10 year: 1.07 (0.94–1.22)	76/83

Johansen	2007	Denmark: Date of first employment or to date of death or 1968–2002	First primary cases from Danish utility cohort of 28,224 people employed at least 3 months	O	Expert assessment by job title and task	Cohort specific JEM	Cohort	RR	Breast cancer, women: < 0.09 μT: 1 0.1–0.99 μT: 0.77 (0.56–1.07) ≥ 1.0 μT: 1.04 (0.32–3.34)	All women: 188 Exposure level strata unavailable
									Leukemia, men: < 0.09 μT: 1 0.1–0.99 μT: 0.97 (0.51–1.85) ≥ 1.0 μT: 1.04 (0.53–2.04)	All men: 70 Exposure level strata unavailable
									Brain cancer, men: < 0.09 μT: 1 0.1–0.99 μT: 0.80 (0.47–1.37) ≥ 1.0 μT: 0.69 (0.38–1.25)	All men: 85 Exposure level strata unavailable
									Brain cancer, women: < 0.09 μT: 1 0.1–0.99 μT: 1.37 (0.51–3.69) ≥ 1.0 μT: No cases	All women: 25 Exposure level strata unavailable
Karipidis	2007	Australia: Dxed between 1987 and 1991	414 glioma cases from Melbourne hospitals; 421 age-, sex-, postcode-matched controls from electoral rolls	O	Questionnaire occ hx, IH assessment, ELF-MF JEM	UWash JEM (Bowman et al.)	Case–control (H)	OR	Low grade glioma, cumulative ELF-MF exposure (μT) total exposure quartiles: 0–3: 1 3 < – 6: 0.81 (0.45–1.46) 6 < – 10: 0.51 (0.25–1.04) > 10: 0.74 (0.34–1.61)	All: 110/421 39/93 30/103 18/115 23/110
									High grade glioma, cumulative ELF-MF exposure (μT) total exposure quartiles: 0–3.5: 1 3.5 < – 6: 0.77 (0.50–1.20) 6 < – 11: 0.56 (0.34–0.91) > 11: 0.72 (0.42–1.23)	All: 304/421 79/102 74/107 66/116 85/96
Karipidis	2007	New South Wales: Dxed between 2000 and 2001	Incident cases from cancer registry; age-, sex-, region-matched controls from electoral rolls	O	Questionnaire occ hx, IH assessment, ELF-MF JEM	UWash JEM (Bowman et al.)	Case–control (P)	OR	Non-Hodgkin's lymphoma, cumulative ELF-MF exposure (μT) total exposure quartiles: < 3.92: 1 3.92 – < 6.31: 0.80 (0.57–1.11) 6.31 – < 9.85: 1.29 (0.90–1.86) ≥ 9.85: 1.33 (0.90–1.96)	All: 694/694 170/177 150/197 185/162 189/158

Continued

TABLE 10.1 (*Continued*)

Epidemiologic Studies of Residential and Occupational Exposure to Magnetic Fields and Adult Breast Cancer, Brain Tumors, and Leukemia Published Subsequent to WHO EHC 238

Author	Year	Location and Study Period	Study Base and Subject Identification	Study Type	Exposure Assessment Approach	JEM Used	Study Design	RR Measure	Risk Estimates and Confidence Intervals	Number of Cases/ Controls or Person/Years
Lowenthal	2007	Australia, Tasmania: Dxed between 1972 and 1980	Incident cases; age-, sex-matched controls randomly selected from electoral rolls	R	Questionnaire lifetime residential hx	—	Case–control (P)	OR	Leukemias and lymphomas, ever having lived distance per year of residence: > 300 m: 1 51–300 m: 1.01 (0.98–1.04) 0–50 m: 1.07 (0.99–1.17)	All: 854/854 760/790 75/55 19/9
McElroy	2007	U.S., Three states—Massachusetts, New Hampshire, and Wisconsin	Incident cases from cancer registry; age-, residence-matched controls from driver license rosters and Medicare	O	Questionnaire occ hx, IH assessment exposure ranking	—	Case–control (P)	OR	Breast cancer, all women: Background: 1 Low: 1.05 (0.97–1.13) Medium: 1.11 (0.99–1.26) High: 1.17 (0.90–1.53)	All: 6,213/7,390 2,783/3,420 2,714/3,165 601/678 115/127
Ray	2007	China, Shanghai: Enrolled in breast examination trial between 1989 and 1991	Incident cases from a 267,400 female textile factory workers from 526 factories	O	Record-based occ hx, IH assessment and ELF-MF JEM	Cohort-specific JEM	Case–cohort		Breast cancer, ELF-MF exposures in years: 0: 1.00 < 5: 0.97 (0.73–1.28) 5 to < 10: 1.02 (0.77–1.34) 10 to < 20: 1.09 (0.91–1.30) +20: 0.86 (0.74–1.01)	All: 1,709/3,155 608/1006 103/175 111/166 377/567 510/1241
Röösli	2007	Switzerland: Follow-up between 1972 and 2002	All deaths from a cohort of 20,141 Swiss railway	O	Calculated lifetime ELF-MF exposures	—	Cohort	HR	Per increase by 10 µT-years: Leukemia: 1.02 (0.98–1.07) Lymphoid leukemia: 0.98 (0.92–1.04) Myeloid leukemia: 1.06 (0.99–1.14) Per increase by 10 µT-years: Hodgkin's: 1.09 (1.00–1.19) Non-Hodgkin's: 0.99 (0.94–1.05) Per increase by 10 µT-years Brain tumor: 0.94 (0.88–1.01)	 66 36 23 15 46 38

Coble	2009	U.S.: Cities of Boston, Pittsburgh, and Phoenix: Three brain tumor treatment hospitals	Cases age 18 years or more recently Dxed; age-, gender-, hospital-, race/ethnicity-, residential distance-matched controls from patient registers	O	Questionnaire occ hx, ELF-MF JEM	Bowman et al.	Case–control (H)	OR	Gliomas, cumulative exposures > 1.5 mG (mG-years): 0: 1 > 0–45: 0.8 (0.6–1.1) > 45: 0.8 (0.5–1.2) Meningiomas, cumulative exposures > 1.5 mG (mG-years): 0: 1 > 0–45: 1.0 (0.7–1.6) >45: 1.0 (0.6–1.8)	All: 489/799 107/159 276/490 91/131 All: 197/791 39/159 121/490 32/131
Wong	2010	China, Shanghai: Dxed between 2003 and 2008	Incident cases from 25 hospitals; age-, gender-matched hospital controls	R	Questionnaire	—	Case–control (H)	OR	Living within 100 m of HV power line: NHLN-Total: 1.05 (0.83–133)	All: 649/1,298 154/289
Baldi	2011	France, Gironde region: Identified between 1999 and 2001	Incident CNS cases; age-, sex-, residence-matched controls	O/R	Questionnaire resid and occ hx, IH assessment exposure ranking	—	Case–control (P)	OR	Occupational ELF-MF exposed: All brain: 1.59 (0.97–2.61) Gliomas: 1.20 (0.66–2.17) Meningiomas: 3.02 (1.10–8.25) Acoustic neurinomas: 1.23 (0.26–5.75) Environmental exposure, proximity to power lines: All brain: 1.51 (0.74–3.07) Gliomas: 0.66 (0.21–2.07) Meningiomas: 2.99 (0.86–10.40) Acoustic neurinomas:3.23 (0.28–36.62)	All: 221/442 38/52 21/36 13/12 4/7 All: 221/442 16/20 5/12 7/6 2/1
Marcilio	2011	Brazil, municipalities including Sao Paulo; deaths from 2001 to 2005	All deaths in 39 municipalities; age-, sex-, residence-matched controls from death certificates	R	Proximity to power lines, Calculated ELF-MF based on hx line data	—	Case–control (P)	OR	Leukemia: ≤ 0.1 μT: 1 >0.1–0.3 μT: 1.34 (0.65–2.73) > 0.3 μT: 1.61 (0.91–2.86) Brain cancer: ≤ 0.1 μT: 1 >0.1–0.3 μT: 1.13 (0.50–2.30) > 0.3 μT: 1.16 (0.6–2.07)	All: 1,857/4,706 1,818/4629 15/31 24/46 All: 2,357/4,706 2,323/4,629 12/31 22/46

Continued

TABLE 10.1 (*Continued*)

Epidemiologic Studies of Residential and Occupational Exposure to Magnetic Fields and Adult Breast Cancer, Brain Tumors, and Leukemia Published Subsequent to WHO EHC 238

Author	Year	Location and Study Period	Study Base and Subject Identification	Study Type	Exposure Assessment Approach	JEM Used	Study Design	RR Measure	Risk Estimates and Confidence Intervals	Number of Cases/Controls or Person/Years
Elliott	2012	England and Wales; Dxed between 1974 and 2008	Cases aged 15–74 years; 79,507 year-, region-matched controls selected from National Cancer Registry	R	Questionnaire and calculated ELF-MF	—	Case–control (P)	OR	Breast cancer, women:	All: 29,202/29,202
									< 100 nT: 1.00	27,130/27,229
									100–199 nT: 1.23 (0.95–1.59)	136/106
									200–299 nT: 0.72 (0.51–1.03)	57/72
									300–399 nT: 0.77 (0.49–1.23)	34/41
									400–999 nT: 1.27 (0.94–1.70)	104/82
									≥ 1000 nT: 0.87 (0.60–1.26)	55/61
									Brain/CNS:	All: 6,781/20,343
									< 100 nT: 1.00	6,354/19,146
									100–199 nT: 1.10 (0.76–1.59)	44/104
									200–299 nT: 0.98 (0.53–1.78)	15/45
									300–399 nT: 1.36 (0.66–2.81)	12/25
									400–999 nT: 0.92 (0.54–1.55)	20/61
									≥ 1000 nT: 1.02 (0.47–2.22)	9/28
									Leukemia:	All: 7,823/23,469
									< 100 nT: 1.00	7,362/22,082
									100–199 nT: 1.19 (0.83–1.70)	44/106
									200–299 nT: 1.22 (0.74–2.02)	22/51
									300–399 nT: 1.19 (0.60–2.36)	12/29
									400–999 nT: 0.91 (0.57–1.46)	23/72
									≥ 1000 nT: 1.03 (0.57–1.86)	16/40
Sorahan	2012	England and Wales; employment between 1973 and 1982	Cancer morbidity among cohort of 81,842 electric utility workers	O	Employment in generation and transmission of electricity	—	Cohort		Breast cancer:	
									Males: 116 (81–162)	35/30.1 (O/E)
									Females: 108 (99–118)	535/494 (O/E)
									Leukemia:	
									Males: 94 (84–104)	333/355.4 (O/E)
									Acute lymphoid leukemia:	
									Males: 76 (36–139)	10/13.2 (O/E)
									Chronic lymphoid leukemia:	
									Males: 104 (88–122)	151/145.1 (O/E)

									Acute myeloid leukemia:	
									Males: 81 (65–100)	86/106.4 (O/E)
									Chronic myeloid leukemia:	
									Males: 110 (79–148)	43/39.3 (O/E)
									Brain:	
									Males: 103 (91–116)	264/256.9 (O/E)
									Females: 63 (35–106)	14/22.1 (O/E)
Kheifets	2008	Update, with new studies published between 1993 and 2007	Brain cancer studies	O	Study-specific	—	Meta-analysis	OR	Brain cancer, including gliomas:	
									All: 1.14 (1.07–1.22)	47 Studies
									New: 1.10 (0.99–1.23)	20 Studies
			Leukemia studies						Leukemia:	
									All: 1.16 (1.11–1.22)	56 Studies
									New: 1.13 (1.07–1.19)	21 Studies
Chen	2010	Studies published between 2000 and 2007	Breast cancer studies	O/R	Study-specific	—	Meta-analysis	OR	Breast cancer:	
									All: 0.99 (0.90–1.09)	15 Studies
									Residential: 1.02 (0.92–1.12)	10 Studies

Note: O, occupational; R, residential or O/R, both; dx/ed, diagnosis/diagnosed; hx, history; RDD, random digit dialing; NSAID, non-steroidal anti-inflammatory drug; HR, hazard ratio; OR, odds ratio; RR, relative risk; SRR, standardized registration ratios; (P), population-based or (H) hospital-based, (IH), industrial hygienist and (JEM), job-exposure matrix.

[a] Most prevalently used medication.

ELF-MF and breast cancer. Results did not differ by the women's menopausal status, and no risk increase was observed for either estrogen receptor-positive or –negative breast cancer.

In a Danish cohort study, a JEM was applied to the employment history of 28,224 electric utility workers from 99 electric utility companies (Johansen et al. 2007). Linking these records to the Danish Cancer Registry did not support an association between ELF-MF and breast cancer (Table 10.1).

In a case–control study, cases (identified from cancer registries among women aged 20–69 years) and age-matched controls (selected from among licensed drivers and Medicare beneficiaries) were interviewed and included in the analyses (McElroy et al. 2007). Women were asked to provide complete work history, including any job held for at least 1 year since the age of 14 years. ELF-MF exposures were assigned by a previously developed exposure scheme for the most representative jobs the women held (Coogan et al. 1996). The relative odds of breast cancer among women with low, medium, and high exposure were 1.05 (95% CI, 0.97–1.13), 1.11 (95% CI, 0.99–1.26), and 1.17 (95% CI, 0.90–1.53), respectively, compared with women with background exposure with a statistically significant trend among all women and among postmenopausal women (McElroy et al. 2007).

No support for an association between female breast cancer and occupational ELF-MF exposure was obtained from a cohort study of 267,400 female textile workers from 526 factories in Shanghai, China (Ray et al. 2007) (Table 10.1).

Sorahan (2012) assessed the incidence of breast cancer in a cohort of generation and transmission workers and compared it to the incidence in the general population of England and Wales, standardizing on sex, age, and calendar period. No evidence of an association between employment in electric utilities and breast cancer among men or women was found.

Leukemia

Leukemia is the malignancy of the hematopoietic system arising in the bone marrow, lymph nodes, or other parts of the lymphoid systems. Depending on the type of progenitor cells from which the leukemia originates, it is broadly grouped into two main categories: myeloid leukemia and lymphoid leukemia. Based on the level of maturity of the leukemia cells and the progression of the disease, leukemia is also grouped as acute leukemia (rapidly progressing and characterized by mostly immature precursor cells) and chronic leukemia (slowly progressing and characterized by more mature cells). The four main types of leukemia, acute lymphocytic leukemia (ALL), chronic lymphocytic leukemia (CLL), acute myeloid leukemia (AML), and chronic myeloid leukemia (CML), are further categorized by cell types and cytogenetic features. Lymphomas, malignancies typically arising from peripheral lymphoid tissues (Hodgkin and non-Hodgkin's lymphoma), are also included here.

For adult leukemia, perhaps the most studied cancer type in ELF-MF epidemiologic studies, the IARC reviewed more than a dozen residential exposure studies. In these primarily case–control studies, exposure assessment mostly relied on residential proximity to power lines, with some including calculated ELF-MF fields, and even fewer studies including spot measurements in the households or data on electric blanket and appliance use. Epidemiologic studies of occupational ELF-MF exposure include cohort studies involving "electrical occupations" with assumed exposure to ELF-MF and no actual exposure assessment, and workers of various industries (such as electric utility and railways) with potential exposure to ELF-MF (see Chapter 3). Many of the latter, most recently conducted studies focusing on specific industries, included extensive, frequently measurement-based exposure assessment. The more than a dozen adult leukemia case–control studies of occupational ELF-MF exposure reviewed by IARC included studies of the general population, and case–control studies nested in industry-based cohorts. A pooled analysis of three electric utility worker studies and a meta-analysis of 38 occupational exposure studies showed a small risk increase for adult leukemia. According to the IARC evaluation, residential exposure studies of adult leukemia provided no basis for an association. In contrast, some of the occupational exposure studies were suggestive of a weak risk increase for adult leukemia. However, no consistent pattern emerged across studies when exposure response was evaluated, and no consistent pattern was identified for specific leukemia subtypes.

For adult leukemia, the WHO EHC examined one residential exposure nested case–control study (Tynes and Haldorsen 2003), one electric blanket use study (Oppenheimer and Preston-Martin 2002), three occupational cohort studies (Rubtsova et al. 1999; Van Wijngaarden et al. 2001b; Hakansson et al. 2002), and four occupational title case–control studies (Bethwaite et al. 2001; Bjork et al. 2001; Oppenheimer and Preston-Martin 2002; Willett et al. 2003) for the 2007 review. Only one of these studies reported an increased leukemia risk with estimated occupational exposure to ELF-MF. The WHO EHC concluded that the recently published adult leukemia studies did not change the earlier IARC conclusion that the evidence for an association between ELF-MF exposure and leukemia is inadequate. The review also included three studies of occupational titles and risk of non-Hodgkin's lymphoma (Cano and Pollan 2001; Fabbro-Peray et al. 2001; Band et al. 2004). Excess risks were reported among electrical engineers, systems analysts/computer programmers, and electrical equipment installing and welding occupations; however, few cases were captured within each of these occupations.

Since the review of WHO EHC, several studies have been published to examine ELF-MF and adult leukemia, including one meta-analysis (Table 10.1).

Residential Studies, 2005–2012

A case–control study of residential proximity to transmission lines in Tasmania included patients with myeloproliferative disorders (MPDs) and lymphoproliferative disorders (LPDs) and their matched controls selected from electoral rolls (Lowenthal et al. 2007). Having lived within 50 m of a transmission line, the OR for leukemia was elevated but imprecise. The OR further increased if exposure only during childhood was considered.

Marcilio et al. (2011) compared all leukemia deaths occurring among those aged ≥40 years in the metropolitan region of São Paolo, Brazil, to controls selected from other causes of deaths with no previously suggested relationship with ELF-MF. They reported moderate associations for those living within 50 m of any power line (OR = 1.47; 95% CI, 0.99–2.18) compared with those living >400 m away, although no exposure response pattern was evident. Moderate associations were also observed between calculated ELF-MF levels and adult leukemia (Table 10.1).

A hospital-based case–control study in China examined risk of non-Hodgkin's lymphoid neoplasms (NHLNs) along with several environmental risk factors, including living within 100 m near a power transmission line (Wong et al. 2010). Exposure information was sought using in-person questionnaires by trained interviewers blinded to case/control status. No associations were observed between living within 100 m of high-voltage transmission lines and NHLN, overall, or for any of the subtypes.

Elliott et al. (2013) also evaluated leukemia risk in their residential ELF-MF case–control study conducted in England and Wales. Residential distance to power lines and calculated

ELF-MF showed no association and no consistent exposure–response relationship with risk of leukemia at the time of diagnosis or 5 years before diagnosis.

Occupational Studies, 2005–2012

Occupational title was used as a proxy for exposure in a case–control study from Germany (Mester et al. 2006). The authors reported elevated ORs for lymphoma among plumbers and welders working for 1–10 years and among electrical workers working for >10 years, compared with the reference group of white collar workers (Table 10.1).

Johansen et al. (2007) also assessed the incidence of leukemia and its relationship to occupational ELF-MF exposure in the cohort of workers at Danish electric utility companies. They reported no increased risk of adult leukemia with occupational exposure to ELF-MF (Table 10.1).

A 2007 case–control study included incident cases of NHL identified from the cancer registry of New South Wales and from frequency-matched controls selected from electoral rolls (Karipidis et al. 2007a). Cumulative occupational exposure to ELF-MF (μT-years) was determined based on detailed work history and linkage of individual jobs to a previously developed JEM. Considering a 5-year exposure lag and compared with the lowest exposure quartile, the adjusted relative risk (RR) in the highest quartile was 1.43 (95% CI, 0.96–2.14).

Röösli et al. (2007) examined leukemia mortality in a cohort of Swiss railway workers. Exposure assessment was based on measurements and modeling for each occupational groups. Occupations were ranked by estimated exposure using several exposure metrics. They reported some evidence of exposure–response association for myeloid leukemia and Hodgkin's disease but not for lymphoid leukemia and non-Hodgkin's disease.

In an update of a previous meta-analysis, Kheifets et al. (2008b) evaluated epidemiologic studies of occupational ELF-MF exposure and adult leukemia. Based on a total of 56 studies providing sufficient data on occupational ELF-MF and adult leukemia, a small risk increase was observed (RR = 1.16; 95% CI, 1.11–1.22). Although some recent studies included more sophisticated exposure assessment methods, the overall risk increase tended to be smaller in those recent studies compared with older studies. The findings were insensitive to study characteristics (e.g., design, outcome measure, exposure assessment method) as well as to various assumptions, weighting schemes, or funding sources. No evidence was found for publication bias, and no overly influential study was identified. The lower RR estimates for more recent studies, taken together with the lack of consistent exposure–response relationship and with the lack of consistent pattern within the leukemia subtype analyses, did not change the conclusions reached by the WHO EHC that the overall evidence is inadequate to link adult leukemia to ELF-MF.

Among generation and transmission workers, Sorahan (2012) assessed the incidence of leukemia comparing with the incidence in the general population of England and Wales. No evidence for an association between employment in electric utilities and leukemia was found (Table 10.1).

Brain Tumors

The majority of nervous system tumors originate in the central nervous system (CNS). Of these CNS tumors, roughly 95% are tumors of the brain, cranial nerves, and cranial meninges. Intracranial CNS tumors are frequently described collectively as brain tumors.

Brain tumors may arise in glial cells; in cells with neuroepithelial origin (including gliomas, glioblastomas, astrocytomas, and oligodendrogliomas); and in tissues surrounding the brain itself, the meninges (meningiomas), or Schwann cells (schwannomas or acoustic neuroma). The degree of malignancy (based on level of cell differentiation, growth pattern, and other anaplastic features) may be expressed by tumor grading. Grade 1 (lowest grade) tumors are well localized, showing no aggressive growth and include more differentiated cells. Grade 4 (highest grade) tumors are highly invasive and include less differentiated, immature cells. Of all brain tumors, gliomas account for a large fraction and, of these, glioblastomas are the most frequent (Bondy et al. 2008; Ohgaki 2009). Few ELF-MF epidemiologic studies have had sufficient power to examine the risk of brain tumors by tumor subtype.

The few epidemiologic studies of residential MF exposure and brain cancer reviewed by IARC, overall, showed no increased risk. ELF-MF exposure in most of these studies was assessed by residential proximity to transmission lines or calculated field. Only one included spot measurements, and two studies investigated electric blanket and other appliance use. The epidemiologic studies of occupational ELF-MF exposure and adult brain cancer, similarly to adult leukemia studies, included several "electrical occupations" cohort studies and workers in industries with potential exposure to ELF-MF. Many case–control studies were also conducted in the general population and within workers of specific industries, such as electric utility and railways. Again, analogous to the adult leukemia results, a pooled analysis of three electric utility studies and a meta-analysis of 29 occupational exposure studies showed a small increased risk of adult brain cancer with occupational ELF-MF exposure. Overall, the IARC conclusion for adult brain cancer was similar to that for adult leukemia. Significant improvements in exposure assessment methods of occupational epidemiologic studies were observed, but uncertainties in exposure assignment remain. Although some studies reported elevated risk of adult brain cancer with occupational exposure, there remains a lack of consistent exposure–response pattern and inconsistencies in risk increases with specific brain cancer subtypes.

The WHO EHC found that new studies, published up to 2005, offered no new evidence for an association between ELF-MF and risk of brain tumors. The additional studies examined included one residential study of electrical appliance use and acoustic neuroma (Kleinerman et al. 2005), five cohort studies of occupational exposure (Rubtsova et al. 1999; Van Wijngaarden et al. 2001b; Hakansson et al. 2002; Navas-Acien et al. 2002; Wesseling et al. 2002), one occupational ELF-MF case–control study (Villeneuve et al. 2002), and three case–control studies of occupational titles without specific focus on ELF-MF (De Roos et al. 2003; Krishnan et al. 2003; Schlehofer et al. 2005). Since the review, many studies on ELF-MF and brain tumors have been published (Table 10.1).

Residential Studies, 2005–2012

A recent population-based case–control study examined EMF (both ELF-MF and radiofrequency [RF]) exposure and risk of brain tumors by subtype (Baldi et al. 2011). Primary brain tumors, including glioma, meningioma, and acoustic neuromas, were identified from the Gironde region in France, with two matched controls selected from electoral rolls. Living within 100 m of a power line was associated with risk of meningioma (OR = 2.99; 95% CI, 0.86–10.4) and acoustic neuroma (OR = 3.23; 95% CI, 0.28–37), but with wide confidence intervals.

In their mortality-based case–control study, Marcilio et al. (2011) also examined residential distance to power lines and brain cancer. Unlike for leukemia, they reported no consistent associations for brain cancer with residential proximity to power lines or calculated ELF-MF fields.

Elliott et al. (2013) included brain and CNS cancers as well in their analysis of residential proximity to power lines in England and Wales. Similarly to breast cancer and leukemia, no risk increase was observed for brain and CNS cancers with either residential distance to power lines or calculated fields.

Occupational Studies, 2005–2012

A population-based case–control study identified acoustic neuroma cases from the Swedish cancer registry and controls from a random sample of the general population (Forssen et al. 2006). Using census-based job titles and exposure levels from sex-specific JEMs (Floderus et al. 1996; Forssen et al. 2004), several exposure metrics were examined, including occupational title, time-weighted average, peak, and rate of change. They observed no evidence for an association between occupational exposure to ELF-MF and acoustic neuroma.

A population-based case–control study examined occupational ELF-MF exposure and risk of glioma including 14 adult cases from Melbourne, Australia, hospitals and controls selected from electoral rolls in the state of Victoria (Karipidis et al. 2007c). For self-reported exposure and exposure based on ELF-MF JEM, they reported an inverse association between exposure and risk of glioma. Based on expert industrial hygienist assessment, however, the OR for glioma was increased in the highest exposure category. They concluded an association between occupational ELF-MF exposure and risk of glioma was not supported.

Johansen et al. (2007) also evaluated the relationship between occupational ELF-MF exposure and brain tumors in their study of Danish electric utility workers. No risk increases were seen among men or women (Table 10.1).

Brain tumor mortality also was examined in the Swiss railway worker cohort (Röösli et al. 2007). They found no evidence for a risk increase or an exposure–response association for brain tumors by using occupational groups as proxies for exposure level.

Kheifets et al. (2008b) also evaluated epidemiologic studies of occupational exposure to ELF-MF and brain cancer in their updated meta-analysis. They reported a small elevation in the RR estimate for brain cancer with occupational ELF-MF exposure (RR = 1.14; 95% CI, 1.07–1.22) based on a total of 47 studies. The results were insensitive to various characteristics of the included studies (such as type of study design, outcome measures, and exposure assessment methods for occupational exposure). Various analytical assumptions, individual studies, weighting schemes, publication bias, or funding source for the individual studies did not appear to have undue influence on the results. Similar to results observed for the leukemia studies, RR estimates in newer brain tumor studies were lower than in older studies. Overall, the more recent studies do not change the conclusion reached by the WHO EHC that the evidence for a potential association between ELF-MF exposure and adult brain cancer is inadequate.

In a hospital-based case–control study of occupational ELF-MF exposure brain tumors, no association was observed between any of the exposure metrics developed, based on detailed job histories and the risk of glioma or meningioma (Coble et al. 2009).

Occupational ELF-MF exposure was also considered in a French case–control study (Baldi et al. 2011). Exposure assessment relied on interview-based job history, a previously developed Swedish JEM, and industrial hygienist evaluation. Lifetime occupational exposure to ELF-MF was not associated with risk of gliomas or acoustic neurinomas. An increased risk was observed for meningioma based on 13 exposed cases.

Sorahan (2012) reported on cancer incidence among U.K. electric utility workers, which included analyses on brain cancer. The observed number of brain cancer cases was close to expectation among males and below expectation among females.

Other Cancers

Mostly in the 1990s, several epidemiologic studies were published focusing on occupational exposure to ELF-MF and male breast cancer, with some showing a positive association. Most of these studies, however, had severe limitations in exposure assessment and were based on small number of exposed cases. The IARC and the WHO EHC evaluations have reviewed numerous epidemiologic studies on a variety of cancer types, some considering residential exposure but most focused on occupational exposure to ELF-MF. The investigated cancers included prostate, pancreatic, lung, kidney, testicular, thyroid, and endometrial cancers and melanoma. Although some of these cancers were associated with exposure to ELF-MF in individual studies, these results were not replicated or remained inconsistent across studies and provided no basis for an association with ELF-MF exposure.

Residential Studies, 2005–2012

Elliott et al. (2013), in their previously described study of residential proximity to power lines and adult cancers, also examined the risk of malignant melanoma. In their assessment, no consistent pattern was observed for risk of malignant melanoma by calculated MF levels or residential distance to power lines.

Occupational Studies, 2005–2012

A hospital-based case–control study was conducted to examine the relationship between occupation and testicular cancer among 229 cases and 800 controls from five French cities (Walschaerts et al. 2007). The OR for welding occupations was 1.49 (95% CI, 0.53–4.15) compared with nonwelding occupations.

Behrens et al. (2010) reported case–control study results on estimated cumulative lifetime occupational exposure to ELF-MF among 298 uveal melanoma patients from several European countries. Controls were selected from variety of sources in the participating countries, including population registries, hospital inpatients, emergency ward patients, and patient lists of general practitioners. They observed an increased risk of uveal melanoma among men and women with dark-colored eyes in the highest percentiles of cumulative exposure to ELF-MF, with ORs of 1.59 (95% CI, 0.56–4.51) and 3.92 (95% CI, 1.29–11.88), respectively.

In the update of the U.K. electricity generation and transmission worker study, Sorahan (2012) examined a wide variety of cancers in addition to the cancers discussed above, including cancer of the eye, stomach, small and large intestines, esophagus, larynx, pleura (mesothelioma), pancreas, skin (melanoma and nonmelanoma), nose and sinus, and small intestines. The patterns of RRs for the individual cancer types appeared consistent, with random variation with similar numbers of elevated and decreased summary relative risks (SRRs). Sorahan (2012) noted significant risk increases for mesothelioma (SRR_{Males} = 331; 95% CI, 303–361), nonmelanoma skin cancers (SRR_{Males} = 107; 95% CI, 104–111), and prostate cancers (SRR_{Males} = 107; 95% CI, 103–111), whereas significant decreases were noted for cancers of the esophagus (SRR_{Males} = 87; 95% CI, 79–96), larynx (SRR_{Males} = 65; 95% CI, 55–77), and the lungs (SRR_{Males} = 82; 95% CI, 78–85).

Electric Fields

Although a vast epidemiologic literature is available investigating the potential effect of ELF-MF on adult cancer development, only a handful of studies examined the role of ELF electric fields (EF). In addition to lack of known metric and mechanism by which ELF-EF may exert a potential effect, a shared difficulty with ELF-MF research, there are several added complexities in ELF-EF exposure assessment. Unlike ELF-MF, ELF-EF are perturbed by the presence of electrically conductive objects, such as vegetation, buildings, and humans, thus making measurements and modeling of ELF-EF exposure significantly more complicated.

Kheifets et al. (2010) reviewed the epidemiologic literature on ELF-EF. Although their review included several childhood cancer studies, only three occupational cohort studies (one among railway workers and two among electric utility workers, with two additional reanalyses from one of the cohorts) and one cancer registry-based case–control study on occupational exposure were identified in relation to adult cancer. An association between occupational exposure to ELF-EF and adult leukemia and non-Hodgkin's lymphoma, but not with brain cancer, was observed in one study population. Another study reported an association with adult brain cancer, but not with leukemia. The other studies showed no association with any of these outcomes. Overall, no consistent pattern for an association with any adult cancer can be discerned from the small number of available studies. Since the 2010 review, no new studies have been published, and there appears to be little momentum, if any, for further research on exposure to ELF-EF and adult cancer.

Methodological Considerations

Epidemiologic studies aim to evaluate potential impact of "exposures" (e.g., environmental stimuli, nutritional components, medication use, genetic characteristics) on health status or health "outcomes" (e.g., mortality, disease incidence). Both exposure and outcome may be defined on a continuum. Exposure assessment, a key step in epidemiologic studies, involves measurement or characterization of various attributes that are proxies for the underlying biologically effective impulse or dose exerting a potential effect (see Chapter 4). As such, most, if not all, exposure variables in epidemiologic studies could be interpreted as proxies for the true primary causal factor, inevitably resulting in exposure misclassification. Methods of exposure assessment evolved extensively in ELF-MF adult cancer epidemiologic studies, most with an underlying assumption that exposures during adulthood are most relevant.

In residential exposure studies, exposure assessment methods included questionnaire-based information on appliance (including electric blanket) use, residential proximity to electric installations (most frequently to high-voltage transmission lines), categorization of homes (e.g., wire configuration codes), or calculated fields in homes based on various parameters (e.g., distance, configuration, load) of nearby transmission lines, short- and long-term measurements in the homes, and personal exposure monitoring. Electric appliance use, relevant for short-term peak exposures, is a very poor predictor for overall average exposure (Mezei et al. 2001). Because appliance use information is typically based on interviews with the study subjects or proxies (e.g., relatives), there is potential for recall bias, as well. Recall bias may be differential if information gathered is related to disease status (see Chapter 4). Recall may also be problematic for certain CNS cancer subtypes because they may affect cognition. Although exposure assessment methods based on residential proximity to high-voltage transmission lines (of these calculated fields being the most reliable metric) may appropriately categorize historical exposure from the power lines themselves, they neglect other exposure sources, such as appliances and those outside the home (e.g., school, work). Measurements, conversely, reflect exposure from all sources but only during the duration of the measurements, thus they may not reflect diurnal, seasonal variations, or environments not included in the measurements. Therefore, integration of time use surveys and application of hierarchical models to account for these exposure sources may be useful in future studies (Greenland 1994).

In occupational studies, job titles are frequently used as exposure categories without any further exposure specification (see Chapter 3). This method results in a very crude classification for distinct exposures as exposure may vary substantially within jobs depending on the specific tasks, environments, and tools used. The quality and source of job title information (e.g., personal or proxy interviews, death certificates) largely influence how well actual exposure is captured. A death certificate may include the last occupation and may not represent lifetime exposure. Reconstruction of lifetime job history based on company records may provide more refined and reliable estimates than job histories constructed based on personal interviews. Quality of information may be further compromised if interviews are conducted with proxies, who may not know the exact nature of work performed, rather than actual study subjects. The resulting exposure misclassification may be differential if information sources vary for cases and controls. For example in the study by Coble et al. (2009), interviews were more likely conducted with proxies for the glioma cases than controls, potentially leading to different levels of exposure misclassification among cases and controls. Job history may be augmented with industrial hygienist assessment or measurement-based JEMs linking jobs to ELF-MF levels. Three of the six case–control studies on occupational exposure and adult cancer discussed above (Karipidis et al. 2007b, 2007c; Coble et al. 2009) used the same population-based JEM to assign exposure levels (Bowman et al. 2007). Although a uniform approach for exposure assignment makes studies more consistent and comparable, differences in actual exposures across populations and geographic locations could arise and result in nonuniform exposure misclassification within the individual studies. Further refinements in JEMs would incorporate tasks, tools, and work environment. These improvements are often more feasible when a JEM is developed within specific industries or companies (Kheifets et al. 2008a).

Inclusion of various exposure surrogates and exposure metrics can assist researchers in assessing consistency of a potential association with the main exposure of interest. Coble et al. (2009), for example, examined several exposure metrics, such as maximum exposure, duration >1.5 mG, lifetime average, and cumulative exposure, in association with brain cancer risk and reported no association with any of those metrics, even though previous studies reported increased brain cancer risk with certain job titles. This argues against an effect of ELF-MF on brain cancer risk. An association with specific job titles could potentially be influenced or biased by other co-exposures encountered during work whereas measured or calculated fields are specific to ELF-MF exposure.

Latency period is the interval between exposure and disease onset. Its actual length is generally unknown but could be up to 20 years or longer for some adult cancers, thus presenting additional challenges for epidemiologists. Determining exposure levels at the etiologically relevant time period remains a key and frequently elusive task in an epidemiologic study. A common practice is to use exposure "windows" or lag periods, considering only certain time intervals when exposure is determined. Forssen et al. (2006) examined exposure within 10 years, and at least 10 years before reference date in their study of acoustic neuroma, whereas Karipidis et al. (2007b) used 5-year exposure lag (i.e., excluding exposure during the past 5 years) in their study of NHL. Lowenthal et al. (2007) investigated early childhood exposure and risk of LPDs and MPDs among adults. Studies that present RR estimates for exposures at different time intervals can contribute to better understanding of disease etiology and where, in the exposure continuum, future research needs to be focused.

With regard to disease outcome, the epidemiologic evidence for a link between exposure and disease is strengthened with use of incidence over prevalence or mortality data. Prevalence of, and mortality from a disease are influenced by disease duration, case fatality, and recovery rate, in addition to disease development. Overall, studies published after 2005 have sought to identify newly diagnosed cases of cancers with few studies using prevalence or mortality. Although not unique to ELF-MF studies, misclassification of adult cancer is also a potential problem. Grouping of heterogeneous cancer subtypes may lead to unclear, inconsistent associations. Histologic verification of cancers, identification of cancer subtypes by genetic markers and use of uniform classification systems can greatly reduce the magnitude of disease misclassification. As an example, some studies have attempted to improve outcome classification by including histopathologic confirmation or expert assessment of disease diagnosis (Wong et al. 2010; Baldi et al. 2011). Other studies have attempted to assess disease risk by examining glioma apart from meningioma (Baldi et al. 2011) or by analyzing the data using grading of the disease (Karipidis et al. 2007c). This strategy could pinpoint risk increases if the risk factor under investigation plays a role in the pathogenesis of only certain subtypes of the disease. Epidemiologic studies are frequently challenged by insufficient numbers of cases in specific subgroups. In such cases, presentation of subtype distribution and discussion of grouping rationale may be informative (Ray et al. 2007).

Selection bias, a systematic error in which the study or analytical sample does not reflect the source population from which cases and controls arise, can also occur in epidemiologic studies (see Chapter 5). Selection bias could be, at least theoretically, eliminated or minimized in a case–control study with well-defined population at risk, complete identification of cases from that population at risk, and with a representative sample of noncases as controls from the same population. This is most frequently accomplished with registry and population-based studies. Of the 11 case–control studies published after 2005, 8 were population based (Mester et al. 2006; Davis and Mirick 2007; Lowenthal et al. 2007; McElroy et al. 2007; Karipidis et al. 2007b; Baldi et al. 2011; Marcilio et al. 2011; Elliott et al. 2013). In hospital-based studies, selection bias may develop if patients of certain rare cancer types are selected as cases and controls are selected from among patients with more trivial diseases. Controls could likely come from local regions as opposed to cases, who may travel from a larger area to specialized centers for diagnosis and treatment. In such instances, controls in a hospital-based study may not fully represent the population at risk for the cases. Nonresponse or missing data could still lead to a form of selection bias, even with population-based designs, if participation is affected by disease and exposure status. For example, in the study of Baldi et al. (2011), although participation rates were similar among cases and controls, the reasons for nonparticipation were different in the two groups; this may lead to selection bias if these reasons are also related to exposure status. In cohort studies, selection bias may stem from differential loss to follow-up or differential censoring from the study. Choice of comparison population may greatly affect results in occupational cohort studies. Internal comparison within the cohort is typically preferred (e.g., Johansen et al. 2007) because it may eliminate or account for the healthy worker effect seen in occupational cohort studies that use external comparison groups (Sorahan 2012).

In studies that solely consider occupational titles as the exposure surrogate, along with uncertain and crude characterization of exposure, the possibility of confounding by other potential co-exposures is also introduced (see Chapter 6). Potential confounding may also be introduced by socioeconomic status and education associated with job titles. Correlated exposures and potential confounders may be accounted for within JEMs. Use of hierarchical models can aid in reducing interdependencies between various exposures and the number of comparisons required in a study (Greenland 2008). No occupational ELF-MF studies have incorporated hierarchical models to date.

Finally, statistical considerations may introduce biases into the best-designed studies. These considerations included sparse data within exposure levels (Greenland et al. 2000), model misspecification, and rationale regarding exposure level cut-points (see Chapter 4). Recent studies have demonstrated the value of understanding changes in estimates with varying assumptions and evaluating uncertainties using sensitivity analyses (Geneletti et al. 2013). Probabilistic sensitivity analyses could also be used to examine the impacts of exposure and outcome misclassification, selection bias, confounding, and random error with assigning a range of probability distributions to various bias parameters (Lash et al. 2009).

Summary

As adult cancer represents a significant societal burden and ELF-MF is a ubiquitous exposure in modern societies, a potential relationship between them, even if weak, may have significant public health impact. Over the past, more than, three decades, a large body of epidemiologic literature emerged examining the relationship between residential and occupational exposure to ELF-MF and adult cancers. The evidence is substantially more limited on ELF-EF exposure. Among both the residential and occupational studies, wide variation in quality, study design, exposure assessment, outcome measure, and other study characteristics exists. Over time, however, significant improvements have advanced the science. More recent studies tended to include histologically verified incident cases, relied on more sophisticated exposure assessment methods, and more of them were designed to primarily focus on the potential carcinogenic effects of ELF-MF exposure as opposed to being an opportunistic observation. In spite of these recent improvements, significant uncertainties remain with most cancer outcomes.

The strongest but overall still weak evidence has emerged in studies of adult brain cancer and adult leukemia. For both of these diseases, although large risk increases could be excluded, ELF-MF exposure appears to be associated with a small risk increase when the entirety of the epidemiologic literature is considered. Some features, however, such as no consistent exposure–response relationship, no clear pattern within subtypes and overall decrease in RR estimates in more recent studies, argue against causality for both adult brain cancer and leukemia. For female breast cancer, the more recently accumulated evidence suggests that ELF-MF exposure is, likely, not related to the disease. For other cancers, the available epidemiologic evidence is not sufficient to draw firm conclusion, as only sporadic data are available based on either single studies or underpowered studies, showing mostly inconsistent results.

References

Baldi, I., Coureau, G., Jaffre, A., Gruber, A., Ducamp, S., Provost, D., et al. 2011. Occupational and residential exposure to electromagnetic fields and risk of brain tumors in adults: a case-control study in Gironde, France. *International Journal of Cancer* 129(6): 1477–84.

Band, P. R., Le, N. D., Fang, R., Deschamps, M., Gallagher, R. P. and Yang, P. 2000. Identification of occupational cancer risks in British Columbia. A population-based case-control study of 995 incident breast cancer cases by menopausal status, controlling for confounding factors. *Journal of Occupational and Environmental Medicine* 42(3): 284–310.

Band, P. R., Le, N. D., Fang, R. and Gallagher, R. 2004. Identification of occupational cancer risks in British Columbia: a population-based case-control study of 769 cases of non-Hodgkin's lymphoma analyzed by histopathology subtypes. *Journal of Occupational and Environmental Medicine* 46(5): 479–89.

Behrens, T., Lynge, E., Cree, I., Sabroe, S., Lutz, J. M., Afonso, N., et al. 2010. Occupational exposure to electromagnetic fields and sex-differential risk of uveal melanoma. *Occupational and Environmental Medicine* 67(11): 751–9.

Bethwaite, P., Cook, A., Kennedy, J. and Pearce, N. 2001. Acute leukemia in electrical workers: a New Zealand case-control study. *Cancer Causes & Control* 12(8): 683–9.

Bjork, J., Albin, M., Welinder, H., Tinnerberg, H., Mauritzson, N., Kauppinen, T., et al. 2001. Are occupational, hobby, or lifestyle exposures associated with Philadelphia chromosome positive chronic myeloid leukaemia? *Occupational and Environmental Medicine* 58(11): 722–7.

Bondy, M. L., Scheurer, M. E., Malmer, B., Barnholtz-Sloan, J. S., Davis, F. G., Il'yasova, D., et al. 2008. Brain tumor epidemiology: consensus from the Brain Tumor Epidemiology Consortium. *Cancer* 113(7 Suppl): 1953–68.

Bowman, J., Touchstone, J. and Yost, M. 2007. A population-based job exposure matrix for power-frequency magnetic fields. *Journal of Occupational and Environmental Hygiene* 4(9): 715–28.

Cano, M. I. and Pollan, M. 2001. Non-Hodgkin's lymphomas and occupation in Sweden. *International Archives of Occupational and Environmental Health* 74(6): 443–9.

Chen, C., Ma, X., Zhong, M. and Yu, Z. 2010. Extremely low-frequency electromagnetic fields exposure and female breast cancer risk: a meta-analysis based on 24,338 cases and 60,628 controls. *Breast Cancer Research and Treatment* 123(2): 569–76.

Coble, J. B., Dosemeci, M., Stewart, P. A., Blair, A., Bowman, J., Fine, H. A., et al. 2009. Occupational exposure to magnetic fields and the risk of brain tumors. *Neuro-Oncology* 11(3): 242–9.

Coogan, P. F., Clapp, R. W., Newcomb, P. A., Wenzl, T. B., Bogdan, G., Mittendorf, R., et al. 1996. Occupational exposure to 60-hertz magnetic fields and risk of breast cancer in women. *Epidemiology* 7(5): 459–64.

Davis, S. and Mirick, D. K. 2007. Residential magnetic fields, medication use, and the risk of breast cancer. *Epidemiology* 18(2): 266–9.

Davis, S., Mirick, D. K. and Stevens, R. G. 2002. Residential magnetic fields and the risk of breast cancer. *American Journal of Epidemiology* 155(5): 446–54.

De Roos, A. J., Stewart, P. A., Linet, M. S., Heineman, E. F., Dosemeci, M., Wilcosky, T., et al. 2003. Occupation and the risk of adult glioma in the United States. *Cancer Causes & Control* 14(2): 139–50.

Elliott, P., Shaddick, G., Douglass, M., de Hoogh, K., Briggs, D. J. and Toledano, M. B. 2013. Adult cancers near high-voltage overhead power lines. *Epidemiology* 24: 184–90.

Fabbro-Peray, P., Daures, J. P. and Rossi, J. F. 2001. Environmental risk factors for non-Hodgkin's lymphoma: a population-based case-control study in Languedoc-Roussillon, France. *Cancer Causes & Control* 12(3): 201–12.

Floderus, B., Persson, T. and Stenlund, C. 1996. Magnetic-field exposures in the workplace: reference distribution and exposures in occupational groups. *International Journal of Occupational and Environmental Health* 2(3): 226–38.

Forssen, U. M., Lonn, S., Ahlbom, A., Savitz, D. A. and Feychting, M. 2006. Occupational magnetic field exposure and the risk of acoustic neuroma. *American Journal of Industrial Medicine* 49(2): 112–18.

Forssen, U. M., Mezei, G., Nise, G. and Feychting, M. 2004. Occupational magnetic field exposure among women in Stockholm County, Sweden. *Occupational and Environmental Medicine* 61(7): 594–602.

Forssen, U. M., Rutqvist, L. E., Ahlbom, A. and Feychting, M. 2005. Occupational magnetic fields and female breast cancer: a case-control study using Swedish population registers and new exposure data. *American Journal of Epidemiology* 161(3): 250–9.

Gardner, K. M., Ou Shu, X., Jin, F., Dai, Q., Ruan, Z., Thompson, S. J., et al. 2002. Occupations and breast cancer risk among Chinese women in urban Shanghai. *American Journal of Industrial Medicine* 42(4): 296–308.

Geneletti, S., Best, N., Toledano, M. B., Elliott, P. and Richardson, S. 2013. Uncovering selection bias in case-control studies using Bayesian post-stratification. *Statistics in Medicine* 32: 2555–70.

Greenland, S. 1994. Hierarchical regression for epidemiologic analyses of multiple exposures. *Environmental Health Perspectives* 102(Suppl 8): 33.

Greenland, S. 2008. Multiple comparisons and association selection in general epidemiology. *International Journal of Epidemiology* 37(3): 430–4.

Greenland, S., Schwartzbaum, J. A. and Finkle, W. D. 2000. Problems due to small samples and sparse data in conditional logistic regression analysis. *American Journal of Epidemiology* 151(5): 531–9.

Hakansson, N., Floderus, B., Gustavsson, P., Johansen, C. and Olsen, J. H. 2002. Cancer incidence and magnetic field exposure in industries using resistance welding in Sweden. *Occupational and Environmental Medicine* 59(7): 481–6.

IARC. 2002. Non-ionizing radiation, Part 1: static and extremely low-frequency (ELF) electric and magnetic fields. *IARC Monographs on the Evaluation of Carcinogenic Risks to Humans* 80: 1–395.

IARC. 2013. *GLOBOCAN Breast Cancer Incidence and Mortality Worldwide in 2008 Summary*. Available from: http://globocan.iarc.fr/factsheets/cancers/breast.asp

Johansen, C. 2004. Electromagnetic fields and health effects—epidemiologic studies of cancer, diseases of the central nervous system and arrhythmia-related heart disease. *Scandinavian Journal of Work, Environment & Health* 30(Suppl 1): 1–30.

Johansen, C., Raaschou Nielsen, O., Olsen, J. H. and Schüz, J. 2007. Risk for leukaemia and brain and breast cancer among Danish utility workers: a second follow-up. *Occupational and Environmental Medicine* 64(11): 782–4.

Karipidis, K., Benke, G., Sim, M., Fritschi, L., Yost, M., Armstrong, B., et al. 2007a. Occupational exposure to power frequency magnetic fields and risk of non-Hodgkin lymphoma. *Occupational and Environmental Medicine* 64(1): 25–9.

Karipidis, K. K., Benke, G., Sim, M. R., Kauppinen, T., Kricker, A., Hughes, A. M., et al. 2007b. Occupational exposure to ionizing and non-ionizing radiation and risk of non-Hodgkin lymphoma. *International Archives of Occupational and Environmental Health* 80(8): 663–70.

Karipidis, K. K., Benke, G., Sim, M. R., Yost, M. and Giles, G. 2007c. Occupational exposure to low frequency magnetic fields and the risk of low grade and high grade glioma. *Cancer Causes & Control* 18(3): 305–13.

Kheifets, L., Bowman, J. D., Checkoway, H., Feychting, M., Harrington, M., Kavet, R., et al. 2008a. Future needs of occupational epidemiology of extremely low frequency (ELF) electric and magnetic fields (EMF): review and recommendations. *Occupational and Environmental Medicine* 66(2): 72–80.

Kheifets, L., Monroe, J., Vergara, X., Mezei, G. and Afifi, A. A. 2008b. Occupational electromagnetic fields and leukemia and brain cancer: an update to two meta-analyses. *Journal of Occupational and Environmental Medicine* 50(6): 677–88.

Kheifets, L., Renew, D., Sias, G. and Swanson, J. 2010. Extremely low frequency electric fields and cancer: assessing the evidence. *Bioelectromagnetics* 31(2): 89–101.

Kleinerman, R. A., Linet, M. S., Hatch, E. E., Tarone, R. E., Black, P. M., Selker, R. G., et al. 2005. Self-reported electrical appliance use and risk of adult brain tumors. *American Journal of Epidemiology* 161(2): 136–46.

Kliukiene, J., Tynes, T. and Andersen, A. 2003. Follow-up of radio and telegraph operators with exposure to electromagnetic fields and risk of breast cancer. *European Journal of Cancer Prevention* 12(4): 301–7.

Kliukiene, J., Tynes, T. and Andersen, A. 2004. Residential and occupational exposures to 50-Hz magnetic fields and breast cancer in women: a population-based study. *American Journal of Epidemiology* 159(9): 852–61.

Koc, M. and Polat, P. 2001. Epidemiology and aetiological factors of male breast cancer: a ten years retrospective study in eastern Turkey. *European Journal of Cancer Prevention* 10(6): 531–4.

Krishnan, G., Felini, M., Carozza, S. E., Miike, R., Chew, T. and Wrensch, M. 2003. Occupation and adult gliomas in the San Francisco Bay Area. *Journal of Occupational and Environmental Medicine* 45(6): 639–47.

Labreche, F., Goldberg, M. S., Valois, M. F., Nadon, L., Richardson, L., Lakhani, R., et al. 2003. Occupational exposures to extremely low frequency magnetic fields and postmenopausal breast cancer. *American Journal of Industrial Medicine* 44(6): 643–52.

Lash, T., Fox, M. P. and Fink, A. K. 2009. *Applying Quantitative Bias Analysis to Epidemiologic Data*. Boston, MA: Springer.

London, S. J., Pogoda, J. M., Hwang, K. L., Langholz, B., Monroe, K. R., Kolonel, L. N., et al. 2003. Residential magnetic field exposure and breast cancer risk: a nested case-control study from a multiethnic cohort in Los Angeles County, California. *American Journal of Epidemiology* 158(10): 969–80.

Lowenthal, R. M., Tuck, D. M. and Bray, I. C. 2007. Residential exposure to electric power transmission lines and risk of lymphoproliferative and myeloproliferative disorders: a case-control study. *Internal Medicine Journal* 37(9): 614–19.

Marcilio, I., Gouveia, N., Pereira Filho, M. L. and Kheifets, L. 2011. Adult mortality from leukemia, brain cancer, amyotrophic lateral sclerosis and magnetic fields from power lines: a case-control study in Brazil. *Revista brasileira de epidemiologia = Brazilian Journal of Epidemiology* 14(4): 580–8.

McElroy, J. A., Egan, K. M., Titus-Ernstoff, L., Anderson, H. A., Trentham-Dietz, A., Hampton, J. M., et al. 2007. Occupational exposure to electromagnetic field and breast cancer risk in a large, population-based, case-control study in the United States. *Journal of Occupational and Environmental Medicine* 49(3): 266–74.

McElroy, J. A., Newcomb, P. A., Remington, P. L., Egan, K. M., Titus-Ernstoff, L., Trentham-Dietz, A., et al. 2001. Electric blanket or mattress cover use and breast cancer incidence in women 50–79 years of age. *Epidemiology* 12(6): 613–17.

Mester, B., Nieters, A., Deeg, E., Elsner, G., Becker, N. and Seidler, A. 2006. Occupation and malignant lymphoma: a population based case control study in Germany. *Occupational and Environmental Medicine* 63(1): 17–26.

Mezei, G., Kheifets, L. I., Nelson, L. M., Mills, K. M., Iriye, R. and Kelsey, J. L. 2001. Household appliance use and residential exposure to 60-Hz magnetic fields. *Journal of Exposure Analysis and Environmental Epidemiology* 11(1): 41–9.

Navas-Acien, A., Pollan, M., Gustavsson, P., Floderus, B., Plato, N. and Dosemeci, M. 2002. Interactive effect of chemical substances and occupational electromagnetic field exposure on the risk of gliomas and meningiomas in Swedish men. *Cancer Epidemiology, Biomarkers & Prevention* 11(12): 1678–83.

Ohgaki, H. 2009. Epidemiology of brain tumors. In *Methods of Molecular Biology, Cancer Epidemiology,* Vol. 472, edited by M. Verma. Totowa, NJ: Humana Press, 323–42.

Oppenheimer, M. and Preston-Martin, S. 2002. Adult onset acute myelogenous leukemia and electromagnetic fields in Los Angeles County: bed-heating and occupational exposures. *Bioelectromagnetics* 23(6): 411–15.

Pollan, M., Gustavsson, P. and Floderus, B. 2001. Breast cancer, occupation, and exposure to electromagnetic fields among Swedish men. *American Journal of Industrial Medicine* 39(3): 276–85.

Ray, R. M., Gao, D. L., Li, W., Wernli, K. J., Astrakianakis, G., Seixas, N. S., et al. 2007. Occupational exposures and breast cancer among women textile workers in Shanghai. *Epidemiology* 18(3): 383–92.

Repacholi, M. 2012. Concern that "EMF" magnetic fields from power lines cause cancer. *The Science of the Total Environment* 426: 454–8.

Röösli, M., Lortscher, M., Egger, M., Pfluger, D., Schreier, N., Lortscher, E., et al. 2007. Leukaemia, brain tumours and exposure to extremely low frequency magnetic fields: cohort study of Swiss railway employees. *Occupational and Environmental Medicine* 64(8): 553–9.

Rubtsova, N. B., Tikhonova, G. I. and Gurvich, E. B. 1999. Study of commercial frequency electromagnetic field effects on human health and their hygienic rating criteria. Paper read at Electromagnetic fields: biological effects and hygienic standardization. *Proceedings of an International Meeting,* Moscow, 18–22 May, 1998, Geneva, Switzerland.

Schlehofer, B., Hettinger, I., Ryan, P., Blettner, M., Preston-Martin, S., Little, J., et al. 2005. Occupational risk factors for low grade and high grade glioma: results from an international case control study of adult brain tumours. *International Journal of Cancer.* 113(1): 116–25.

Schoenfeld, E. R., O'Leary, E. S., Henderson, K., Grimson, R., Kabat, G. C., Ahnn, S., et al. 2003. Electromagnetic fields and breast cancer on Long Island: a case-control study. *American Journal of Epidemiology* 158(1): 47–58.

Schüz, J., Lagorio, S. and Bersani, F. 2009. Electromagnetic fields and epidemiology: an overview inspired by the fourth course at the International School of Bioelectromagnetics. *Bioelectromagnetics* 30(7): 511–24.

Siegel, R., Naishadham, D. and Jemal, A. 2013. Cancer statistics, 2013. *CA: A Cancer Journal for Clinicians* 63(1): 11–30.

Sorahan, T. 2012. Cancer incidence in UK electricity generation and transmission workers, 1973–2008. *Occupational Medicine* 62(7): 496–505.

SSM. 2010. *Recent Research on EMF and Health Risks. Seventh Annual Report from SSM:s Independent Expert Group on Electromagnetic Fields, 2010.* Document Number: 2010:44. Stockholm, Sweden: SSM.

Stevens, R. G. and Davis, S. 1996. The melatonin hypothesis: electric power and breast cancer. *Environmental Health Perspectives* 104(Suppl 1): 135–40.

Teitelbaum, S. L., Britton, J. A., Gammon, M. D., Schoenberg, J. B., Brogan, D. J., Coates, R. J., et al. 2003. Occupation and breast cancer in women 20-44 years of age (United States). *Cancer Causes & Control: CCC* 14(7): 627–37.

Tynes, T. and Haldorsen, T. 2003. Residential and occupational exposure to 50 Hz magnetic fields and hematological cancers in Norway. *Cancer Causes & Control: CCC* 14(8): 715–20.

Van Wijngaarden, E., Nylander-French, L. A., Millikan, R. C., Savitz, D. A. and Loomis, D. 2001b. Population-based case-control study of occupational exposure to electromagnetic fields and breast cancer. *Annals of Epidemiology* 11(5): 297–303.

Van Wijngaarden, E., Savitz, D. A., Kleckner, R. C., Kavet, R. and Loomis, D. 2001b. Mortality patterns by occupation in a cohort of electric utility workers. *American Journal of Industrial Medicine* 40(6): 667–73.

Vijayalaxmi and Prihoda, T. J. 2009. Genetic damage in mammalian somatic cells exposed to extremely low frequency electro-magnetic fields: a meta-analysis of data from 87 publications (1990–2007). *International Journal of Radiation Biology* 85(3): 196–213.

Villeneuve, P. J., Agnew, D. A., Johnson, K. C. and Mao, Y. 2002. Brain cancer and occupational exposure to magnetic fields among men: results from a Canadian population-based case-control study. *International Journal of Epidemiology* 31(1): 210–17.

Walschaerts, M., Muller, A., Auger, J., Bujan, L., Guerin, J. F., Le Lannou, D., et al. 2007. Environmental, occupational and familial risks for testicular cancer: a hospital-based case-control study. *International Journal of Andrology* 30(4): 222–9.

Wertheimer, N. and Leeper, E. 1979. Electrical wiring configurations and childhood cancer. *American Journal of Epidemiology* 109(3): 273–84.

Wesseling, C., Pukkala, E., Neuvonen, K., Kauppinen, T., Boffetta, P. and Partanen, T. 2002. Cancer of the brain and nervous system and occupational exposures in Finnish women. *Journal of Occupational and Environmental Medicine* 44(7): 663–8.

WHO. 2007. *Environmental Health Criteria 238: Extremely Low Frequency Fields*. Geneva, Switzerland: World Health Organization.

Willett, E. V., McKinney, P. A., Fear, N. T., Cartwright, R. A. and Roman, E. 2003. Occupational exposure to electromagnetic fields and acute leukaemia: analysis of a case-control study. *Occupational and Environmental Medicine* 60(8): 577–83.

Wong, O., Harris, F., Wang, Y. and Fu, H. 2010. A hospital-based case-control study of non-Hodgkin lymphoid neoplasms in Shanghai: analysis of personal characteristics, lifestyle, and environmental risk factors by subtypes of the WHO classification. *Journal of Occupational and Environmental Medicine* 52(1): 39–53.

Zhu, K., Hunter, S., Payne-Wilks, K., Roland, C. L. and Forbes, D. S. 2003. Use of electric bedding devices and risk of breast cancer in African-American women. *American Journal of Epidemiology* 158(8): 798–806.

11

Neurodegenerative Diseases and ELF-EMF

Anke Huss and Roel Vermeulen

CONTENTS

Neurodegenerative Diseases

Neurons, also called nerve cells, are cells that transmit and process information. Together with the brain and the spinal cord, they form the nervous system. When neurons lose structure or function, or die, it is called neurodegeneration [1]. If this loss cannot be compensated, a neurodegenerative disease develops [2], causing problems with movement or mental functioning. This chapter focuses on Alzheimer's disease, Parkinson's disease, and amyotrophic lateral sclerosis (ALS), conditions that represent the most common neurodegenerative diseases and those which have been investigated in electromagnetic field (EMF)-related research. Another relatively common neurodegenerative disease, Huntington's disease, is largely hereditary and is not discussed here. Also not summarized in this book chapter were several other neurodegenerative diseases (e.g., lewy body disease, spinal muscular atrophy) that have never been the focus of EMF-related research.

Alzheimer's Disease/Senile Dementia

Dementia involves a gradual deterioration of a person's abilities with regard to memory, attention, language, and thinking. Alzheimer's disease is the most common form of dementia, accounting for around 70% of all dementia cases [3]. Alzheimer's disease has been associated with the extracellular accumulation of the protein beta-amyloid, called plaques, and the intracellular accumulation of the protein tau, called tangles [4]. The accumulation of the proteins interferes with the neuron-to-neuron communication of synapses, the connectors between neurons, and contributes to cell death [5]. Alzheimer's disease eventually leads to death. Incidence of Alzheimer's disease increases with age, peaking after age 75 [5]. A small percentage of Alzheimer's disease is caused by genetic mutations, but the majority occur as a sporadic form, likely as a complex interaction between genetic and environmental factors [6].

Parkinson's Disease/Parkinsonism

Parkinsonism is an umbrella term for several conditions that share symptoms with Parkinson's disease. Parkinson's disease is a motor system disorder, the result of the loss of brain cells that produce dopamine. Parkinson's disease is characterized by tremors (shaking), bradykinesia (slowness of movement), rigidity (stiffness of limbs and trunk), or postural instability (impaired balance and coordination) [7]. Parkinson's disease is a progressive disease that can lead to severe impairment and disability. It is the second most frequent neurodegenerative disease after Alzheimer's disease [8]. Average age of onset of Parkinson's disease is around age 50–60 years [9]. Around 90% of the cases are estimated to be sporadic (i.e., not hereditary).

ALS/Motor Neuron Disease

Motor neurons are nerve cells in the brain, brainstem, and spinal cord that control the voluntary muscles of the body. Motor neuron diseases are characterized by a gradual degeneration and death of motor neurons [10]. In ALS, motor neurons degenerate, causing progressive muscle atrophy, weakness, and spasticity [11]. This leads to paralysis and premature death, in most cases within 5 years of disease onset. Incidence of ALS increases with age, peaking approximately between 55 and 75 years. Some percentage of the cases inherit ALS, but no genetic component is apparent in the majority of cases [12].

Why Research Neurodegenerative Diseases in Association with Magnetic Field Exposures?

To date, there is no known mechanism by which magnetic field exposures could cause neurodegenerative diseases. However, what Alzheimer's disease, Parkinson's disease, and ALS have in common is that they are chronic, progressive, and, to date, are without a cure. In addition, they pose a large burden on patients and caregivers. If indeed the majority of Alzheimer's disease, Parkinson's disease, and ALS cases are sporadic, this implies that environmental (i.e., nongenetic) factors play a large role in the development of these diseases. Understanding modifiable risk factors could therefore provide opportunities for prevention.

Some early case reports associated ALS with trauma, and a report in 1931 linked ALS-like symptoms to electrical trauma [13]. Electrical injury can involve cardiac arrest; thermal burns; and muscle, nerve, and tissue destruction. The extent of the injuries depends on current, voltage, tissue resistance, the pathway of the current through the body, and the duration of the contact [14]. Depending on the type of accident, harm from an electrical injury can also include neurological symptoms. Several early case reports, however, described signs of ALS that had developed outside the pathway of the electrical current, which spoke against a direct effect of the current. A case–control study published in 1964 reported that persons with ALS had more occupational contact with electricity compared with controls, and linked the disease to frequent, but minor electrical shocks that had not involved hospitalization [15]. This report was followed up in a case–control study in 1986 that observed an association of ALS with shocks producing unconsciousness [16]. In addition to electrical shocks, in this study, occupations "at risk of electrical exposure" were also evaluated. Such occupations included, for example, electrical or electronic engineers, technicians or mechanics, power station operators, or conductors. People who were diagnosed with ALS were more likely to have worked in one of these "electrically related" occupations.

After these initial publications, research interest subsequently shifted to include other neurodegenerative diseases such as Alzheimer's disease [17,18] or Parkinson's disease [19]. In addition, magnetic field exposures became the main research focus once measurement devices able to log this exposure became available (somewhere around 1990).

Issues in Exposure Assessment

Magnetic field measurements, or any other workplace measurement for that matter, are time intensive and thus costly to do. In addition, if exposure has to be assessed retrospectively, measurements might not be possible. Although some studies performed measurement surveys with the aim to assess magnetic field exposure levels and to use the data for subsequent health analyses, others have used (available) measurement data to develop so-called job–exposure matrices (JEMs), a cross-tabulation of average exposure levels per job. A JEM can subsequently be used in an epidemiological study to assess the association between occupational exposure and disease. This approach has been widely applied (see Chapters 3, 4, and 7).

A few studies have relied on self-reports of exposure, but because magnetic fields represent an exposure that cannot be seen or felt (at least not at the low levels encountered in the majority of jobs), such an information would be likely very unreliable.

Furthermore, because no mechanisms are known for the possible association between electrical occupations-related exposures and neurodegenerative diseases, it has been unclear what exposure should be estimated (magnetic or electric fields, electrical shocks, nuisance shocks, or contact currents). The same holds for the definition of the exposure metric, such as ever having been exposed, maximum exposure, cumulative exposure, and so forth.

Errors may also have been introduced into analyses of studies depending on how study participants' type of occupation was determined. For example, in many studies occupation at only one point in time was used to assess exposure levels, for example, at baseline of the cohort, based on longest held occupation, or current occupation. This would mean misclassifying exposures that might have occurred in occupations at earlier or later time points. In case–control studies, reported occupations might potentially suffer from recall bias, if study participants are aware of the study hypothesis (see Chapter 4). In addition, recall of study participants may be affected especially when analyzing Alzheimer's disease where it is not uncommon for researchers to rely on information provided by "the most knowledgeable informant," often these are family members. In other cases, studies have used occupational information as provided on death certificates, likely further reducing the accuracy of the occupational information compared with family member information.

Issues in Outcome Assessment

The ability to diagnose neurodegenerative diseases has improved over time, especially with the introduction of magnetic resonance imaging (MRI) and biomarkers. However, because patients are mostly seen as outpatients, and because disease registries for neurological diseases do not exist in most countries, inclusion of newly diagnosed cases in studies has represented a challenge. Several researchers have therefore used mortality registries to identify cases with neurodegenerative diseases. Death certificates are however inherently less reliable compared with, for example, hospital or clinical records, leading to underreporting of the disease. Underreporting can be expected especially for diseases that are not necessarily fatal, such as Alzheimer's and Parkinson's diseases. In addition, Alzheimer's disease and Parkinson's disease, in particular, have a slow disease progression; thus, a large uncertainty about the time of onset. Knowing the time of onset, however, might be important to be able to evaluate relevant time periods of exposure.

Overview of Literature on Occupational Magnetic Field Exposure Studies

Here, we provide an overview of studies analyzing the effect of occupational magnetic field exposures on Alzheimer's disease, Parkinson's disease, and ALS. Several previous reviews have noted that the individual studies report different (heterogenous) results that could not easily be explained with methodological aspects of the different studies (e.g., design, or participation rates). Reviews to date, however, have rarely focused on possible differences in the methods and quality of exposure assessment, and how these may relate to heterogeneity in study outcomes. For this overview, we therefore checked whether study characteristics, in particular differences in exposure assessment of the individual studies, helped in explaining why study results had been divergent. In particular, we assessed whether a quantitative assessment of the exposure was performed, if exposure was assessed based on self-report, at how many time points occupation was assessed and so on. We did this by performing a meta-analysis, a statistical synthesis of results of a series of studies.

The results of our overview include studies that were published until February 2013. Studies are shown that reported risks for workers exposed to "high" versus "low" magnetic field exposure. Within studies where measurements were performed, or where JEMs were used, the microtesla reported here corresponds to the lower cutoffs of the (average) exposure level to which workers were exposed. Because some studies reported risks for "electrical workers" rather than for exposure of magnetic fields, these are presented separately. Studies based on death certificates [20–23], which are not suitable to correctly assign the occupation of persons potentially exposed to magnetic fields [24], and studies that used self-reported exposure [25,26] were also treated as a separate group.

Studies differed in many aspects: although some performed case–control studies in the general population, others performed cohort studies in workers that included many higher exposed persons such as railway or utility workers [19,27–31]. Characteristics of all studies are listed in Table 11.1.

TABLE 11.1
Study Characteristics

	Outcome	Outcome: Source of Information	Population	Exposure	Exposure Information Source	Time Point of Exposure Assessment
Buckley 1983 [20]	ALS	Death certificates	Deceased from England and Wales	Electrical and electronics workers	Death certificates	Unclear
Deapen 1986 [16]	ALS	Physician diagnosis	Cases via ALS society, controls were neighbors, workmates, or acquaintances of cases	Occupational magnetic field (MF), self-reported electric shocks with unconsciousness	Interviews/ questionnaire	Occupation 3 years before diagnosis
Gunnarsson 1991 [32]	ALS	Death certificates	Swedish population	Occupation: electricity work	Census	Occupation at census
Gunnarsson 1992 [25]	ALS	Medical records	Cases from departments of neurology and internal medicine, general population controls	Occupation: electricity work, self-reported exposure	Interviews/ questionnaire	Occupational history
Sobel 1995 [17]	Alzheimer's disease (AD)	Hospital records	Cases were diagnosed at AD hospital, controls were demented and nondemented patients without AD	Occupational MF	Interviews/ questionnaire	Primary occupation
Sobel 1996 [18]	AD	Hospital records	Cases were diagnosed at AD hospital, controls were demented and nondemented patients without AD	Occupational MF	Interviews/ questionnaire	Primary occupation
Davanipour 1997 [33]	ALS	Diagnostic criteria	Clinic-based cases, blood and nonblood relatives as controls	Occupational MF	Interviews/ questionnaire	Occupational history

Continued

TABLE 11.1 (*Continued*)

Study Characteristics

	Outcome	Outcome: Source of Information	Population	Exposure	Exposure Information Source	Time Point of Exposure Assessment
Savitz 1998 a [23]	AD, Parkinson's disease (PD), ALS	Death certificates	Deceased that had occupational information on death certificate from 25 U.S. states	Occupational MF	Death certificates	Primary occupation
Savitz 1998 b [19]	AD, PD, ALS	Death certificates	Electric utility workers	Occupational MF	Occupational records	Occupational history
Johansen 1998 [29]	ALS	Death certificates	Utility companies	Occupation	Occupational records	Occupation at baseline/census
Feychting 1998 [34]	AD	Screening	Twins	Occupational MF, electrical occupations	Interviews/ Questionnaire	Last and primary occupation
Graves 1999 [35]	AD	Diagnostic criteria	Cases enrolled in Health Maintenance Organization clinics, controls were nondemented patients of same program	Occupational MF	Interviews/ Questionnaire	Occupational history
Johansen 2000 [36]	AD, PD, ALS	Hospital records	Utility companies	Occupational MF	Occupational records	Occupation at baseline/census
Noonan 2002 [21]	AD, PD, ALS	Death certificates	Deceased aged at least 60 years from Colorado	Occupational MF, electrical occupations	Death certificates	Primary occupation
Harmanci 2003 [37]	AD	Diagnostic criteria	Population aged 70 or older of Kadikoy, Turkey	Occupational MF	Interviews/ questionnaire	Primary occupation
Håkansson 2003 [27]	AD, PD, ALS	Death certificates	Industry cohort of engineering workers	Occupational MF	Census	Occupation at baseline/census
Feychting 2003 [38]	AD, PD	Death certificates	Economically active Swedish population at census	Occupational MF	Census	Occupation at baseline/census

Qiu 2004 [39]	AD	Diagnostic criteria	Community based cohort of persons aged 75 and older, Stockholm, Sweden	Occupational MF	Interviews/ questionnaire	Occupational history
Park 2004 [22]	AD, PD, ALS	Death certificates	Deceased from 22 U.S. states	Occupational MF, occupation	Death certificates	Primary occupation
Seidler 2006 [40]	AD	Medical records	Cases were patients from general practitioner practices in Frankfurt, Germany; controls were population based	Occupational MF; occupation	Interviews/ questionnaire	Occupational history
Sorahan 2007 [31]	AD, PD, ALS	Death certificates	Electricity generation and transmission workers	Occupational MF	Occupational records in addition to location information for which modeled exposure was used	Occupational history
Davanipour 2007 [41]	AD	Diagnostic criteria	Cases were diagnosed at AD hospital, controls were demented and nondemented patients without AD	Occupational MF	Interviews/ questionnaire	Last occupation
Röösli 2007 [30]	AD, PD, ALS	Death certificates	Swiss railway employees	Occupational MF	Occupational records	Occupational history
Andel 2010 [42]	AD	Screening instrument	Twins	Occupational MF	Interviews/ questionnaire	Primary occupation
Fang 2009 [26]	ALS	Diagnostic criteria	Cases were from two major referral centers in New England; general population controls	Self-reported	Interviews/ questionnaire	Ever 10 times or more exposed
Parlett 2011 [43]	ALS	Death certificates	National longitudinal mortality study based on census information, United States	Occupational MF	Census	Occupation at baseline/census

Whether a study participant had a neurodegenerative disease was either assessed by death certificates [19,21–23,27,29–34], screening, or doctors' diagnoses [16–18,35–41]. If death certificates were used, applied International Classification of Disease (ICD) codes for the outcomes were usually ICD-9 331.0 and ICD-10 G30 for Alzheimer's disease (no code existed in ICD 8); ICD-8 348, ICD-9 335.2, and ICD-10 G12.2 for ALS; and ICD-8 342, ICD-9 332, and ICD-10 G20 (also G21 and G25.9 in Röösli et al. [30]) for Parkinson's disease. If morbidity was analyzed [17,18,34,35,37,39–42], then Alzheimer's disease was generally based on a physician's diagnosis using the National Institute of Neurological and Communicative Disorders and Stroke/Alzheimer's Disease and Related Disorders Association (NINCDS/ADRA) criteria [44]. For ALS, only three earlier studies [16,25,33] identified cases based on physician's diagnoses; and for Parkinson's disease, only one study used hospital discharge registration [36].

Studies evaluated the primary occupation [17,18,34,37,42], the last occupation [43,34], or cumulative exposure over the respective occupational history of the worker [19,30,31,35,39,40]. Occupations were determined from interviews or questionnaires [16–18,25,32–35,37,39,40,42], from information provided in censuses [27,34,43], from company records [19,28–31,36], or from death certificates [20–23]. Job titles were then used to identify those persons likely exposed to magnetic fields, with the exception of one study that used a combination of job categories and modeled site-specific exposure [31]. An early study on magnetic field exposure and ALS [16] identified a group of "electrical occupations" and compared persons in such an occupation with those with all other occupations. In subsequent studies, this original list was slightly expanded and average magnetic field exposure levels were assigned to these jobs. Percentages of persons classified as exposed varied in these studies between 2% [16] and 8% [41]. Later, JEMs were developed from measurements or in combination with expert estimations and were used to assign average exposure values to the occupations. In some of the population-based studies that used JEMs, exposure assessment was (likely) less specific than in the previous studies, with about 22% [35] to 33% [39] of the study participants classified as exposed. This is much less specific than the 2%–8% of the previous studies. A high specificity of the exposure assessment is especially important to detect potential effects of rare exposures [45]. Under the assumption that different general populations should have essentially similar exposure distributions, the reported exposure prevalence would be expected to reflect the specificity of the exposure assignment, and a higher specificity would be expected to result in less exposure misclassification. Therefore, as an additional analysis, a meta-analysis was included that assesses whether the proportion of persons classified as high-exposed was related to reported risk estimates in population-based studies.

Alzheimer's Disease

Figure 11.1 presents studies assessing risk of Alzheimer's disease grouped by the type of exposure assessment. For those studies that assessed magnetic field exposure quantitatively, studies are presented sorted by the reported exposure levels. These studies suggested about a 50% increase in risk for those high versus low exposed. Only two studies addressed risk of Alzheimer's disease in electrical workers, and there was no association [19,40]. Studies that assessed occupation from death certificates reported risks close to unity for those exposed.

Although the forest plot looks suggestive of an increase in risk of Alzheimer's disease with higher average magnetic field exposures at work, this association was not statistically significant. Risk estimates tended to be somewhat lower in studies when exposure had been

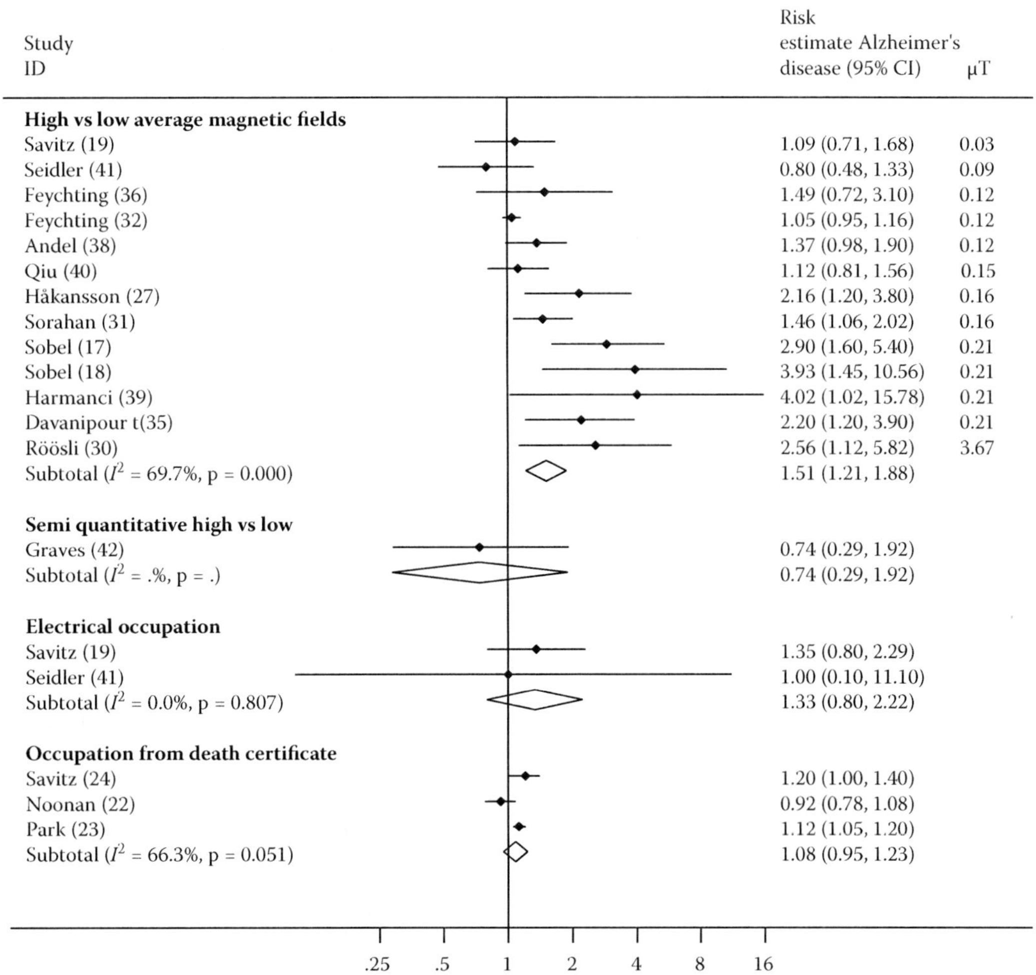

FIGURE 11.1
Risk estimates of studies assessing risk of Alzheimer's disease in association with occupational exposure to magnetic fields, comparing high to low exposure. Please note that the risk estimate of the study by Seidler et al. [40] has very wide confidence intervals, because it is based on only two cases.

assessed for the full occupational history, compared with only assessing jobs at one point in time. There was no evidence that risk estimates differed depending on whether the outcome had been assessed from death certificates compared with screening or doctors' diagnoses.

Within 10 population-based studies, there was a linear relationship with presumed specificity of the exposure cutoff: the smaller the percentage of high-exposed workers in population-based studies, the higher the risk estimate (Figure 11.2), and this effect was statistically significant ($p = 0.004$).

Parkinson's Disease

Nine studies reported on the association between occupational magnetic field exposures and Parkinson's disease. An overview is given in Figure 11.3. Overall, occupational exposure to magnetic fields did not seem to be associated with Parkinson's disease,

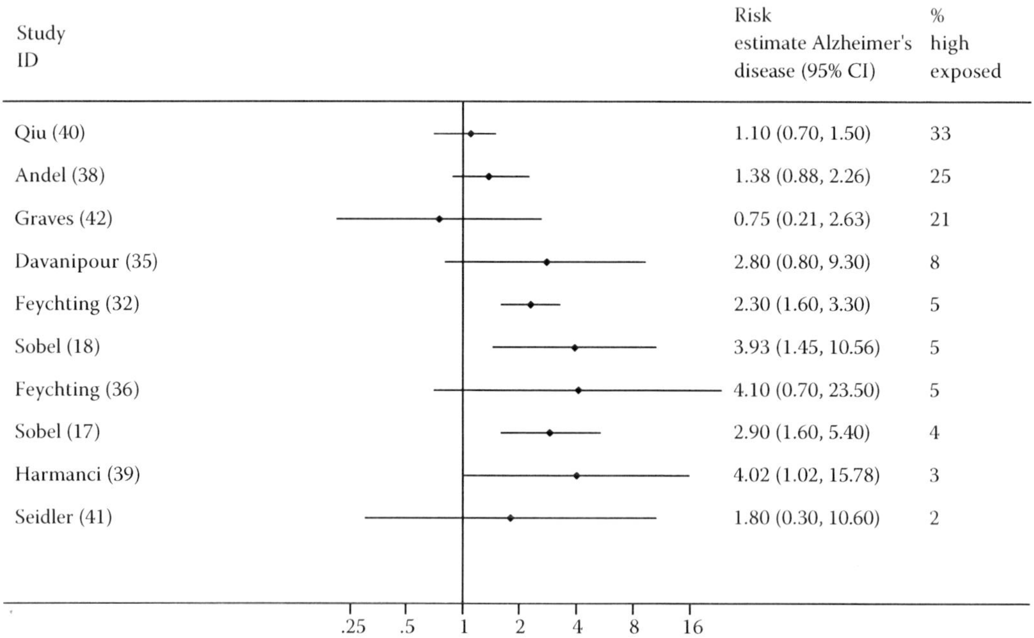

FIGURE 11.2
Alzheimer's disease and reported risk estimates of highest as well as longest exposure duration, ordered by the percentage of persons classified as exposed. Percentage of exposed in Feychting, 1998, approximated from Feychting, 2003.

with summary risk estimates consistently around unity. Of all studies, only one study [36] analyzed hospital discharge records and not mortality, but this study also provided no evidence for an increased risk of Parkinson's disease in those exposed.

Amyotrophic Lateral Sclerosis

Of those studies that compared workers with high versus low magnetic field exposure, only a slightly increased risk was observed (Figure 11.4), but study results were very heterogeneous. Of note, increased risks were only identified in worker cohorts with a high proportion of exposed persons [19,29–31], with population-based studies reporting risks around unity [33,38,43].

More consistently, increased risks of ALS was reported in studies analyzing "electricity workers" [32], "electricity work" [25], or for workers with an "electrical occupation" [16,19,38], with less heterogeneity between studies. Two studies [19,38] evaluated both average magnetic field exposure as well as electrical occupations. In these studies, an association emerged for electrical occupations, but not for magnetic field exposure. Studies evaluating self-reported exposure as well as studies that assessed occupation from death certificates yielded risk estimates around unity.

There was no evidence that risk estimates increased linearly with increasing magnetic field exposures, the specificity of exposure assessment in population-based studies (based on only five studies), whether a full occupational history was collected or not, or whether the outcome was based on screening, doctors' diagnoses, or death certificates.

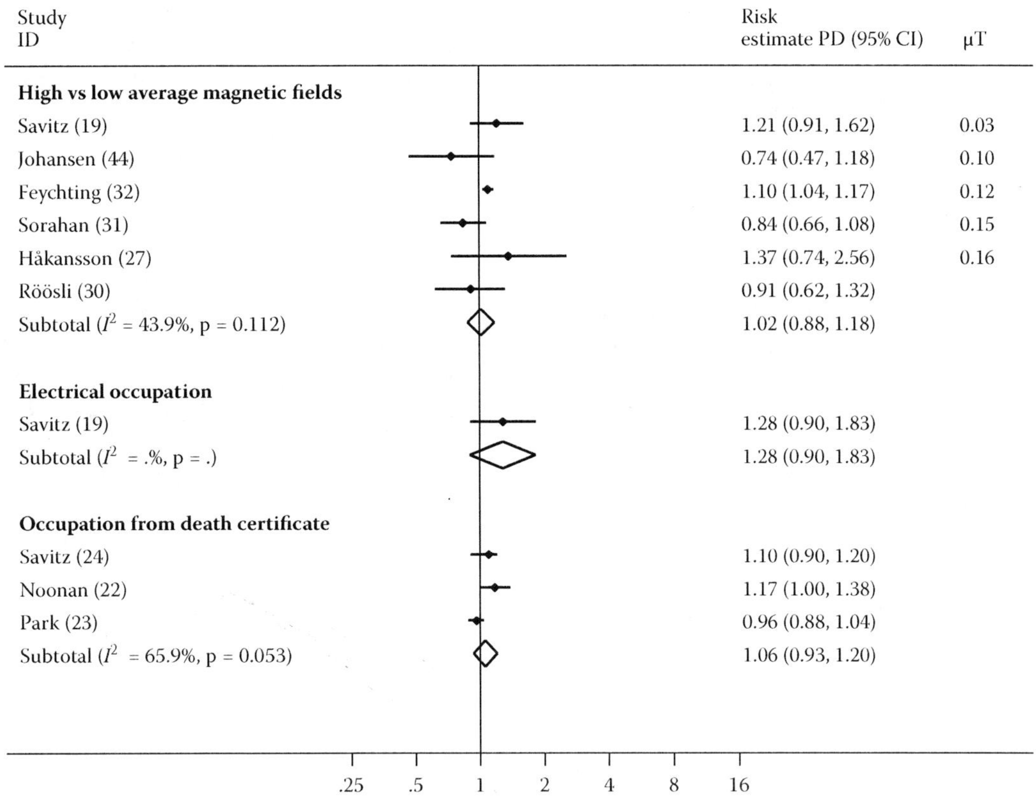

FIGURE 11.3
Risk estimates of studies assessing risk of Parkinson's disease in association with occupational exposure to magnetic fields, comparing high to low exposure categories.

Overview of Literature on Residential Magnetic Field Exposure Studies

Only very few studies have addressed residential exposures to magnetic fields. Elevated residential magnetic field exposures can arise, for example, if persons live close to overhead power lines. A very small Italian study in 345 residents exposed to (modeled) magnetic fields from power lines did not observe any cases of Alzheimer's disease or ALS [46]. A Brazilian study analyzed 367 deaths from ALS, but of these only one case was exposed to residential exposure levels of >0.3 μT [47]. No elevated risk of cognitive impairment emerged from an analysis of Taiwanese persons living within 50 or 100 m of a power line [48]. In contrast, a census study of the full Swiss population reported an increased risk of death in persons with Alzheimer's disease who had lived for at least 10 years within a distance of 50 m of a 220- or 380-kV power line [49]. However, no ALS case had lived within 50 m of a high-voltage power line, and no increased risk was found for the occurrence of Parkinson's disease [49]. More recently, a Danish case–control study performed a similar analysis addressing incidence of Alzheimer's disease, ALS, and Parkinson's disease [50]. The study did not report increased risks for people who had lived close (<50 m) to a power line. The study was, however, limited in addressing risks for persons who

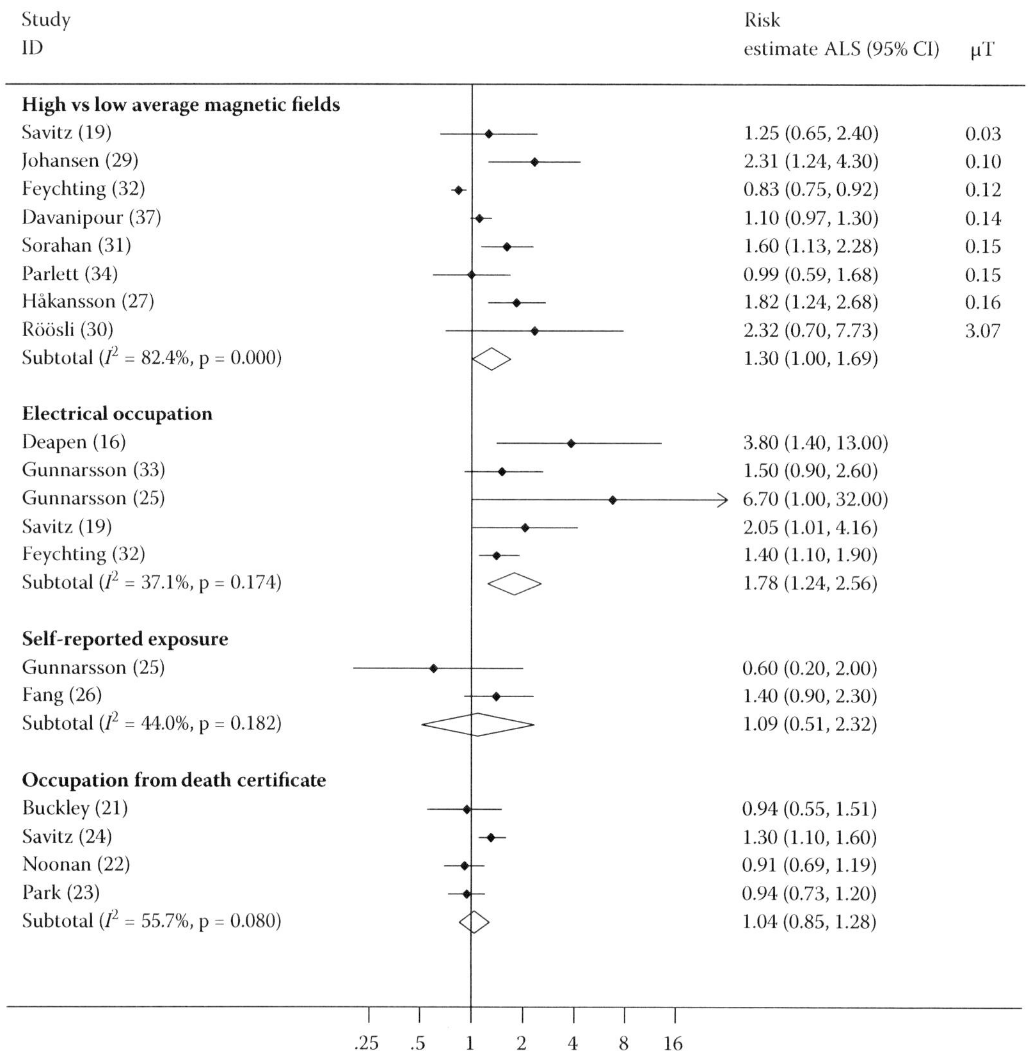

FIGURE 11.4
Risk estimates of studies assessing risk of ALS in association with occupational exposure to magnetic fields, comparing high to low exposure categories.

lived close to those lines that are likely to generate stronger exposures (220- or 400-kV lines): the majority of persons living close to a power line had lived in the vicinity of a 132- or 150-kV line [50].

Critical Evaluation

Previous reviews have commented on the difference in results between studies and have evaluated several aspects of underlying heterogeneity, in particular study design (e.g., cohort study or case–control), type of outcome assessment (e.g., clinical examination or death certificates), and also exposure assessment (e.g., whether or not quantitative exposure information was reported, or whether workers in a study had been exposed to at least specific exposure levels (e.g., >0.2 μT)). Previous meta-analyses on Alzheimer's disease and ALS concluded that studies remained heterogenous in their results and that future studies should improve, in particular, on the exposure assessment. In this chapter, we put focus on the exposure assessment: studies that had applied a quantitative exposure assessment were analyzed separately from those assessing groups of electrical workers and from studies expected to have a higher risk of exposure misclassification (i.e., self-reported exposure or occupational information extracted from death certificates).

Some degree of exposure misclassification will have been at play in all presented studies, where a variety of methods were used to assign exposure levels to job titles. For example, although some studies performed measurements in their study population, others assigned exposure levels using measurement-based JEMs, used combinations of measurements and expert opinion, or relied on experts only. Given that absolute levels of magnetic field exposure reported in the individual studies were assessed with a variety of methods, they may not be directly comparable across studies. An additional calculation was therefore included in this chapter in an attempt to assess whether risk estimates differed according to the percentage of workers classified as being high exposed. For Alzheimer's disease, higher risks were identified in those studies where fewer workers were classified as exposed. This approach, however, assumes a similar exposure distribution of high-exposed workers across countries and decades.

Different methods were also applied to identify jobs held by workers: occupations were assessed from company records, self-reports, mixed self-reports with information provided by next of kin, or death certificates. In addition, studies evaluated the full occupational history, assessed occupations at baseline of the study, or the primary or last occupation that was held by a person. Within those studies that did not evaluate the full occupational history, the question arises whether all relevant magnetic field occupational exposures during the life course were captured. For example, a population-based study in Swedish twins found that 31%–36% of the population reported different occupations for primary (longest held) and last job [34]. A similar percentage for job changes was reported in one Swedish region between the 1960 and 1970 census [32]. Exposure misclassification due to job changes would matter less if skilled (and possibly higher exposed) workers were less likely to be changing jobs. For some studies, this might be the case, especially in those that restricted their population to some industries or to electric occupations. For example, in the Swiss train driver study, only 2%–3% of train drivers and shunting yard engineers had changed their job over about a decade [30]; and in the Danish utility worker study, only about 1% of workers had changed occupation over the assessed period [29].

Two of four studies that evaluated the relevance of when persons were exposed during their life course on Alzheimer's disease risk reported that later exposures were more relevant than earlier exposures. If indeed timing of exposure is an important factor, this could potentially be an underlying reason for our observation that studies using the full occupational history reported, on average, lower relative risks of Alzheimer's disease compared with studies using only the primary or the last job. No effect of timing, however, was observed for studies on ALS.

Initial interest in the association between ALS and magnetic field exposure developed from the question of whether people who experience electrical trauma are at increased risk of disease [15]. In line with this question, most previous reviews have stressed that the increased risk of ALS in electrical occupations could also be due to electric shocks and not necessarily to magnetic field exposure. Electric shocks, however, occur by accident and therefore cannot be measured. Recently, attempts were made to identify occupations where workers are at higher risk of electric shock, using registered occupational electrical injuries [51,52]. One of the studies reported a high concordance of about 67% where occupations at risk for electrical injury were also more likely to occur in those occupations with higher magnetic field exposures [51]. A recent Danish study assessed neurological diseases in a cohort of workers who had survived electrical injuries and observed increased risks for some conditions such as migraine and vertigo. However, magnetic field exposures for these workers were not taken into account [53]. Because, to date, no previous study has analyzed both exposures at the same time, it has not been possible to disentangle the potential effects of these exposures. In addition, workers in electrical occupations will likely also be exposed to other exposures associated with the use of electricity (such as electric fields, imperceptible contact currents, or nuisance shocks [54]) or other occupational exposures to chemicals. Exposure to, for example, metals and solvents might also occur more frequently in some occupations with higher magnetic field exposure. These exposures have been associated with increased risks of neurodegenerative diseases [6], but to our knowledge, they have not been accounted for in the previous studies. Studies addressing residential exposures could serve as an interesting addition to evaluate this question, because persons living, for example, close to power lines would be exposed to magnetic fields (and electric fields when outside of their homes) but would be unlikely to have an increased risk of electrical shocks or some of the other chemical exposures. Previous studies, however, have not been conclusive.

The assessment of the outcomes also differed between the studies, for example, some used hospital discharge or GP diagnoses, and others used information provided on death certificates. While ALS is likely well captured on a death certificate [21,55], other diseases such as Alzheimer's or Parkinson's disease would be expected to be strongly underreported. Underreporting as such would lead to a loss of power in the analysis. Bias would arise if this underreporting was associated with levels of exposure to magnetic fields. This overview, however, did not show that results differed significantly depending on whether or not the outcome was assessed from death certificates.

In summary, the overview presented in this chapter indicates slightly elevated risks of Alzheimer's disease for persons occupationally exposed to magnetic fields. Future studies should improve on magnetic field exposure assessment, in particular on quantitative measures of exposure. In addition, temporal effects of exposure should be evaluated to be able to identify etiologically relevant time windows—in particular the indication of some studies that late exposures might be more important than earlier exposures deserves some attention. For Parkinson's disease, studies so far have not indicated that exposed workers were at higher risk of this disease. Studies analyzing ALS showed higher risks for workers in electrical occupations. Whether ELF magnetic fields are the relevant exposure, or whether they are just a proxy for another exposure such as electrical injuries or other occupational exposures, remains an open research question. The recently developed shock JEMs [51,52] provide a possibility to evaluate this question.

References

1. Przedborski S, Vila M, Jackson-Lewis V. Series introduction: neurodegeneration: what is it and where are we? *J Clin Invest* 2003;111(1):3–10.
2. Mattsson MO, Simkó M. Is there a relation between extremely low frequency magnetic field exposure, inflammation and neurodegenerative diseases? A review of in vivo and in vitro experimental evidence. *Toxicology* 2012;301(1–3):1–12.

3. Reitz C, Brayne C, Mayeux R. Epidemiology of Alzheimer disease. *Nat Rev Neurol* 2011;7(3):137–52.
4. Mayeux R, Stern Y. Epidemiology of Alzheimer disease. *Cold Spring Harb Perspect Med* 2012;2(8):a006239.
5. Alzheimer's Association. 2012 Alzheimer's disease facts and figures. *Alzheimers Dement* 2012;8(2):131–68.
6. Migliore L, Coppedè F. Genetics, environmental factors and the emerging role of epigenetics in neurodegenerative diseases. *Mutat Res* 2009;667(1):82–97.
7. von Campenhausen S, Bornschein B, Wick R, Bötzel K, Sampaio C, Poewe W, et al. Prevalence and incidence of Parkinson's disease in Europe. *Eur Neuropsychopharmacol* 2005;15(4):473–90.
8. de Lau LML, Breteler M. Epidemiology of Parkinson's disease. *Lancet Neurol* 2006;5(6):525–35.
9. Bertram L, Tanzi RE. The genetic epidemiology of neurodegenerative disease. *J Clin Invest* 2005;115(6):1449–57.
10. Bossy-Wetzel E, Schwarzenbacher R, Lipton SA. Molecular pathways to neurodegeneration. *Nat Med* 2004;(10 Suppl):S2–9.
11. Rothstein JD. Current hypotheses for the underlying biology of amyotrophic lateral sclerosis. *Ann Neurol* 2009;65(S1):S3–9.
12. Majoor-Krakauer D, Willems P, Hofman A. Genetic epidemiology of amyotrophic lateral sclerosis. *Clin Genet* 2003;63(2):83–101.
13. Mitchell J, Beaumont A. Motor neuron disorders with other diseases and exposures. *Handbook Clin Neurophysiol* 2004;4:451–66.
14. Dzhokic G, Jovchevska J, Dika A. Electrical injuries: etiology, pathophysiology and mechanism of injury. *Mac J Med Sci* 2008;1(2):54–8.
15. Haynal A, Regli F. Amyotrophic lateral sclerosis associated with accumulated electric injury. *Confin Neurol* 1964;24:189–98.
16. Deapen DM, Henderson BE. A case-control study of amyotrophic lateral sclerosis. *Am J Epidemiol* 1986;123(5):790–9.
17. Sobel E, Davanipour Z, Sulkava R, Erkinjuntti T, Wikstrom J, Henderson VW, et al. Occupations with exposure to electromagnetic fields: a possible risk factor for Alzheimer's disease. *Am J Epidemiol* 1995;142(5):515–24.
18. Sobel E, Dunn M, Davanipour Z, Qian Z, Chui H. Elevated risk of Alzheimer's disease among workers with likely electromagnetic field exposure. *Neurology* 1996;47(6):1477–81.
19. Savitz DA, Checkoway H, Loomis DP. Magnetic field exposure and neurodegenerative disease mortality among electric utility workers. *Epidemiology* 1998;9(4):398–404.
20. Buckley J, Warlow C, Smith P, Hilton-Jones D, Irvine S, Tew J. Motor neuron disease in England and Wales, 1959–1979. *J Neurol Neurosurg Psychiatr* 1983;46(3):197–205.
21. Noonan CW, Reif JS, Yost M, Touchstone J. Occupational exposure to magnetic fields in case-referent studies of neurodegenerative diseases. *Scand J Work Environ Health* 2002;28(1):42–8.
22. Park RM, Schulte PA, Bowman JD, Walker JT, Bondy SC, Yost MG, et al. Potential occupational risks for neurodegenerative diseases. *Am J Ind Med* 2005;48(1):63–77.
23. Savitz DA, Loomis DP, Tse CKJ. Electrical occupations and neurodegenerative disease: analysis of US mortality data. *Arch Environ Health* 1998;53(1):71–4.
24. Andrews KW, Savitz DA. Accuracy of industry and occupation on death certificates of electric utility workers: implications for epidemiologic studies of magnetic fields and cancer. *Bioelectromagnetics* 1999;20(8):512–18.
25. Gunnarsson LG, Bodin L, Söderfeldt B, Axelson O. A case-control study of motor neurone disease: its relation to heritability, and occupational exposures, particularly to solvents. *Br J Ind Med* 1992;49(11):791–8.
26. Fang F, Quinlan P, Ye W, Barber MK, Umbach DM, Sandler DP, et al. Workplace exposures and the risk of amyotrophic lateral sclerosis. *Environ Health Perspect* 2009;117(9):1387.
27. Håkansson N, Gustavsson P, Johansen C, Floderus B. Neurodegenerative diseases in welders and other workers exposed to high levels of magnetic fields. *Epidemiology* 2003;14(4):420–6.

28. Johansen C, Koch-Henriksen N, Rasmussen S, Olsen JH. Multiple sclerosis among utility workers. *Neurology* 1999;52(6):1279–82.
29. Johansen C, Olsen JH. Mortality from amyotrophic lateral sclerosis, other chronic disorders, and electric shocks among utility workers. *Am J Epidemiol* 1998;148(4):362–8.
30. Röösli M, Lörtscher M, Egger M, Pfluger D, Schreier N, Lörtscher E, et al. Mortality from neurodegenerative disease and exposure to extremely low-frequency magnetic fields: 31 years of observations on Swiss railway employees. *Neuroepidemiology* 2007;28(4):197–206.
31. Sorahan T, Kheifets L. Mortality from Alzheimer's, motor neuron and Parkinson's disease in relation to magnetic field exposure: findings from the study of UK electricity generation and transmission workers, 1973–2004. *Occup Environ Med* 2007;64(12):820–6.
32. Gunnarsson LG, Lindberg G, Söderfeldt B, Axelson O. Amyotrophic lateral sclerosis in Sweden in relation to occupation. *Acta Neurol Scand* 1991;83(6):394–8.
33. Davanipour Z, Sobel E, Bowman J, Qian Z, Will A. Amyotrophic lateral sclerosis and occupational exposure to electromagnetic fields. *Bioelectromagnetics* 1997;18(1):28–35.
34. Feychting M, Pedersen NL, Svedberg P, Floderus B, Gatz M. Dementia and occupational exposure to magnetic fields. *Scand J Work Environ Health* 1998;24:46–53.
35. Graves AB, Rosner D, Echeverria D, Yost M, Larson EB. Occupational exposure to electromagnetic fields and Alzheimer disease. *Alzheimer Dis Assoc Disord* 1999;13(3):165.
36. Johansen C. Exposure to electromagnetic fields and risk of central nervous system disease in utility workers. *Epidemiology* 2000;11(5):539–43.
37. Harmanci H, Emre M, Gurvit H, Bilgic B, Hanagasi H, Gurol E, et al. Risk factors for Alzheimer disease: a population-based case-control study in Istanbul, Turkey. *Alzheimer Dis Assoc Disord* 2003;17(3):139–45.
38. Feychting M, Jonsson F, Pedersen NL, Ahlbom A. Occupational magnetic field exposure and neurodegenerative disease. *Epidemiology* 2003;14(4):413–19.
39. Qiu C, Fratiglioni L, Karp A, Winblad B, Bellander T. Occupational exposure to electromagnetic fields and risk of Alzheimer's disease. *Epidemiology* 2004;15(6):687–94.
40. Seidler A, Geller P, Nienhaus A, Bernhardt T, Ruppe I, Eggert S, et al. Occupational exposure to low frequency magnetic fields and dementia: a case–control study. *Occup Environ Med* 2007;64(2):108–14.
41. Davanipour Z, Tseng CC, Lee PJ, Sobel E. A case-control study of occupational magnetic field exposure and Alzheimer's disease: results from the California Alzheimer's Disease Diagnosis and Treatment Centers. *BMC Neurol* 2007;7(1):13.
42. Andel R, Crowe M, Feychting M, Pedersen NL, Fratiglioni L, Johansson B, et al. Work-related exposure to extremely low-frequency magnetic fields and dementia: results from the population-based study of dementia in Swedish twins. *J Gerontol A Biol Sci Med Sci* 2010;65(11):1220.
43. Parlett LE, Bowman JD, van Wijngaarden E. Evaluation of occupational exposure to magnetic fields and motor neuron disease mortality in a population-based cohort. *J Occup Environ Med* 2011;53(12):1447.
44. McKhann G, Drachman D, Folstein M, Katzman R, Price D, Stadlan EM. Clinical diagnosis of Alzheimer's disease report of the NINCDS-ADRDA Work Group under the auspices of Department of Health and Human Services Task Force on Alzheimer's Disease. *Neurology* 1984;34(7):939–44.
45. Blair A, Stewart P, Lubin JH, Forastiere F. Methodological issues regarding confounding and exposure misclassification in epidemiological studies of occupational exposures. *Am J Ind Med* 2007;50(3):199–207.
46. Fazzo L, Tancioni V, Polichetti A, Iavarone I, Vanacore N, Papini P, et al. Morbidity experience in populations residentially exposed to 50 hz magnetic fields: methodology and preliminary findings of a cohort study. *Int J Occup Environ Health* 2009;15(2):133–42.
47. Marcilio I, Gouveia N, Pereira Filho ML, Kheifets L. Adult mortality from leukemia, brain cancer, amyotrophic lateral sclerosis and magnetic fields from power lines: a case-control study in Brazil. *Rev Bras Epidemiol* 2011;14(4):580–8.
48. Li CY, Sung FC, Wu SC. Risk of cognitive impairment in relation to elevated exposure to electromagnetic fields. *J Occup Environ Med* 2002;44(1):66–72.

49. Huss A, Spoerri A, Egger M, Röösli M. Residence near power lines and mortality from neurodegenerative diseases: longitudinal study of the Swiss population. *Am J Epidemiol* 2009;169(2):167–75.
50. Frei P, Poulsen AH, Mezei G, Pedersen C, Salem LC, Johansen C, et al. Residential distance to high-voltage power lines and risk of neurodegenerative diseases: a Danish population-based case-control study. *Am J Epidemiol* 2013;177(9):970–8.
51. Huss A, Vermeulen R, Bowman JD, Kheifets L, Kromhout H. Electric shocks at work in Europe: development of a job exposure matrix. *Occup Environ Med* 2013;70(4):261–7.
52. Vergara XP, Kheifets L, Silva M, Bracken TD, Yost M. New electric-shock job exposure matrix. *Am J Ind Med* 2012;55(3):232–40.
53. Grell K, Meersohn A, Schüz J, Johansen C. Risk of neurological diseases among survivors of electric shocks: a nationwide cohort study, Denmark, 1968–2008. *Bioelectromagnetics* 2012;33(6):459–65.
54. Bracken TD, Kavet R, Patterson RM, Fordyce TA. An integrated job exposure matrix for electrical exposures of utility workers. *J Occup Environ Hyg* 2009;6(8):499–509.
55. Chiò A, Magnani C, Oddenino E, Tolardo G, Schiffer D. Accuracy of death certificate diagnosis of amyotrophic lateral sclerosis. *J Epidemiol Community Health* 1992;46(5):517–18.

12

Brain Tumors and Mobile Phone Use: The Cohort Approach

Aslak Harbo Poulsen and Patrizia Frei

CONTENTS

Introduction

A range of smaller studies on use of mobile phones and risk of central nervous system (CNS) tumors have been published. Apart from one American cohort study that ended after only 1 year for legal reasons (Rothman et al. 1996; Dreyer et al. 1999), these studies have been case–control studies limited by small numbers, short follow-up, or methodological limitations (Baan et al. 2011). In 2011, the International Agency for Research on Cancer (IARC) classified radiofrequency electromagnetic fields as "possibly carcinogenic to humans" (Baan et al. 2011). This classification was primarily based on results from two questionnaire-based case–control studies, namely, the studies done in Sweden by Hardell and colleagues (2011) and the international INTERPHONE study (Cardis et al. 2007). Both of these studies on mobile phones and brain tumors have been ongoing for many years. In general, the interpretation of the case–control studies is severely hampered by their susceptibility to biases due to differential participation, recall, and reporting, depending on case status. These problems have been further aggravated by the need to use proxy reports for deceased cases. Overall, Interphone found no clear evidence of increased risk of CNS tumors in mobile phone users; there were however increased odds ratios (ORs) for both glioma, 1.40 (95% confidence intervals [CI], 1.03–1.89) (INTERPHONE Study Group 2010), and vestibular schwannoma (also called acoustic neuroma), 1.32 (0.88–1.97) (INTERPHONE Study Group 2011) among subjects within the highest decile of cumulative self-reported use (>1640 min). However, several factors speak against a causal interpretation (INTERPHONE Study Group 2010; Swerdlow et al. 2011): implausible levels of daily use were reported in this stratum, there was no suggestion of dose response over the preceding categories of exposure, and the ninth decile was even among the lowest ORs observed. The Hardell studies (Hardell et al. 2011) have reported elevated risks, already, after a few years of phone use; these results are however incompatible with incidence trends (Deltour et al. 2009, 2012; Inskip et al. 2010; Little et al. 2012). Incidence studies, however, can only discern effects that are strong enough to be seen on population level against a backdrop of other factors influencing incidence. The ideal study would be a prospective cohort study with elaborate and repeated or continuous exposure assessment that would avoid or minimize the issues with incidence and case–control studies (WHO 2006; SCENIHR 2009b). Data are being collected for such a study (Schüz et al. 2011a); the results are however still some time in the future. In the meantime, an existing entirely register-based cohort of subscription holders in Denmark until 1995 has been used for investigation of a range of outcomes (Johansen et al. 2001; Schüz et al. 2006b, 2009, 2011b; Frei et al. 2011; Poulsen et al. 2012, 2013). Much concern has, however, been expressed about both actual and conceived shortcomings of this study (e.g., Ahlbom et al. 2007; Khurana 2011; Philips and Lamburn 2011; Söderqvist et al. 2012). A recently published British study has used an ongoing cohort; the million women study, where participants have been asked basic questions on their use of mobile phones (Benson et al. 2013). An overview of these two studies, with a discussion of their strengths and weaknesses, is given here.

Methods

Danish Subscriber Cohort

The Danish cohort has been used to investigate intracranial brain tumors in several studies (Johansen et al. 2001; Schüz et al. 2006b, 2011b; Frei et al. 2011); the focus here is given to the latest studies, where follow-up is the longest. The cohort was established from records of all (723,421) mobile phone numbers active from 1982, when the analog Nordic Mobile Phone Telephony (NMT) cellular service was put into operation, to the end of 1995. The data as well as partial funding were provided by the mobile phone network operators active in Denmark at that time (later publications have not received industry support).

Of the received records, 282,408, primarily corporate subscriptions, were deleted because an individual subscription holder could not be identified. For the remaining records, the national personal identification number (Pedersen 2011) could be established, allowing deletion of duplicate records and ineligible subscription. The final data set constituted 420,086 individual subscribers. Cancer cases were identified from the virtually complete, nationwide Danish Cancer Registry (Storm et al. 1997; Gjerstorff 2011) with additional details on vestibular schwannoma from a clinical database (Stangerup et al. 2010). The study was entirely register based; the only loss to follow-up has been due to emigration (<2.2%), and 53 persons who explicitly requested to be excluded upon learning of the study from their operator.

The two first publications on CNS tumors from the Danish cohort, with follow-up through 1996 (Johansen et al. 2001) and 2002 (Schüz et al. 2006b), calculated standardized incidence ratios. Methods have been updated in the two most recent CNS studies based on the Danish cohort that have investigated vestibular schwannoma specifically through 2006 (vestibular schwannoma study), and updated the analysis on all CNS tumors with follow-up through 2007 (CNS tumor study). For subjects with first subscription before 1987, the year of first subscription was changed to 1987 because truly handheld phones only became available in that year. Log linear Poisson regression models were used to estimate incidence rate ratios (IRRs) with adjustment for age (in 5-year groups), calendar period (in 6-year groups), education, and income. These time-varying socioeconomic indicators were obtained by restricting the study population to members of a nationwide cohort on social inequality and cancer (CANULI) (Dalton et al. 2008), that is, to all Danes who were not immigrants or descendants, born in 1925 or later, and resident in Denmark at the start of cancer follow-up. The CNS tumor study had follow-up from age 30 or 1990, which ever came last, through 2007; the vestibular schwannoma study had follow-up from age 30 years or 1998, which ever came last, through 2006. In both studies, eligible subjects who started using phones before age 30 years would then enter the study at age 30 with their accumulated exposure level. After these restrictions, the cohort size was further reduced; in the CNS tumor study, cohort size was reduced to 358,403 subscription holders.

Both studies allowed focus on comparatively long-term exposure groups. In the vestibular schwannoma study, subjects with ≥11 years of exposure were compared with all subjects with no or shorter exposure because previous studies of this slow-growing tumor did not suggest an increased risk with exposure durations of ≤10 years. In the CNS tumor study, subjects were counted as exposed 1 year after acquiring the first subscription, and the duration of exposure was then stratified with the longest exposure group having ≥13 years since first subscription. In the study on vestibular schwannoma, additional analysis was done on tumor size, and laterality and the age-specific incidence rates of vestibular schwannoma in subscribers and nonsubscribers were compared.

Million Women Study

During 1996–2001, all women aged 50–64 years in the United Kingdom were routinely invited to participate in a national breast cancer screening program. At 66 of about 100 national screening centers; the women were invited to participate in a prospective cohort with a primary focus on hormone replacement therapy (Million Women Study Collaborative Group 1999; Million Women Study Collaborators 2002). Of all women invited, about 50% returned complete consent and questionnaire data (Million Women Study Collaborative Group 1999), yielding a final cohort of about 1.3 million women (Benson et al. 2013). A follow-up questionnaire sent out between 1999 and 2005 included mobile phone questions; 866,525 women responded to this follow-up questionnaire and 791,710 remained after exclusion of 14,387 who received a questionnaire version without these questions; 11,981 who did not answer the questions; 48,531 with an existing cancer diagnosis (except nonmelanoma skin cancer); and six cases of neurofibromatosis (Benson et al. 2013). The women were followed for death and cancer through 2009 via record linkage to the National Health Service (NHS) central register.

Relative risks (RRs) were calculated in Cox models, using only data from the 1999–2005 questionnaire, with age as underlying time scale, and adjusting for baseline data on the following: socioeconomic status, geographical region, age at baseline (in 3-year groups), height, body mass index (BMI), smoking, alcohol intake, exercise level, and use of hormone replacement therapy. For all analyses, never users were the reference, and ever users were stratified by frequency of use (daily, less than daily) and duration (<5, 5–9, and ≥10 years), with exposure levels incrementing from the baseline levels over the follow-up period. Two subanalyses were performed: one subanalysis excluded the first 3 years of follow-up because as-yet-undiagnosed CNS tumors conceivably could influence responses at baseline; and the other subanalysis excluded participants recruited in 1999 and 2000, when the rate of uptake was particularly high among the recruited women and many nonusers may, therefore, have become users shortly after filling in the questionnaire. In addition, the annual incidence rates for vestibular schwannoma during 1998–2008 were investigated using National Statistics data.

Results

Meningioma

Neither the Danish nor the U.K. study found elevated risk for meningioma (Table 12.1). After ≥10 years of subscribing, the IRRs in the Danish study were 0.90 (0.57–1.42; n = 21) for men and 0.93 (0.46–1.87; n = 8) for women. In the British study, the corresponding RR was 1.10 (0.66–1.84; n = 8).

Glioma

The Danish CNS tumor study found no increased risk of glioma in male or female mobile phone subscribers overall or after ≥10 years since first subscription (Table 12.1). The IRR after ≥13 years was 0.98 (0.70–1.36; n = 37) in men, where numbers allowed the analysis. In the U.K. study, the estimates were <1 both overall and in women with ≥10 years since first use (Table 12.1), and the RR was similar also in women who were daily users at baseline, 0.80 (0.56–1.14).

For men in the Danish CNS tumor study, restricting analysis to glioma of the temporal lobe, the area that receives the highest exposure when a phone is held to the ear (Cardis et al. 2008), showed an IRR of 1.13 (0.86–1.48), and the lowest risk was seen among subjects with the longest exposure, that is, ≥10 years of subscribing, where the IRR was 0.81 (0.50–1.32). Within the pooled group of glioma in "other and unspecified" anatomical sites of the brain (IRR = 1.35 [1.05–1.74]), the highest IRR (2.58, 1.08–6.15; n = 8) was seen for glioma of the cerebral ventricles.

Pituitary Gland

The U.K. study presented separate results for pituitary tumors, with an overall RR of 1.52 (0.99–2.33) and a similar level among long-term users (Table 12.1). The RR for women with <5 years of phone use was 2.31 (1.31–4.06), but there was no increased risk in the strata with longer exposure duration.

Vestibular Schwannoma

In the Danish vestibular schwannoma study, subscribing for ≥11 years was not associated with an increased risk of vestibular schwannoma (IRR = 0.87 [0.52–1.46]) in men.

TABLE 12.1

RRs and 95% CI for Tumor Types of the CNS among Mobile Phone Users in Two Cohorts

Study	Site and Sex	Ever Use*			≥10 Years of Use		
		Cases	RR	95% CI	Cases	RR	95% CI
Danish Subscriber Cohort: Frei et al. 2011[a]		≥1 years of use vs. <1 year of use			≥10 years of use vs. <1 year of use		
	CNS, male	714	1.02	(0.94–1.10)	276	1.06	(0.94–1.20)
	CNS, female	132	1.02	(0.86–1.22)	40	1.03	(0.75–1.40)
	Glioma, male	324	1.08	(0.96–1.22)	117	1.04	(0.85–1.26)
	Glioma, female	32	0.98	(0.69–1.40)	10	1.04	(0.56–1.95)
	Meningioma, male	50	0.78	(0.58–1.05)	21	0.90	(0.57–1.42)
	Meningioma, female	30	1.02	(0.71–1.47)	8	0.93	(0.46–1.87)
	Other/unspecified, male	162	1.12	(0.95–1.33)	65	1.19	(0.92–1.55)
	Other/unspecified, female	35	1.19	(0.85–1.67)	12	1.27	(0.72–2.25)
Danish Subscriber Cohort: Schüz et al. 2011[b]					≥11 years of use vs. <11 years of use		
	Vestibular schwannoma, male				15	0.87	(0.52–1.46)
U.K. Million women study: Benson et al. 2013[b]		Ever use vs. never use			≥10 years of use vs. never use		
	CNS	754	1.01	(0.90–1.14)	103	1.02	(0.81–1.27)
	Glioma	334	0.91	(0.76–1.08)	40	0.78	(0.55–1.10)
	Meningioma	149	1.05	(0.81–1.38)	20	1.10	(0.66–1.84)
	Vestibular schwannoma	67	1.44	(0.91–2.28)	8	2.46	(1.07–5.64)
	Pituitary tumors	77	1.52	(0.99–2.33)	11	1.61	(0.78–3.35)

[a] The Danish cohort: follow-up: 1990–2007(CNS) and 1998–2006 (VS); adjustment: age, sex, calendar period, income, and education. Exposure from mobile phone subscription records for 1987–1995.

[b] U.K. cohort: follow-up: 1999–2009; adjustment: age at diagnosis, socioeconomic status, geographical region, age at baseline, height, BMI, smoking, alcohol intake, exercise level, and use of hormone replacement therapy. Exposure from baseline questionnaire collected for 1999–2005.

There was no evidence for more vestibular schwannoma on the right side of the head. The tumor size in long-term subscribers did not differ from that seen in the reference population, although based on small numbers with applicable data on tumor spread ($n = 8$). In women, no cases were observed (1.6 expected). For women in the U.K. study, the RR for vestibular schwannoma in ever mobile phone users was 1.44 (0.91–2.28; $n = 67$); similar estimates were seen for both daily (RR = 1.30 [0.61–3.07]; $n = 8$) and less frequent users (RR = 1.37 [0.61–3.07]; $n = 59$). When stratifying ever users by years since self-reported first use, the estimates increased with increasing exposure duration, reaching 2.46 (1.07–5.64; $n = 8$) among subjects with ≥10 years since first self-reported use. This estimate was 1.98 (0.70–5.59; $n = 5$) in a subanalysis excluding responders from 1999 and 2000 who may well have become users in more recent years.

Strengths and Limitations of the Cohorts

Strengths

The Danish cohort and the Hardell studies are of about equal size in terms of number of cases with ≥10 years of exposure, whereas the U.K. cohort is considerably smaller, and the INTERPHONE study has twice as many cases of glioma and about fivefold more cases of meningioma and vestibular schwannoma. The two cohort studies are, however, alone in having prospectively collected exposure data, free from bias related to future case status. Both cohorts have virtually complete case ascertainment from high-quality registers and supplement each other well. The Danish cohort is nationwide with 3.8 million exposed person-years, has virtually complete and long follow-up, and has an exposure metric that is crude but unbiased and includes subanalysis by anatomical site. The only available confounders are time-varying socioeconomic indicators and there are very few women in the cohort. In contrast, the U.K. study only contains women and has more precise data regarding whether a phone is used together with a range of potential confounder data from baseline. Even though the study has similar number of person-years as the Danish study, it has fewer long-term users.

Limitations

The Danish cohort was augmented with data on socioeconomic status that was associated with likelihood of obtaining a phone in the exposure period covered by the Danish data. This improvement, however, necessitated reduction of the cohort: Cases diagnosed in subjects born before 1925 (i.e., with age at start of follow-up >65 years in the CNS tumor study and >73 years in the vestibular schwannoma study) were lost; however, even in 1995, mobile phone penetration was very low in this age group (Poulsen et al. 2012). Also, the few subscribers diagnosed or otherwise censored, before aged 30 years were lost; however, all others who acquired a subscription at a young age entered the study at age 30 years with their accumulated exposure. Finally, immigrants and their descendants were excluded; incidence of CNS tumors does, however, not differ by ethnicity in this population (Schmidt et al. 2008), and it seems highly implausible that any association of CNS tumors and mobile phones should be different in this subpopulation. Within these restrictions, that were independent of exposure and case status, the study remained nationwide, and the number of exposed cases was sufficient to allow detailed investigations by tumor type and location.

Participation bias is possible in the U.K. study because it depends on active participation. However, even though the response rate for the original cohort was about 50%, the response rate for the follow-up questionnaire, including questions on mobile phones that formed the baseline for the present study, was better than 98%. And a subanalysis excluding the first 3 years of follow-up to account for any effect of prodromal symptoms of undiagnosed tumors on participation did not yield materially different results.

Lack of Exposure Details

A major limitation of the two cohorts is the lack of exposure details: (1) There is no information on call technology or conditions that might influence output power and thereby exposure. (2) Exposure modifiers such as hands-free devices or the preferred ear used when using a mobile phone cannot be taken into account. (3) The only available dose measure is time since first subscription or use, and although the U.K. study identifies daily users at baseline, neither study allows detailed investigation of dose–response relationships or risks restricted to heavy users. The increased ORs reported in the Hardell studies were, however, also detectable when analyzed solely by years since first use (Hardell et al. 2011).

Other Sources of Exposure

The register data used in the Danish cohort do not provide information on other sources of radiofrequency (RF) exposure, and the U.K. questionnaire only asked about mobile phones.

Potential misclassification of exposure due to car and bag phones, available in Denmark since 1982, that only give minimal cranial exposure, has been minimized in the latest studies from the Danish subscriber cohort, by left-truncating exposure data to 1987 (Frei et al. 2011; Schüz et al. 2011b) when truly handheld mobile phones became available in Denmark. Car and bag phones are not an issue in the U.K. study due to the questionnaire approach.

Regarding cordless phones that operate in the same frequency range as digital mobile phones, although typically at substantially lower power loads than GSM and NMT phones (Andersen et al. 2010), the INTERPHONE study did not suggest any risk increase associated with their use (Lönn et al. 2005; Schüz et al. 2006a). Hardell et al. (2011) have reported similar risk estimates for cordless phones and mobile phones; it is, however, noteworthy that they also found the increased risk when looking only at mobile phones. Information on cordless phones would, of course, have been an interesting addition to the two cohorts. With regard to misclassification of exposure the concern is, however, whether people with a mobile phone were, in general, less likely to use a cordless phone. This seems unlikely because the expensive minute charge on the mobile phone would motivate owners of mobile phones to use landline phones around the house and office and not rely on the convenience offered by their mobile phone.

Corporate Users/Lenders Borrowers

The Danish subscriber cohort, which does not have self-reported data, is afflicted by exposure misclassification (Schüz and Johansen 2007) from subscription holders not using their mobile phone and thus erroneously being classified as exposed or vice versa. Because mobile phone subscriptions during this period were very expensive, misclassification from children having subscriptions registered to a parent is unlikely. Conversely, the expense of subscriptions meant that many of the subscriptions of this period were paid by companies for their employees, and some (but not all) of these subscriptions will have been registered to the company rather than to the individual user, making the actual user unidentifiable from the available data. In total, 285,000 prescription records had to be excluded. This number represents an unknown but smaller number of unidentified eligible subscribers, because subscriptions that might have been held simultaneously or consecutively by the same user could not be identified. However, even if each deleted subscription is assumed to represent an unidentified user, the total number remains minute in proportion to the about four million adult Danes without a recorded subscription by the end of 1995.

Late Users

Both cohorts establish exposure at a single point in time, either when first getting a subscription or at the enrollment questionnaire. It is therefore possible that exposure status changes later.

In spite of the combined adjustment for age at diagnosis and age at baseline, it is possible for a woman in the U.K. cohort, not exposed at baseline but acquiring a phone shortly after, to accumulate up to two more years of exposure than an exposed person in the same strata who only had a short usage history before enrollment. This must however be a rare occurrence, and even though data on a subset of cohort members found that half of the nonusers at baseline had started using mobile phones in 2009 (Benson et al. 2013), the uptake of mobile phones in this cohort of middle-aged women was comparatively slow, and risk estimates did not increase in a subanalysis excluding subjects recruited before 2001, that is, when the issue was eliminated in the 10+ years of exposure category and reduced in the shorter exposure categories.

In the Danish cohort, people taking out their first subscription after 1995, when the available subscription records ended, had to be classified as unexposed. Members of the reference population who acquired a phone in 1996 may have accumulated almost as long exposure as those cohort members who got their phone in 1995. Such persons will, however, only constitute a small proportion of the reference population, and the average exposure level in the reference population will be much shorter. Particularly, the applied statistical adjustment for calendar period and time since first subscription ensures that subjects are only compared within the same strata. In the CNS tumor study, for example, this ensured that cohort members with <5 years since first subscription will never be compared with the population after 2001 when such exposure was common. It is also noteworthy that with regard to the long-term estimates of 10+ or 13+ years of use, there is well-defined exposure contrast between cohort members and reference population and neither the Danish subscriber cohort, Interphone, nor the investigations of incidence trends have identified increased risk associated with shorter usage.

In addition, early users were more exposed by their mobile phones than later users because the early phones relied on NMT followed by GSM technology, a technology having much higher emissions than the UMTS phones available since the early 2000s (Andersen et al. 2010). Furthermore, the average annual out-going traffic per subscription was about 1400 min per subscription until 1992, decreasing to about 900 min per subscription from 1995 to 2002 (National IT and Telecom Agency, Denmark 2003), after which use has been steadily increasing, reaching 1450 min per subscription in 2007 (National IT and Telecom Agency, Denmark 2003, 2010). This development presumably reflects that most early users had sufficient need for a mobile phone to justify the expense of the subscription, whereas later subscribers include also many sporadic users.

Finally, all studies on the cohort have investigated long-term user categories that could only include subscribers who acquired their phone one or more years before 1995, thereby increasing contrast, without finding indications of increased risk. Also, the risk estimates were unchanged in a subanalysis in the latest paper on CNS tumors, where exposure misclassification from latter users was eliminated by censoring follow-up in 1996. Similarly, the previous publications on the cohort, where the potential exposure range in the reference population was shorter, did not suggest an increased risk (Schüz et al. 2006b).

In spite of the obvious limitations of the cohorts, they are free from the directional biases that have been demonstrated to affect case–control studies (Vrijheid et al. 2006, 2009a, 2009b). The limited exposure detail and the fact that there are also exposed subjects in the reference population does however mean that the risk estimates will be driven toward the null, and concern has been raised with regard to the Danish cohort that if the total magnitude of exposure misclassification is such that the cohort is unable to detect any differences at all. Apart from the arguments given previously, there is also firsthand evidence that this is not the case: The average early mobile phone user was likely to be more affluent and in Denmark smoking is less prevalent among affluent males, whereas such an association is not seen in women; and indeed in the Danish cohort, there are reduced risks of smoking related cancers among males but not among females (Schüz et al. 2006b; Frei et al. 2011). The fact that the cohort can detect this very indirect association illustrates the cohort's ability to detect associations of appreciable size.

Summary

Compatible with most previous studies (SCENIHR 2009a; EFHRAN 2010; Baan et al. 2011), neither the U.K. nor the Danish study found any indication of an increased risk of meningioma. Similarly, neither study showed indications of an increased risk of glioma in mobile phone users, overall or among long-term subscribers. The picture was the same for the most exposed brain lobe. The finding of no clear association is in accordance with incidence trends (Deltour et al. 2012; Little et al. 2012), but in contrast to the studies by Hardell et al. (2011) that suggested an increased risk of glioma even after few years of mobile phone use.

The Danish subscriber cohort observed no increased risk of vestibular schwannoma among subjects with ≥11 years of mobile phone use, nor were the vestibular schwannoma tumors in mobile phone users larger or more likely to be situated on the right side of the head, reported in the prospective cohort study COSMOS, to be the preferred side of the head (Schüz et al. 2011a). The lack of association within the available time frame agrees with a comprehensive meta-analysis (not including the U.K. study) that did not find an increased risk, although there was marked heterogeneity (Repacholi et al. 2012) with two studies by Hardell et al. (2002, 2006) reporting increased risks.

The U.K. study found an increasing RR of vestibular schwannoma with increasing time since first use of a mobile phone. This was however based on relatively small numbers, and there was no difference in risk estimate among daily users and less frequent users. Also, there was no indication of increasing incidence rates of vestibular schwannoma in the United Kingdom during 1998–2008, as would be expected if use of mobile phones was associated with a pronounced risk increase (Benson et al. 2013). It must therefore be considered whether the mobile phone-using women in the cohort differ from the nonusers in a way that might be associated with an increased risk of vestibular schwannoma and that was not sufficiently covered by the baseline confounder data. Also, the media attention given to a potential association of mobile phones may have led to a surveillance bias, where hearing loss, a symptom of vestibular schwannoma, may have been more meticulously investigated in mobile phone users, leading to an increased likelihood of tumors being found.

The Danish study did not suggest an association of mobile phones use with vestibular schwannoma, and although the annual incidence of vestibular schwannoma in Denmark increased from 1976 to 2004, this is ascribed to the advent of computed tomography (CT) and magnetic resonance imaging (MRI) scanners. And in more recent years, there has not been an increase in Denmark (Stangerup et al. 2004, 2010; Stangerup and Caye-Thomasen 2012) and in the United Kingdom; the annual incidence rates have been constant from 1998 to 2008 (Benson et al. 2013). So, even though the U.K. study found some indication of an association for vestibular schwannoma fitting with the Hardell studies, there are at present too many uncertainties relating to these results to allow any inference about a causal association. However, the exposure detail available in either cohort study does not allow identification of specific user segments such as heavy users, and therefore precludes identification of risks restricted to small user segments. And even if the Danish cohort offers the longest follow-up data available, follow-up time may still be too short for slow-growing tumors such as vestibular schwannoma.

References

Ahlbom A, Feychting M, Cardis E, Elliott P. (2007). Re: cellular telephone use and cancer risk: update of a nationwide Danish cohort study. *J Natl Cancer Inst* **99**: 655–656.

Andersen JB, Mogensen PE, Pedersen GF. (2010). Power variations of wireless communication systems. *Bioelectromagnetics* **31**: 302–310.

Baan R, Grosse Y, Lauby-Secretan B, El Ghissassi F, Bouvard V, Benbrahim-Tallaa L, et al. (2011). Carcinogenicity of radiofrequency electromagnetic fields. *Lancet Oncol* **12**: 624–626.

Benson VS, Pirie K, Schüz J, Reeves GK, Beral V, Green J. (2013). Mobile phone use and risk of brain neoplasms and other cancers: prospective study. *Int J Epidemiol* **42**: 792–802.

Cardis E, Deltour I, Mann S, Moissonnier M, Taki M, Varsier N, et al. (2008). Distribution of RF energy emitted by mobile phones in anatomical structures of the brain. *Phys Med Biol* **53**: 2771–2783.

Cardis E, Richardson L, Deltour I, Armstrong B, Feychting M, Johansen C, et al. (2007). The INTERPHONE study: design, epidemiological methods, and description of the study population. *Eur J Epidemiol* **22**: 647–664.

Dalton SO, Steding-Jessen M, Gislum M, Frederiksen K, Engholm G, Schüz J. (2008). Social inequality and incidence of and survival from cancer in a population-based study in Denmark, 1994–2003: background, aims, material and methods. *Eur J Cancer* **44**: 1938–1949.

Deltour I, Auvinen A, Feychting M, Johansen C, Klaeboe L, Sankila R, et al. (2012). Mobile phone use and incidence of glioma in the Nordic countries 1979–2008: consistency check. *Epidemiology* **23**: 301–307.

Deltour I, Johansen C, Auvinen A, Feychting M, Klaeboe L, Schüz J. (2009). Time trends in brain tumor incidence rates in Denmark, Finland, Norway, and Sweden, 1974–2003. *J Natl Cancer Inst* **101**: 1721–1724.

Dreyer NA, Loughlin JE, Rothman KJ. (1999). Cause-specific mortality in cellular telephone users. *JAMA* **282**: 1814–1816.

EFHRAN. (2010). *Risk Analysis of Human Exposure to Electromagnetic Fields*. European Commission: Luxembourg, pp. 1–28.

Frei P, Poulsen AH, Johansen C, Olsen JH, Steding-Jessen M, Schüz J. (2011). Use of mobile phones and risk of brain tumours: update of Danish cohort study. *BMJ* **343**: d6387.

Gjerstorff ML. (2011). The Danish cancer registry. *Scand J Public Health* **39**: 42–45.

Hardell L, Carlberg M, Hansson Mild K. (2011). Pooled analysis of case-control studies on malignant brain tumours and the use of mobile and cordless phones including living and deceased subjects. *Int J Oncol* **38**: 1465–1474.

Hardell L, Carlberg M, Mild KH. (2006). Case-control study of the association between the use of cellular and cordless telephones and malignant brain tumors diagnosed during 2000–2003. *Environ Res* **100**: 232–241.

Hardell L, Mild KH, Carlberg M. (2002). Case-control study on the use of cellular and cordless phones and the risk for malignant brain tumours. *Int J Radiat Biol* **78**: 931–936.

Inskip PD, Hoover RN, Devesa SS. (2010). Brain cancer incidence trends in relation to cellular telephone use in the United States. *Neuro Oncol* **12**: 1147–1151.

INTERPHONE Study Group. (2010). Brain tumour risk in relation to mobile telephone use: results of the INTERPHONE international case-control study. *Int J Epidemiol* **39**: 675–694.

INTERPHONE Study Group. (2011). Acoustic neuroma risk in relation to mobile telephone use: results of the INTERPHONE international case-control study. *Cancer Epidemiol* **35**: 453–464.

Johansen C, Boice J, Jr, McLaughlin J, Olsen J. (2001). Cellular telephones and cancer—a nationwide cohort study in Denmark. *J Natl Cancer Inst* **93**: 203–207.

Khurana VG. (2011). Questions about selection, exposure, and tumour incidence. *BMJ* **343**: d7893.

Little MP, Rajaraman P, Curtis RE, Devesa SS, Inskip PD, Check DP, et al. (2012). Mobile phone use and glioma risk: comparison of epidemiological study results with incidence trends in the United States. *BMJ* **344**: e1147.

Lönn S, Ahlbom A, Hall P, Feychting M. (2005). Long-term mobile phone use and brain tumor risk. *Am J Epidemiol* **161**: 526–535.

Million Women Study Collaborative Group. (1999). The Million Women Study: design and characteristics of the study population. *Breast Cancer Res* **1**: 73–80.

Million Women Study Collaborators. (2002). Patterns of use of hormone replacement therapy in one million women in Britain, 1996–2000. *BJOG* **109**: 1319–1330.

National IT and Telecom Agency – Denmark. (2003). *Tele Yearbook—2002*. pp. 1–39.

National IT and Telecom Agency – Denmark. (2010). *Telecommunication Markets in the Nordic Countries 2010*. Copenhagen, Denmark: National IT and Telecom Agency; pp. 1–28.

Pedersen CB. (2011). The Danish civil registration system. *Scand J Public Health* **39**: 22–25.

Philips A, Lamburn G. (2011). Updated study contains poor science and should be disregarded. *BMJ* **343**: d7899.

Poulsen AH, Friis S, Johansen C, Jensen A, Frei P, Kjaer SK, et al. (2013). Mobile phone use and the risk of skin cancer: a nationwide cohort study in Denmark. *Am J Epidemiol* **178**: 190–197.

Poulsen AH, Stenager E, Johansen C, Bentzen J, Friis S, Schüz J. (2012). Mobile phones and multiple sclerosis—a nationwide cohort study in Denmark. *PLoS One* **7**: e34453.

Repacholi MH, Lerchl A, Röösli M, Sienkiewicz Z, Auvinen A, Breckenkamp J, et al. (2012). Systematic review of wireless phone use and brain cancer and other head tumors. *Bioelectromagnetics* **33**: 187–206.

Rothman KJ, Loughlin JE, Funch DP, Dreyer NA. (1996). Overall mortality of cellular telephone customers. *Epidemiology* **7**: 303–305.

SCENIHR. (2009a). *Health Effects of Exposure to EMF*. Scientific Committee on Emerging and Newly Identified Health Risks. Bruxelles: European Commission, pp. 1–83.

SCENIHR. (2009b). *Research Needs and Methodology to Address the Remaining Knowledge Gaps on the Potential Health Effects of EMF*. Scientific Committee on Emerging and Newly Identified Health Risks. Bruxelles: European Commission, pp. 1–28.

Schmidt LS, Nielsen H, Schmiedel S, Johansen C. (2008). Social inequality and incidence of and survival from tumours of the central nervous system in a population-based study in Denmark, 1994–2003. *Eur J Cancer* **44**: 2050–2057.

Schüz J, Bohler E, Berg G, Schlehofer B, Hettinger I, Schlaefer K, et al. (2006a). Cellular phones, cordless phones, and the risks of glioma and meningioma (INTERPHONE Study Group, Germany). *Am J Epidemiol* **163**: 512–520.

Schüz J, Elliott P, Auvinen A, Kromhout H, Poulsen AH, Johansen C, et al. (2011a). An international prospective cohort study of mobile phone users and health (Cosmos): design considerations and enrolment. *Cancer Epidemiol* **35**: 37–43.

Schüz J, Jacobsen R, Olsen JH, Boice JD, Jr, McLaughlin JK, Johansen C. (2006b). Cellular telephone use and cancer risk: update of a nationwide Danish cohort. *J Natl Cancer Inst* **98**: 1707–1713.

Schüz J, Johansen C. (2007). A comparison of self-reported cellular telephone use with subscriber data: agreement between the two methods and implications for risk estimation. *Bioelectromagnetics* **28**: 130–136.

Schüz J, Steding-Jessen M, Hansen S, Stangerup SE, Caye-Thomasen P, Poulsen AH, et al. (2011b). Long-term mobile phone use and the risk of vestibular schwannoma: a Danish nationwide cohort study. *Am J Epidemiol* **174**: 416–422.

Schüz J, Waldemar G, Olsen JH, Johansen C. (2009). Risks for central nervous system diseases among mobile phone subscribers: a Danish retrospective cohort study. *PLoS One* **4**: e4389.

Söderqvist F, Carlberg M, Hardell L. (2012). Review of four publications on the Danish cohort study on mobile phone subscribers and risk of brain tumors. *Rev Environ Health* **27**: 51–58.

Stangerup SE, Caye-Thomasen P. (2012). Epidemiology and natural history of vestibular schwannomas. *Otolaryngol Clin North Am* **45**: 257–68, vii.

Stangerup SE, Tos M, Caye-Thomasen P, Tos T, Klokker M, Thomsen J. (2004). Increasing annual incidence of vestibular schwannoma and age at diagnosis. *J Laryngol Otol* **118**: 622–627.

Stangerup SE, Tos M, Thomsen J, Caye-Thomasen P. (2010). True incidence of vestibular schwannoma? *Neurosurgery* **67**: 1335–1340.

Storm HH, Michelsen EV, Clemmensen IH, Pihl J. (1997). The Danish cancer registry—history, content, quality and use. *Dan Med Bull* **44**: 535–539.

Swerdlow AJ, Feychting M, Green AC, Leeka Kheifets LK, Savitz DA. (2011). Mobile phones, brain tumors, and the INTERPHONE study: where are we now? *Environ Health Perspect* **119**: 1534–1538.

Vrijheid M, Armstrong BK, Bedard D, Brown J, Deltour I, Iavarone I, et al. (2009a). Recall bias in the assessment of exposure to mobile phones. *J Expo Sci Environ Epidemiol* **19**: 369–381.

Vrijheid M, Cardis E, Armstrong BK, Auvinen A, Berg G, Blaasaas KG, et al. (2006). Validation of short term recall of mobile phone use for the INTERPHONE study. *Occup Environ Med* **63**: 237–243.

Vrijheid M, Richardson L, Armstrong BK, Auvinen A, Berg G, Carroll M, et al. (2009b). Quantifying the impact of selection bias caused by nonparticipation in a case-control study of mobile phone use. *Ann Epidemiol* **19**: 33–41.

WHO. (2006). *WHO Research Agenda for Radiofrequency Fields*. Geneva: WHO.

13

Brain Tumors and Mobile Phone Use: The Case–Control Approach

Chelsea Eastman Langer and Elisabeth Cardis

CONTENTS

An Overview: Role of Case–Control Studies

Potential health effects of radiofrequency (RF) have been reviewed in recent years by several national and international organizations (AFSSET 2009; SCENIHR 2009; CCARS 2011; Swerdlow et al. 2011; Sienkiewicz et al. 2012; IARC 2013). The most recent and comprehensive evaluations, conducted by the International Agency for Research on Cancer (IARC) Monographs and by EFRHAN, classified RF from portable phones as "possibly carcinogenic to humans (2B)" (Baan et al. 2011; Sienkiewicz et al. 2012; IARC 2013) based on limited evidence in humans. The evidence came from observations of positive associations between glioma and acoustic neuroma and exposure to radiofrequency electromagnetic fields (RF-EMFs) from wireless phones in large-scale case–control studies.

Case–control studies on the health effects of RF from mobile phones are very complex, particularly due to difficulties in exposure assessment and due to the rapid changes in phone use and in communication systems occurring within the span of the studies. Furthermore, like all epidemiological study designs, they are subject to methodological limitations that complicate interpretation of results. For case–control studies, the main issues of concern are potential errors (systematic and random) in recall and selection of population due to participation.

In its evaluation, the IARC Monographs Working Group on RF recognized that the positive associations observed in recent large-scale epidemiological studies are susceptible to bias—due to recall error (see Chapter 4) and selection for participation (see Chapter 5)—but concluded that the findings could not be dismissed as reflecting bias alone and that a causal interpretation between mobile phone RF-EMF exposure and glioma is possible (IARC 2013).

Brain and Central Nervous System Tumors

A primary brain tumor originates in the brain; however, not all such tumors are malignant (Ohgaki and Kleihues 2005). Although some of the published studies have considered brain tumors (or brain and central nervous system [CNS] tumors) as a whole, several more recent studies have focused on tumors occurring in the areas of the brain that receive much of the RF energy emitted by mobile phones: gliomas, a broad category of mainly malignant brain tumors that occur in the glial cells; and meningiomas and acoustic neuromas (vestibular schwannoma), generally benign brain tumors arising in the meninges and the auditory vestibular nerve, respectively. In some case–control studies (discussed below), researchers have also focused more specifically on gliomas and meningiomas arising in parts of the brain where absorption of RF energy from mobile phones is highest (Cardis et al. 2008). Table 13.1 presents a brief overview and summary of results for studies discussed in this chapter.

Studies of Risk of Brain Tumors in General

Two early case–control studies investigating the possible association of mobile phone use and the risk of brain tumors included brain tumors without distinguishing between specific diagnoses.

In one of the first case–control studies studying this topic, adult patients (aged 18–80 years) with a primary brain cancer diagnosed between 1994 and 1998 were recruited from five U.S. medical centers (Muscat et al. 2000). Analyses included 469 cases and 422 controls matched by hospital, and frequency matched by age, sex, race, and month of admission. The multivariate odds ratio (OR) for ever having been a regular user (defined as ever having a subscription to a mobile telephone service) was 0.8 (95% confidence intervals [CI], 0.6–1.2). Considering three exposure measures (hours per month, years of use, and lifetime cumulative hours) showed no association between mobile phone use and risk of brain cancer, although few participants (14% of cases and 18% of controls) had ever used a mobile phone regularly and only 17 cases had used a phone for ≥4 years.

In the same timeframe as the study conducted by Muscat and colleagues, 782 brain tumor patients were enrolled in a separate case–control study in three cities in the United States (Inskip et al. 2001). The 799 controls were hospital-based (admitted to same hospitals as cases, for a variety of nonmalignant conditions) and frequency matched according to hospital, age, sex, race, and proximity of residence to hospital. In keeping with the findings of Muscat et al. (2000), Inskip et al. (2001) found no statistically significant association between mobile phone and risk of brain tumors (the OR for ever having been a regular mobile phone user was 0.8 [0.6–1.1]). Regular use was infrequent among study subjects (18% among cases and 22% among controls; only 22 brain tumor cases had used a phone for ≥5 years), and the authors acknowledged that their data were insufficient to evaluate risks among long-term, heavy users, and for long induction periods.

TABLE 13.1

Case–Control Studies Investigating Mobile Phone Use and Brain Tumor Risk

				Years of Use		Cumulative Call Time	
Study	Diagnostic Period	No. of Cases	% "Regular" Mobile Phone Use	OR (95% CI)	n Cases, Long-Term Users	OR (95% CI)	n Cases, Heavy Users
Muscat et al. 2000	1994–1998	All malignant brain tumors (n = 469; age range, 18–80 years)	Cases: 14% Controls: 18%	≥4.0 years 0.7 (0.4–1.4)	17	>480 hr 0.7 (0.3–1.4)	14
Inskip et al. 2001	1994–1998	All (n = 782; age range, 18–90 years) Glioma (n = 489) Meningioma (n = 197) Acoustic neuroma (n = 96)	Cases: 18% Controls: 22%	≥5.0 years Glioma: 0.6 (0.3–1.4) Meningioma: 0.9 (0.3–2.7) Acoustic neuroma: 1.9 (0.6–5.9)	 11 6 5	>500 hr Glioma: 0.5 (0.2–1.3) Meningioma: 0.7 (0.2–2.4) Acoustic neuroma: 0.4 (0.0–3.3)	 11 6 1
Hardell et al. 2011	1997–2003	Malignant brain tumor cases (n = 1251; age range, 20–80 years) Glioma (n = 1148)	Cases: 46% Controls: 39%	>10-year latency All: 2.5 (1.8–3.3) Glioma: 2.5 (1.8–3.3)	 134 123	>2000 hr All: 3.0 (1.9–4.8) Glioma: 3.2 (2.0–5.1)	 61 58
Hardell et al. 2013a	2007–2009	Malignant brain tumor cases (n = 593; age range, 18–75 years)	Cases: 92% Controls: 89%	>10–15 years: 1.3 (0.8–2.2) >15–20 years: 1.5 (0.8–2.6) >20–25 years: 1.9 (1.1–3.5) >25 years: 2.9 (1.4–5.8)	163 76 48 30	>2376 hr 2.8 (1.6–4.8)	137
Carlberg et al. 2013	2007–2009	Meningioma (n = 709; age range, 18–75 years)	Cases: 84% Controls: 89%	>10–15 years: 1.0 (0.7–1.4) >15–20 years: 1.0 (0.6–1.5) >20–25 years: 0.8 (0.5–1.4) >25 years: 1.2 (0.6–2.3)	185 78 29 16	>2376 hr 1.3 (0.8–1.9)	84

Continued

TABLE 13.1 ***(Continued)***

Case–Control Studies Investigating Mobile Phone Use and Brain Tumor Risk

				Years of Use		Cumulative Call Time	
Study	**Diagnostic Period**	**No. of Cases**	**% "Regular" Mobile Phone Use**	**OR (95% CI)**	**n Cases, Long-Term Users**	**OR (95% CI)**	**n Cases, Heavy Users**
INTERPHONE Study Group 2010	2000–2004	Glioma (n = 2708) Meningioma (n = 2409; age range, 30–59 years)	Glioma: 62% Meningioma: 52% Controls: 60%	≥10 years Glioma: 0.98 (0.76–1.26)	252	≥1640 hr Glioma: 1.40 (1.03–1.89)	210
				Meningioma: 0.83 (0.61–1.14)	110	Meningioma: 1.15 (0.81–1.62)	130
INTERPHONE Study Group 2011	2000–2004	Acoustic neuroma (n = 1105; age range, 30–59 years)	Cases: 58% Controls: 61%	1-year analysis: ≥10 years 0.76 (0.52–1.11)	68	1-year analysis: ≥1640 hr	77
				5-year analysis: ≥10 years 0.83 (0.58–1.19)	68	1.32 (0.88–1.97)	36
						5-year analysis: ≥1640 hr 2.79 (1.51–5.16)	
Hardell et al. 2013b	1997–2003, 2007–2009	Acoustic neuroma (n = 316; age range, 18–80 years)	Cases: 63% Controls: 61%	>10–15 years: 2.1 (1.3–3.5)	34	>1486 hr 2.6 (1.5–4.4)	30
				>15–20 years: 2.1 (1.02–4.2)	12		
				>20 years: 4.5 (2.1–9.5)	12		
Sato et al. 2011[a]	2000–2006	Acoustic neuroma (n = 787; all ages)	1-year analysis: 55% 5-year analysis: 28%	>10 years 1-year analysis: 1.62 (0.79–4.77)	12	>20 min/day 1-year analysis: 2.74 (1.18–7.85)	23
				5-year analysis: 1.00 (0.59–3.23)	6	5-year analysis: 3.08 (1.47–7.41)	33
Aydin et al. 2011	2004–2008	Brain tumor cases (n = 352; age 7–19 years)	Cases: 55% Controls 51%	>5 years 1.26 (0.70–2.28)	46	>144 hr 1.55 (0.86–2.82)	49

[a] Case-case study: RRs are reported here, not OR.

Studies of Glioma

In most of the case–control studies that analyzed glioma specifically, no increased risk was observed overall in association with regular mobile phone use (Table 13.1). However, an increased risk of glioma was observed at the highest exposure levels in the largest case–control studies.

Of the 782 brain tumor patients included in the study by Inskip et al. (2001), 489 were glioma patients. Similar to the overall findings, Inskip et al. (2001) found no statistically significant association between mobile phone and risk of glioma in any of the categories. Only 11 cases reported ≥5 years of use, however; hence, the study has little power to detect a possible association.

In contrast, when Hardell et al. (2011) pooled data from two case–control studies of malignant brain tumors in Sweden (including cases diagnosed between 1997 and 2003), an association was found between amount and duration of mobile phone use for all malignant brain tumors as well as for glioma specifically (Table 13.1) (Hardell et al. 2011). The highest ORs were found for astrocytoma (n = 952), a common type of glioma; the OR for ≥10 years of mobile phone use was 2.7 (1.9–3.7); in comparison, the OR for use of cordless phones for ≥10 years was 1.8 (1.2–2.9). Mobile and cordless phone use appeared to be uncorrelated and adjustment for both variables at the same time had little effect on the ORs. The risk for astrocytoma appeared to be the highest in the group with first use of a wireless phone before the age of 20 years. More recently, Hardell and colleagues studied adult malignant brain tumor cases (primarily gliomas) diagnosed during 2007–2009 and observed an estimated OR of 2.9 (1.4–5.8) for the longest mobile phone users—25 years or more (Hardell et al. 2013a). In contrast to their previous work, the association between the risk of malignant brain tumor and 10–20 years of mobile phone use was not statistically significant. Whereas Hardell's results suggest that mobile phone use may increase the risk of glioma, the similarity of ORs for mobile and cordless phones is unexpected given the difference in average output power of these phone types.

The largest study to date on this topic, INTERPHONE, included 2708 glioma cases (in addition to meningioma and acoustic neuroma cases; discussed below) diagnosed between 2000 and 2004 in 13 countries by using a common protocol (INTERPHONE Study Group 2010b). A reduced OR related to ever having been a regular mobile phone user was seen for glioma (0.81 [0.70–0.94]), possibly reflecting participation bias or other methodological limitations. No elevated OR was observed ≥10 years after first phone use (0.98 [0.76–1.26]). Furthermore, ORs were below 1.0 for all deciles of lifetime number of phone calls and nine deciles of cumulative call time. Although there was no trend of increasing ORs with increasing cumulative call time or cumulative number of calls, a higher (and statistically significant) OR was seen in the highest decile of recalled cumulative call time, ≥1640 hours: 1.40 (1.03–1.89) for glioma. These ORs for glioma were higher for tumors in the temporal lobe (1.87, [1.1–3.2]) than in other lobes of the brain, and in subjects who reported usual phone use on the same side of the head as their tumor (1.96, [1.2–3.2]) than on the opposite side. The INTERPHONE Study Group concluded that although no overall association was observed, there were suggestions of an increased risk of glioma at the highest exposure levels although biases and error prevent a causal interpretation.

Meningioma

Analyses by Inskip et al. (2001) included 197 meningioma cases and their controls. Similar to the findings for glioma, Inskip et al. (2001) did not find an association between mobile phone and risk of meningioma. Only six cases reported ≥5 years of use, however; hence, the study has little power to detect a possible association.

The INTERPHONE analyses included 2409 meningioma cases and their matched controls. As with glioma, a reduced OR related to ever having been a regular mobile phone user was seen for meningioma (0.79 [0.68–0.91]). No elevated OR was observed ≥10 years after first phone use (meningioma: 0.83; [0.61–1.14]); furthermore, ORs were below 1.0 for all deciles of lifetime number of phone calls and nine deciles of cumulative call time, possibly reflecting participation bias or other methodological limitation. There was no trend of increasing ORs with increasing cumulative call time or cumulative number of calls, although a higher OR was seen in the highest decile of recalled cumulative call time: 1.15 (0.81–1.62) for meningioma (INTERPHONE Study Group 2010). This OR was higher when analyses were restricted to subjects who reported usual phone use on the same side of the head as their tumor 1.45 (0.80–2.61) but not when restricting analyses to tumors in the temporal lobe.

In the most recent Hardell study (Carlberg et al. 2013), no statistically significant association was seen between years of use of mobile phones and risk of meningioma, although an indication of an increased risk was seen in the group with highest cumulative use.

Acoustic Neuroma

The first study on acoustic neuroma (also called vestibular schwannoma) by Inskip et al. (2001) included 96 acoustic neuroma cases and their controls. Their study found an OR of 1.9 (0.6–5.9) among people who had used a mobile phone for ≥5 years and an OR of 0.4 (0.0–3.3) among subjects who had talked on a mobile phone for >500 hr. The wide confidence intervals are likely a result of the small number of cases; only one case reported >500 hr of use, and five cases used a mobile phone for ≥5 years.

The INTERPHONE study included 1105 patients with newly diagnosed acoustic neuroma and 2145 controls (INTERPHONE Study Group 2011). In the primary analysis, exposure time was censored at 1 year before the reference date (date of diagnosis for cases and date of diagnosis of the matched case for controls); analyses censoring exposure at 5 years before the reference date were also done to allow for a possible longer latency period. As in analyses of glioma and meningioma, most ORs were <1, possibly reflecting selection bias or other methodological limitations. The OR was 0.85 (0.69–1.04) for regular use and 0.76 (0.52–1.11) for use ≥10 years in the past. There was no trend of increasing ORs with increasing cumulative call time or cumulative number of calls, with the lowest OR of 0.48 (0.30–0.78) being observed in the ninth decile of cumulative call time. In the 10th decile (≥1640 hr) of cumulative call time, the OR was 1.32 (0.88–1.97); it was 2.79 (1.51–5.16) in analyses censoring use 5 years before the reference date. These ORs were 2.33 (1.23–4.40) and 3.53 (1.59–7.82), respectively, when analyses were restricted to subjects reporting using the phone on the side of the head where the tumor developed. As for glioma and meningioma, the elevated ORs observed at the highest level of cumulative call time could be due to chance, reporting bias or a causal effect. As acoustic neuroma is usually a slowly growing tumor, the interval between introduction of mobile phones and occurrence of the tumor might have been too short to observe an effect, if there was one.

A case–case study was conducted in Japan, including 787 acoustic neuroma cases, to avoid potential selection biases related to low response rates among controls (see section "Selection Bias") (Sato et al. 2011). The study investigated laterality of mobile phone use. The estimated risk ratios were 1.08 (0.93–1.28) for regular mobile phone use at 1 year before diagnosis and 1.14 (0.96–1.40) at 5 years before diagnosis. A significantly increased risk was observed for self-reported mobile phone use for on average >20 min/day (based on 23 cases).

When Hardell et al. pooled data from their two case–control studies of acoustic neuromas in Sweden (including cases diagnosed 1997–2003 and 2007–2009), an association was found between amount and duration of mobile phone use and risk of acoustic neuroma (Hardell 2013b). The OR for ≥20 years of mobile phone use was 4.5 (2.1–9.5); by comparison, the OR for use of cordless phones for ≥10 years was 6.5 (1.7–26). Although these results indicate mobile phone use may increase the risk of acoustic neuroma, it is surprising to observe similar ORs for mobile and cordless phones given the difference in average output power of these phone types.

Studies in Children and Adolescents

CEFALO, a multicenter case–control study of brain tumors in young people (aged 7–19 years), was conducted in Denmark, Sweden, Norway, and Switzerland and included all cases diagnosed between 2004 and 2008 (Aydin et al. 2011). Analyses included 352 cases and 646 population controls matched by age, sex, and geographical region. The OR for having been a regular mobile phone user (55% of cases and 51% of controls) was 1.36 (0.92–2.02). As in the studies discussed above, there was no association with duration or amount of use. In a small subset of study participants for whom operator recorded data were available, brain tumor risk was related to the time elapsed since the mobile phone subscription was started, but not to amount of use. The subjects in this study were young (the median age at diagnosis was 13 years), and the study included very few long-term (46 cases with more than 5 years of use) or heavy users (median lifetime use of 35 hr).

Although none of these three studies showed evidence of an association, there were very few long-term or heavy users included in the studies; hence, they provide no information about the existence or not, of an association between mobile phone use and risk of brain tumors.

Potential Sources of Bias in Case–Control Studies

There are inherent limitations in case–controls studies, specifically the potential for selection and recall bias, discussed below.

Selection Bias

Selection bias (see Chapter 5) may occur if individuals have unequal probabilities of being included in the study sample according to the exposure, outcome of interest, or both (Szklo and Nieto 2007). This is a particular concern in studies of mobile phone use where (with the notable exception of the Hardell studies; Hardell et al. 2011), refusal rates in controls tend to be quite high (30% in INTERPHONE, with an additional 15% untraceable; INTERPHONE Study Group 2010). One way of gauging if selection bias has occurred is to administer a short questionnaire to subjects who refuse to participate in the full study; in this way, it is possible to estimate whether differences between participants and nonparticipants exist. This was done in the INTERPHONE study where, based on the responses of those who agreed to complete the nonresponse questionnaire (NRQ), it appears that mobile phone users were more likely to participate in the study (i.e., 56% of controls who answered the nonresponse questionnaire were regular mobile phone users compared with 69% of participating controls; similar results for cases; Vrijheid et al. 2006). In addition, fewer nonrespondents had started using a mobile phone in 1998 or earlier compared with participants (controls, 41% vs. 52%; cases, 41% vs. 54%). Thus, the high refusal rates among controls, combined with unrepresentativeness of participants, may explain ORs <1 reported in several studies. In INTERPHONE, based on results of the NRQ and a series of plausible phone use scenarios among other nonparticipants, it is estimated that ORs may be underestimated by a factor of 10%–20% (Vrijheid et al. 2009a), thus explaining, in part, the large number of ORs <1 in that study. Other factors that may contribute to the reduced ORs include a protective effect of mobile phones (unlikely as the definition of "regular mobile phone use" is very broad), intrinsic differences between exposed and nonexposed subjects that have not been measured or controlled (a means for correcting this is to conduct analyses restricted only to exposed subjects as was done in INTERPHONE (INTERPHONE Study Group 2010), confounding (see Chapter 6); recall bias (discussed below and in Chapter 4), and reverse causation (brain tumor cases less likely to become mobile phone users as a result of their symptoms, although this would be expected to be more important for recent use than for longer term use), or a combination.

Recall Bias

Recall bias, an example of information bias, arises when the study subject's memory is inaccurate regarding an exposure of interest, and may thus result in misclassifying the subject's exposure status (Szklo and Nieto 2007). In case–control studies researching mobile phone use, investigators have two primary options to check whether there is recall bias: (1) comparing self-reported phone use to mobile phone operator records and (2) distributing software-modified phones (SMPs). SMPs are regular mobile phones with modified software to record the date, start time, and duration of each call. Both INTERPHONE and CEFALO attempted to characterize and quantify recall errors (Vrijheid et al. 2006, 2009b; Aydin et al. 2011). To investigate the accuracy of self-reported phone use, two validation substudies were conducted in some of the INTERPHONE centers. Among healthy volunteers using SMPs, phone use in the past year was reported with substantial random error; with over- and underestimation both being frequent (Vrijheid et al. 2006). Errors were larger for duration of calls than for number of calls, and phone use was underestimated by light users and overestimated by heavy users. In another substudy, records of mobile phone use up to 6 years previously were obtained for some participants in three INTERPHONE centers, allowing comparison of the interview responses with the records (Vrijheid et al. 2009b). Overall, there was little evidence that recall quality differed between cases and controls, but there was some indication of greater overreporting by cases than by controls for periods 3–5 years before interview. These substudies provide no information regarding differential reporting error for periods greater than 5 years before interview.

In INTERPHONE and the Hardell studies, ipsilateral ORs were almost always greater than contralateral ORs. This may reflect a true association or laterality recall bias. This was examined in some detail in INTERPHONE; for glioma, a trend toward a stronger effect of ipsilateral use relative to contralateral use with increasing exposure was observed for cumulative number of calls as well as, when excluding the lowest cumulative call time category, with increasing cumulative call time. The observation of an unlikely ipsilateral effect in the lowest cumulative call time category, however, suggests that cases might have overreported use on the side of the tumor. There is, however, evidence of lack of such reporting bias from an INTERPHONE substudy. In three centers, participants were asked at the end of their interview to put a mobile phone to their ear as if answering a call (INTERPHONE Study Group 2010a). The concordance between the reported side of use of the phone and the side where it was held was lower for cases (72% glioma cases, 66% meningioma) than controls (95%). The greater degree of concordance among controls suggests differential reporting quality. Among cases, however, there was as much discrepancy in the contralateral direction (52 instances) as in the ipsilateral direction (48 instances). Thus, it is possible that the ipsilateral effect is a true effect, and is due to reporting bias or is a mixture of both.

An important result in the INTERPHONE analysis is the finding of a higher OR among heavy users for tumors in the temporal lobe, where most of the RF energy is absorbed. Although laterality analyses may be biased by the respondent's knowledge of the side of the tumor, results for tumors in different lobes are probably less susceptible to reporting bias.

These sources of bias, inherent in case–control studies, complicate the interpretation of results.

Exposure Assessment

Assessing RF exposure from mobile phones is extremely complex: exposure varies by mobile phone make and model; how and where the subject holds the phone; if the subject uses a hands-free kit or speakerphone; network; and generation (see Chapter 8). Most studies rely on self-reported information on mobile phone use as exposure proxies, namely, self-reported years of use and cumulative call time (generally based on average call time multiplied by average number of calls). Although these provide better estimates of true exposure to mobile phones than a dichotomous estimate such as "ever" or "never" mobile phone user, they are subject to biases and errors as discussed above. Attempts to refine exposure assessment are discussed below.

To explore the observed associations taking into account the localized nature of RF energy absorption when using a mobile phone, further analyses of subgroups of INTERPHONE countries were conducted (Larjavaara et al. 2011; Cardis et al. 2011a, 2011b). A case-only analysis of data from seven INTERPHONE countries (Denmark, Finland, Germany, Italy, Norway, Sweden, United Kingdom-South) was conducted to evaluate whether gliomas occur preferentially in the areas of the brain having the highest exposure to RF fields, based on estimated distance from the center of their tumor to a hypothetical phone axis (Larjavaara et al. 2011). No difference was found between tumors with a center within 5 cm of the phone line and tumors with a center further than 5 cm, either in terms of ever having used a mobile phone regularly or duration of phone use. Complementary case-specular analyses were also conducted, in which the distance from the center of the tumor to the phone axis was compared between the cases and their "specular" controls (for each case, the location of the specular tumor was obtained as a mirror image in two dimensions—within the same brain hemisphere—of the location of the original tumor). In these analyses, an OR of 2.00 (0.68–5.85) was observed among long-term users (≥10 years) based on small numbers of cases.

The main parameters thought to influence absorption of RF energy in the brain from mobile phone use were also investigated (Cardis et al. 2011a), based on information from the INTERPHONE questionnaire; network operators; laboratory measurements; and SMPs issued to a subset of study participants. An algorithm was developed to evaluate the total cumulative RF energy (in joules per kilogram), or dose, absorbed at a particular location in the brain. The main determinants of absorbed energy were the communication system and frequency band, location in the brain, and the amount and duration of mobile phone use. Although there was substantial agreement between categorization of subjects by cumulative absorbed energy and by cumulative call time (the exposure variable used in the main INTERPHONE analyses and in many other epidemiological studies), misclassification appeared non-negligible, particularly at higher frequency bands.

The above-mentioned algorithm was applied to INTERPHONE Study subjects in five countries (Australia, Canada, France, Israel, and New Zealand) (Cardis et al. 2011b). An increased risk of glioma was seen in individuals at the highest quintile of absorbed dose, although reduced risks were seen in the four lower quintiles. When risk was examined as a function of absorbed dose received in different time windows before diagnosis, an increasing trend was observed with increasing absorbed dose for exposures ≥7 years in the past, but not for more recent use.

Complementary case-case analyses in which laterality of phone use was not considered to avoid a possible laterality recall bias also indicated an increased risk in the most exposed region of the brain, based on small numbers of subjects, compared with other areas among long-term users. Patterns of risk for meningioma in relation to absorbed dose were similar, although increases in risk were much smaller than for glioma, and not statistically significant. These results may suggest an increased risk of glioma in the most exposed area of the brain among long-term and heavy users of mobile phones. However, the exposure algorithm still relies on the bias-susceptible questionnaire data, and as pointed out by the authors, there are uncertainties associated with tumor center localization, estimation of absorbed dose, and sample size. These results require replication in an independent, and preferably improved, setting before they could be taken to indicate a cause–effect relationship.

The reasons for the differences in the results of the two studies of independent subsets of INTERPHONE countries are unclear. However, there are differences in the detail of exposure assessment. For the case-case analyses based only on location of the tumor, the most exposed area of the brain was defined from analyses of results of experiments on the spatial distribution of the specific energy absorption rate (SAR) on phantoms for >100 phone models (Cardis et al. 2008, 2011b), whereas Larjavaara et al. (2011) calculated distance from the center of the tumor to a hypothetical phone axis. Different approaches were also used to define the center of the tumor in both studies, and formal comparisons of the approaches are needed to assist in the interpretation of the results.

Potential Confounders

Most of the studies reviewed collected information on potential confounders (see Chapter 6) of an association between mobile phone use and brain tumor risk. Higher socioeconomic status (SES) has been associated with a higher risk of brain cancer in some but not all relevant studies (Chakrabarti et al. 2005; Schmidt et al. 2008), and with mobile phone use, particularly when the technology was new (Schuz et al. 2006). Many studies adjusted for education level in analyses, but this is an imperfect indicator of SES. Otherwise, there are few well-established risk factors for brain tumors.

The INTERPHONE study conducted thorough sensitivity analyses to determine factors that may have biased the results, including study center, mention of mobile phones in the introductory letter to subjects; centers with a hospital-based design or particularly low participation rates; respondents whose interviews were considered by the interviewer to be of poor quality; subjects for whom proxies provided the responses; subjects who reported implausibly high amounts of mobile phone use; use of matching and conditional analysis; and adjustment for possible confounders. For glioma, results from the sensitivity analyses were generally similar to those from the primary analysis. Most ORs were well within the 95% CI of the OR from the main analysis. When subjects with high reported use were included, but with use truncated at 5 hr/day, the OR was hardly affected. When subjects who reported >5 hr call time per day were excluded altogether, on the premise that such responses were unreliable, the OR decreased from 1.39 to 1.27 (0.92–1.75).

Strengths of Case–Control Studies

The primary advantage of case–control studies compared with cohort studies (see Chapter 12) is there is no need to follow up a large population for a long time, thus optimizing speed and efficiency (Gordis 2004). This is especially important for studying a rare disease, such as brain tumors. For example, INTERPHONE included cases covering a source population of approximately 48.8 million people in 13 countries (Cardis et al. 2007), whereas a recent update of the Danish cohort study included >350,000 mobile phone subscription holders (Frei et al. 2011). Due to the large population covered, INTERPHONE included >6000 cases (glioma, meningioma, and acoustic neuroma); the Danish cohort study, in contrast, included 356 gliomas and 80 meningiomas. Exposure assessment is also more in-depth in case–control studies compared with cohort studies, because there are fewer subjects for whom exposure needs to be estimated. Frei and colleagues did not contact participants: all subjects' information was gathered from registries or other public sources. In contrast, the case–control studies discussed above administer questionnaires with detailed mobile phone use, including laterality, use of hands-free kits/speakerphone, and amount and duration of use.

Conclusions

Case–control studies investigating the health effects of mobile phones are extremely complex. Methodological difficulties and limitations, particularly possible selection bias and recall errors, and difficulties and differences in EMF exposure assessment, complicate the interpretation of results, as inevitably happens in other study designs, be they cohort or ecological studies (see Chapter 14), that have their own, different inherent limitations.

Despite these limitations, however, the case–control study design provides considerable advantages for the study of rare diseases and relatively small risks compared with other epidemiological study designs. Case–control studies to date have included a substantially higher number of brain tumor patients compared with cohort studies with much more detailed exposure information. Case–control studies with detailed substudies to investigate the potential for biases and errors therefore have considerably greater statistical power for the study of the potential effect of RF from mobile phones on the risk of brain tumors.

Although mobile phone use has been very prevalent in Europe since the mid- to late 1990s, heavy mobile phone use is still a relatively recent phenomenon: the median monthly use reported in INTERPHONE controls interviewed between 2000 and 2004 was of the order of 2 hr; in CEFALO, the median years of use was 2.7 hr, although it is not unusual today for individuals, particularly young people, to use their phone 1 hr or more a day. If RF exposure from mobile phones does increase the risk of brain tumors as suggested by some of the results of the INTERPHONE study, its potential impact on cancer trends is likely not to be appreciable yet, if excess risk only manifests more than a decade after phone use begins and if mobile phone use only affects a small proportion of cases, in the most heavily exposed areas of the brain, or a subset of brain tumors. Clearly, however, continued monitoring of trends is needed and may be a very important tool in the future (see Chapter 14).

In the meantime, although results of INTERPHONE and other studies are subject to potential bias, they suggest that a causal role of RF in brain tumorigenesis is possible. Although research continues through further case–control studies such as Mobi-Kids (www.mbkds.com) and other study designs, simple means (use of hands-free kits or speaker mode of the phone and use of data instead of voice) can be adopted to reduce exposure and, if a risk is confirmed in the future, to limit the potential public health impact of mobile phones.

References

AFSSET. (2009). *Mise à jour de l'expertise relative aux radiofréquences*. Paris: Agence Française de Sécurité, Sanitaire de l'Environnement et du Travail.

Aydin, D., Feychting, M., Schüz, J., Tynes, T., Andersen, T.V., Schmidt, L.S., et al. (2011). Mobile phone use and brain tumors in children and adolescents: a multicenter case–control study. *J Natl Cancer Inst 103*, 1264–1276.

Baan, R., Grosse, Y., Lauby-Secretan, B., El Ghissassi, F., Bouvard, V., Benbrahim-Tallaa, L., et al. (2011). Carcinogenicity of radiofrequency electromagnetic fields. *Lancet Oncol 12*, 624–626.

Cardis, E., Armstrong, B.K., Bowman, J.D., Giles, G.G., Hours, M., Krewski, D., et al. (2011a). Risk of brain tumours in relation to estimated RF dose from mobile phones: results from five Interphone countries. *Occup Environ Med 68*, 631–640.

Cardis, E., Deltour, I., Mann, S., Moissonnier, M., Taki, M., Varsier, N., et al. (2008). Distribution of RF energy emitted by mobile phones in anatomical structures of the brain. *Phys Med Biol 53*, 2771–2783.

Cardis, E., Richardson, L., Deltour, I., Armstrong, B., Feychting, M., Johansen, C., et al. (2007). The INTERPHONE study: design, epidemiological methods, and description of the study population. *Eur J Epidemiol 22*, 647–664.

Cardis, E., Varsier, N., Bowman, J.D., Deltour, I., Figuerola, J., Mann, S., et al. (2011b). Estimation of RF energy absorbed in the brain from mobile phones in the Interphone study. *Occup Environ Med 68*, 686–693.

Carlberg, M., Söderqvist, F., Hansson Mild, K. and Hardell, L. (2013). Meningioma patients diagnosed 2007–2009 and the association with use of mobile and cordless phones: a case–control study. *Environ Health 12*, 60.

CCARS. (2011). *Informe sobre radiofrecuencias y salud (2009–2010)*. Madrid, Spain: CCARS.

Chakrabarti, I., Cockburn, M., Cozen, W., Wang, Y.P. and Preston-Martin, S. (2005). A population-based description of glioblastoma multiforme in Los Angeles County, 1974–1999. *Cancer 104*, 2798–2806.

Frei, P., Poulsen, A.H., Johansen, C., Olsen, J.H., Steding-Jessen, M. and Schuz, J. (2011). Use of mobile phones and risk of brain tumours: update of Danish cohort study. *BMJ 343*, d6387–d6387.

Gordis, L. (2004). *Epidemiology*. Philadelphia, PA. Elsevier Saunders.

Hardell, L. (2013b). Pooled analysis of case-control studies on acoustic neuroma diagnosed 1997–2003 and 2007–2009 and use of mobile and cordless phones. *Int J Oncol 43*, 1036–1044.

Hardell, L., Carlberg, M. and Hansson Mild, K. (2011). Pooled analysis of case-control studies on malignant brain tumours and the use of mobile and cordless phones including living and deceased subjects. *Int J Oncol 38*, 1465–1474.

Hardell, L., Carlberg, M., Söderqvist, F. and Mild, K.H. (2013a). Case-control study of the association between malignant brain tumours diagnosed between 2007 and 2009 and mobile and cordless phone use. *Int J Oncol 43*, 1833–1845.

IARC (International Agency for Research on Cancer). (2013). *IARC Monographs on the Evaluation of Carcinogenic Risks to Humans, Vol. 102. Non-Ionizing Radiation, Part II: Radiofrequency Electromagnetic Fields [Includes Mobile Telephones]*. Lyon: IARC.

Inskip, P.D., Tarone, R.E., Hatch, E.E., Wilcosky, T.C., Shapiro, W.R., Selker, R.G., et al. (2001). Cellular-telephone use and brain tumors. *N Engl J Med 344*, 79–86.

INTERPHONE Study Group. (2010a). *Annex 2—Brain Tumour Risk in Relation to Mobile Telephone Use: Results of the INTERPHONE International Case-Control Study*. The Interphone Study Group.

INTERPHONE Study Group. (2010b). Brain tumour risk in relation to mobile telephone use: results of the INTERPHONE international case–control study. *Int J Epidemiol 39*, 675–694.

INTERPHONE Study Group. (2011). Acoustic neuroma risk in relation to mobile telephone use: results of the INTERPHONE international case-control study. *Cancer Epidemiol 35*, 453–464.

Larjavaara, S., Schüz, J., Swerdlow, A., Feychting, M., Johansen, C., Lagorio, S., et al. (2011). Location of gliomas in relation to mobile telephone use: a case-case and case-specular analysis. *Am J Epidemiol 174*, 2–11.

Muscat, J.E., Malkin, M.G., Thompson, S., Shore, R.E., Stellman, S.D., McRee, D., et al. (2000). Handheld cellular telephone use and risk of brain cancer. *JAMA 284*, 3001–3007.

Ohgaki, H. and Kleihues, P. (2005). Epidemiology and etiology of gliomas. *Acta Neuropathol (Berl.) 109*, 93–108.

Sato, Y., Akiba, S., Kubo, O. and Yamaguchi, N. (2011). A case-case study of mobile phone use and acoustic neuroma risk in Japan. *Bioelectromagnetics 32*, 85–93.

SCENIHR. (2009). *Health Effects of EMF*. SCENIHR.

Schmidt, L.S., Nielsen, H., Schmiedel, S. and Johansen, C. (2008). Social inequality and incidence of and survival from tumours of the central nervous system in a population-based study in Denmark, 1994–2003. *Eur J Cancer 44*, 2050–2057.

Schuz, J., Jacobsen, R., Olsen, J.H., Boice, J.D., McLaughlin, J.K. and Johansen, C. (2006). Cellular telephone use and cancer risk: update of a nationwide Danish cohort. *J Natl Cancer Inst 98*, 1707–1713.

Sienkiewicz, Z., Schüz, J., Poulsen, A.H. and Cardis, E. (2012). *EFHRAN Report on Risk Analysis of Human Exposure to Electromagnetic Fields (revised)*. Deliverable Report D2. European Health Risk Assessment Network on Electromagnetic Fields Exposure.

Swerdlow, A.J., Feychting, M., Green, A.C., Kheifets, L., Savitz, D.A. and International Commission for Non-Ionizing Radiation Protection Standing Committee on Epidemiology. (2011). Mobile phones, brain tumors, and the Interphone study: where are we now? *Environ Health Perspect 119*, 1534–1538.

Szklo, M. and Nieto, F.J. (2007). *Epidemiology: Beyond the Basics*. Bartlett Publishers: Sudbury, MA.

Vrijheid, M., Armstrong, B.K., Bédard, D., Brown, J., Deltour, I., Iavarone, I., et al. (2009a). Recall bias in the assessment of exposure to mobile phones. *J Exp Sci Environ Epidemiol 19*, 369–381.

Vrijheid, M., Cardis, E., Armstrong, B.K., Auvinen, A., Berg, G., Blaasaas, K.G., et al. (2006). Validation of short term recall of mobile phone use for the Interphone study. *Occup Environ Med 63*, 237–243.

Vrijheid, M., Richardson, L., Armstrong, B.K., Auvinen, A., Berg, G., Carroll, M., et al. (2009b). Quantifying the impact of selection bias caused by nonparticipation in a case-control study of mobile phone use. *Ann Epidemiol 19*, 33–41.

14

Brain Tumors and Mobile Phone Use: The Ecological Approach

Peter D. Inskip and Mark P. Little

CONTENTS

Introduction

Consideration of a possible causal relationship between mobile phone use and brain cancer risk leads naturally to the question, "Has the incidence of brain cancer increased since the introduction and explosive growth in use of mobile phones?" There would, of course, be a lag of some extent, albeit unknown, depending on the time course of the pathogenesis of the cancer, and its detection, diagnosis, and documentation in a population-based cancer registry. Nonetheless, the increase in use of mobile phones has been so dramatic, and the prevalence of use in the general population is now so high, that it is implausible that any substantial increase in brain cancer risk would not eventually be reflected in population incidence rates.

Ecological studies, in which the unit of observation is a group of people rather than individual people, are typically considered to be inferior to analytic epidemiological studies, such as cohort or case–control studies, for evaluating etiological hypotheses (Rothman 1986; Ahlbom and Feychting 2011). Measures of exposure and disease at the population level are averages that would tend to dampen an association, if one exists. Use of surrogate measures or indicators of exposure also would lessen the ability to detect an association. The unavailability of data for individuals compromises the ability to control for potential confounding factors and opens the door for biased and invalid inferences. For these reasons, ecological studies are considered to be weak and more prone to error compared with analytic studies.

In the matter of mobile phones and brain cancer, however, population-level data have an important role to play in the overall interpretation. This is particularly true in light of methodological limitations of case–control studies on this subject that have been published to date (see Chapters 2 and 13). In earlier chapters of this volume, these limitations, including possible differential participation by level of mobile phone use (see Chapter 5) and questionable accuracies of self-reported exposure histories (see Chapters 4 and 8), are discussed in detail. To this day, considerable diferences of opinion exist regarding interpretation of results from the largest and most expensive case–control study to date, namely, the INTERPHONE study (Swerdlow et al. 2011).

The relevance of time trends in brain cancer incidence to assessment of possible risks due to mobile phone use is enhanced by several factors. The prevalence of use of mobile phones in developed countries increased from nearly zero to something approaching 100% over a period of approximately two decades. Given the magnitude and abruptness of this increase, any appreciable effect on the risk of brain cancer should be reflected in incidence rates, unless the induction period is very long. The availability of high-quality cancer registries dating back to before mobile phones were introduced provides the opportunity to address this question at relatively low cost. As of the early 1990s, when use of mobile phones started to grow rapidly, brain cancer incidence rates in the United States and Nordic countries had been relatively stable (Deltour et al. 2009; Inskip et al. 2010). This stable baseline facilitates detection of a possible increase related to mobile phone use, because it suggests that secular trends in the prevalence of other etiological factors were not concurrently underway. Patterns of change in incidence rates over time—or the lack thereof—set boundaries on the plausible magnitude of effect due to mobile phone use (Swerdlow et al. 2011).

Issues regarding time trends in incidence of benign intracranial tumors of the nervous system (meningioma and acoustic neuroma) are somewhat different from those for glioma, the most common form of brain cancer; for this reason, malignant and benign tumors are considered separately. Much less information is available concerning benign tumors.

Malignant Brain Tumors

There have been a series of papers from northern European countries and the United States concerning time trends in brain cancer incidence and mortality before and after the introduction of mobile phones. The earlier papers compared brain cancer incidence or mortality with some indicator of prevalence of use of mobile phones, such as yearly number of subscribers, in an informal way (Röösli et al. 2007; Deltour et al. 2009; Inskip et al. 2010; Ahlbom and Feychting 2011; de Vocht et al. 2011a). Two more recent papers addressed the relationship using a more formal, analytical approach (Deltour et al. 2012; Little et al. 2012).

Analysis of time trends in brain tumor mortality in Switzerland from 1969 to 2002 showed generally increasing mortality rates from 1969 to 1986 and stable rates from 1987 to 2002 (Röösli et al. 2007) (Table 14.1). Persons >74 years old and, to a lesser extent, aged 60–74 years, were the exception in showing rising mortality rates even after 1986. These are not the ages corresponding to most frequent use of mobile phones. For males aged 45–59 years, a group with higher frequency mobile phone use, the annual change in the brain tumor mortality rate was −0.3% (95% confidence intervals [CI], −1.7 to 1.1). The authors calculated predicted increases in brain tumor mortality on the assumption of a causal association with mobile phone use and with varying hypothetical relative risks and time lags (induction times). A causal relation with very short induction time (1 year) was not consistent with the observed data; however, mobile phone use in Switzerland did not increase markedly until 1996, and population surveillance through 2002 may have been too short of a time to detect an effect with a more plausible induction time.

The incidence of glioma, the most common form of brain cancer, among adult men aged 20–79 years in Denmark, Finland, Norway, and Sweden increased by 0.5%/year (0.2–0.8)

TABLE 14.1

Annual Percentage Change (APC) in Brain Tumor Incidence and Mortality from Studies Conducted in Europe and the United States

Location	Reference	Period	Outcome	Rate Measure	Age Range (years)	Males		Females	
						APC	95% CI	APC	95% CI
Switzerland	Röösli et al. (2007)	1987–2002	Brain cancer	Mortality	0–14	0.7	−3.9, 5.3	−0.7	−6.3, 4.9
					15–29	−2.2	−6.2, 1.8	−0.4	−5.7, 5.0
					30–44	−0.3	−2.4, 1.9	0.8	−1.9, 3.6
					45–59	−0.3	−1.7, 1.1	−0.4	−2.2, 1.3
					60–74	1.2	0.0, 2.4	0.4	−0.9, 1.7
					75+	1.9	0.1, 3.7	3.6	1.9, 5.3
Denmark, Finland, Norway, Sweden	Deltour et al. (2012)	1979–2008	Glioma	Incidence	20–79	0.4	0.1, 0.6	0.3	0.1, 0.5
		1987–2008	Glioma	Incidence	20–39	−0.7	−1.4, 0.1	NA	NA
United States	Inskip et al. (2010)[a]	1992–2006	Brain cancer	Incidence	< 20	−0.1	−1.7, 1.5	−0.8	−2.4, 0.9
					20–29	−0.1	−1.7, 1.4	4.3	1.9, 6.7
					30–39	−1.0	−2.5, 0.5	−0.0	−1.7, 1.6
					40–49	−0.8	−2.0, 0.5	0.3	−1.2, 1.8
					50–64	−0.4	−1.1, 0.4	−0.7	−1.5, 0.1
					65+	−0.3	−1.2, 0.5	0.2	−0.8, 1.1
United States	Kohler et al. (2011)[b]	1987–2007	Glioma[c]	Incidence	All ages#	−0.7	NA	−0.3	NA
Denmark, Finland, Norway, Sweden	Deltour et al. (2009)	1974–2003	Meningioma	Incidence	20–79	0.8	0.4, 1.3	2.9	2.2, 3.7
Denmark	Larjavaara et al. (2011)	1987–2007	Acoustic neuroma	Incidence	All ages	5.3	2.7, 7.9	4.5	2.2, 7.0
Finland	Larjavaara et al. (2011)	1987–2007	Acoustic neuroma	Incidence	All ages	−0.16	−2.5, 2.3	−0.70	−2.8, 1.4
Norway	Larjavaara et al. (2011)	1987–2007	Acoustic neuroma	Incidence	All ages	5.5	2.8, 8.3	4.7	2.1, 7.5
Sweden	Larjavaara et al. (2011)	1987–2007	Acoustic neuroma	Incidence	All ages	1.5	−0.56, 3.6	0.44	−1.7, 2.7

Note: Some values have been rounded from the original published values. APC, annual percentage change in rate; NA, not available.

[a] Based on whites only.

[b] Based on all races.

[c] Glioma and other neuroepithelial cancers.

from 1974 to 2003; however, the increase occurred in the 1970s and 1980s, well before mobile phones were introduced and in wide use (Deltour et al. 2009). Mobile phones were not in widespread use in these countries until the early 1990s. Among young men aged 20–39 years, the incidence increased 3.8%/year from 1974 through mid-year 1987 (2.6–4.9) and decreased 1.1%/year thereafter (–1.9 to –0.3). The reason(s) for the increase in the early years are unknown but may be related to improvements in diagnosis. Clearly, because of the period when this took place, mobile phones were not a factor. Among women aged 20–79 years, there was no significant trend in incidence of glioma over the time period from 1974 to 2003. An updated analysis extended the time trend analysis for these Nordic countries through 2008 (Deltour et al. 2012). Between 1979 and 2008, the annual percentage of change in incidence rates was 0.4% (0.1–0.6) among men and 0.3% (0.1–0.5) among women aged 20–79 years. For both sexes, the incidence of glioma increased slightly among those aged 60–79 years. Among men aged 20–39 years, the incidence rate decreased 0.7%/year (–1.4 to 0.1) from 1987 to 2008. In an analysis focusing on children (aged 5–19 years) from Nordic countries, Aydin et al. (2012) demonstrated relatively stable incidence rates for childhood brain tumors from 1990 to 2009, with no evidence of a secular trend.

The number of mobile phone subscribers in the United States increased from 5.3 million in 1990 to 279 million in 2008 (CTIA 2009). Trends in the incidence of primary brain cancer among whites were evaluated separately for the periods 1977–1991 and 1992–2006 based on data from the original nine Surveillance, Epidemiology and End Results (SEER) cancer registries (Inskip et al. 2010). The earlier time interval precedes the era of widespread mobile phone use but includes years when computerized tomography (CT) and magnetic resonance imaging (MRI) machines were introduced, and increasing in use in the medical care system. Among persons younger than 30 years and those 65 years or older, highly significant increasing trends in brain cancer incidence were seen between 1977 and 1991 (Figure 14.1). With the exception of the 20- to 29-year-old age group, incidence trends from 1992 to 2006 were downward or flat. Among women aged 20–29 years, there was a statistically significant increasing trend between 1992 and 2006; no such trend was seen for males. The increasing trend in 20- to 29-year-old women was due to rising incidence of frontal lobe tumors. No increases were apparent for temporal or parietal lobe cancers, or cancers of the cerebellum, that involve parts of the brain that would typically be more highly exposed to radiofrequency (RF) radiation from mobile phones (Cardis et al. 2008). Incidence rates for frontal lobe cancers also rose among 20- to 29-year-old males, but the increase began earlier than among females, and before mobile phone use was highly prevalent. Among 20- to 29-year-old women, no trend was apparent for temporal or parietal lobe tumors, nor for tumors of poorly specified location. The trend for glioblastoma, the most common type of brain cancer in adults, was similar to that for all types of brain cancer combined. In summary, the dramatic increase in prevalence of use of mobile phones in the United States was not followed by a detectable increase in brain cancer incidence based on follow-up through 2006.

Little et al. (2012) extended the surveillance based on SEER from 1992 to 2008, in this case using 12 SEER cancer registries rather than the original 9 registries. The standardized incidence rate for glioma among non-Hispanic white people was approximately constant from 1992 to 2008 (annual percentage change, –0.02%/year [–0.28 to 0.25]). Positive trends were seen for temporal lobe cancers (0.73%/year [0.23–1.23]) and brain cancers of other specified location (0.79%/year [0.40–1.19]), whereas a significant decreasing trend was seen for brain cancers of poorly specified location (–2.35%/year [–2.81 to 1.89]). Neither temporal lobe cancers nor cancers of other specified sites showed an acceleration of the incidence trend after 1996 compared with 1992–1996. The observed trends might reflect, at least in part, improved specification of tumor location over time.

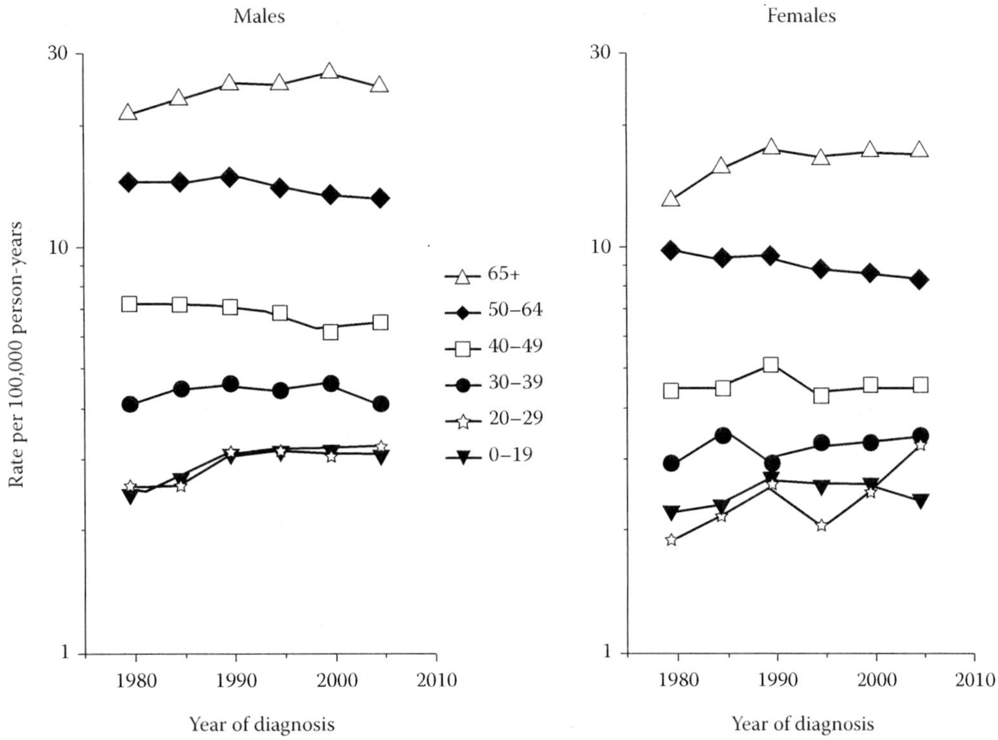

FIGURE 14.1
Brain cancer incidence trends among whites by age in the United States, 1977–1981 to 2002–2006, based on SEER 9 cancer registry data. (From Inskip P. D. et al., *Neuro Oncol* 12, 1147–51, 2010.)

Mobile phone ownership in England increased from 17% of the population in 1996–1997 to 65% in 2001–2002 (Shephard 2007); yet, an analysis of brain cancer incidence rates in England showed no significant time trends between 1998 and 2007 for either sex or any age group (de Vocht et al. 2011a). Analysis by anatomical subsite indicated an increase in temporal lobe tumors in men and women and an increase in frontal lobe tumors in males. The finding for temporal lobe tumors is of interest insofar as part of this lobe would be exposed to RF radiation from a mobile phone held next to the ear (Cardis et al. 2008). However, further analysis showed that an increase in the incidence of temporal lobe tumors preceded widespread use of mobile phones, and a subsequent increase in prevalence of use of mobile phones was not accompanied by increasing incidence of temporal lobe cancers (de Vocht et al. 2011b). Furthermore, the incidence of cancer of the cerebellum, another part of the brain with potential for higher exposure to RF radiation from mobile phones, decreased from 1998 to 2007 (de Vocht et al. 2011a). The frequency of cancers of overlapping lesions of the brain decreased during the same time period, again raising the possibility that the apparent increase in incidence of temporal lobe cancers was an artifact of more precise specification of tumor location.

Hand-held mobile phones entered the marketplace in Sweden in 1987 and quickly came into wide use; by 2002, 87% of adults were mobile phone users (Ahlbom and Feychting 2011). This dramatic increase in use was not accompanied by an increase in the incidence of glioma that remained approximately constant from 1970 to 2009 (Ahlbom and Feychting 2011). This study should have been able to detect a moderate increase in risk associated with mobile phone use with a latency of 10 years. That it did not provides a degree of reassurance.

More detailed analyses are required to assess the consistency between observed brain cancer incidence trends and published results from case–control studies in a formal way. Using data on the prevalence of mobile phone use among controls from Nordic countries in the INTERPHONE case–control study, Deltour et al. (2012) compared observed incidence rates with expected rates assuming a range of relative risks (RRs) and latency (induction) periods. The investigators conducted a simulation study in which they generated multiple hypothetical data sets under the assumption that there is a risk related to mobile phone use. Simulations were based on different combinations of RRs and lag periods. Then, pretending that the true value of the RR was unknown, they used the proportion of 10,000 simulated data sets with statistically significant changes in incidence (or, equivalently, number of cases, as the population size was fixed) as estimates of the probability of detecting different RRs. Selected results are shown in Figure 14.2 and are based on data from Table 2 in their paper. The authors conducted the analyses separately for all users and for heavy users, defined as having >1640 lifetime hours of use. Results shown are based on males aged 40–59 years and using incidence rates for Nordic countries between 1979 and 2008. Observed incidence rates for the 40- to 59-year-old age group were stable over

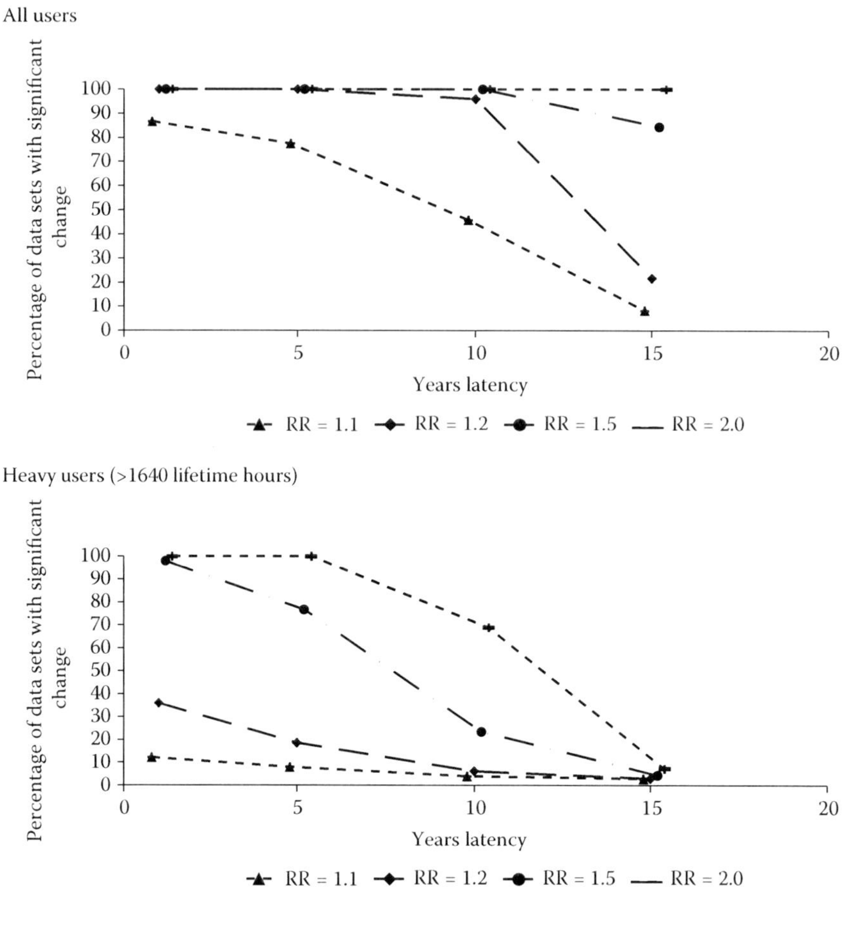

FIGURE 14.2
Proportion of 10,000 simulations showing statistically significant increases or decreases in the incidence rate of glioma among men aged 40–59 years in Denmark, Finland, Norway, and Sweden under various scenarios of RR and latency. (Based on data from Deltour I. et al., *Epidemiology* 23, 301–7, 2012.)

this period. The shorter the induction (lag) period and the greater the RR, the greater the probability of detecting an effect of mobile phone use. Based on analyses that consider all users to be exposed, and for induction periods <10 years, true RRs as low as 1.2 would be expected to have resulted in detectable increases in population incidence rates. If the RR is ≥1.5, an increase in the incidence rate should have been detected even if the induction period is 15 years. If one assumes that an effect is confined to heavy users, defined as >1640 hr of lifetime use, and that the true RR is 2.0, the effect would be expected to be detectable in population incidence rates if the induction period is as long as 10 years.

These hypothetical RRs and induction periods can be compared with actual values reported for case–control studies in the literature. In a study conducted in Sweden, Hardell et al. (2011a) reported RRs for glioma of 1.3 for >5–10-year latency and 2.5 for latency >10 years. These values are not compatible with the stable glioma incidence rates seen among young and middle-aged persons in Nordic countries through 2008 (Deltour et al. 2012).

In a similar analysis based on incidence rates for glioma in the United States for 1992–2008, Little et al. (2012) compared the observed incidence rates with the expected rates, assuming different RRs associated with mobile phone use and varying latency periods. Temporal patterns of prevalence of per capita mobile phone use in the U.S. general population were estimated based on subscriber data (CTIA 2011). This analysis focused on estimation of incidence rates in the different scenarios rather than on the ability to detect a change in incidence, that is, the approach was based on estimation rather than statistical testing. Selected results are shown in Figure 14.3. The observed age-standardized incidence rate of glioma based on data from SEER cancer registries was approximately stable over this time period (annual percentage change, −0.02%/year [−0.28–0.25]). Even with a latency period of 10 years, a true RR of 1.5 should be associated with a nearly 11% increase in the incidence rate. With shorter latency periods or larger RRs, the expected increase in incidence was greater. In general, the incidence of glioma was predicted to increase by at least 20% with a short latency period (≤5 years) or large RRs (≥2.0).

Little et al. (2012) also assessed the compatibility of the observed incidence rates from SEER with what would be expected if RRs and observed latency distributions among controls from two case–control studies (The INTERPHONE Study Group 2010; Hardell et al. 2011a) pertained to the U.S. population. Results from the Swedish study (Hardell et al. 2011a) would have predicted a 44.5% higher incidence rate for the U.S. population than was actually observed (25.5/100,000 people per year vs. 17.7/100,000 per year), a difference that was highly significant. The observed incidence rate was more consistent with resuts from the INTERPHONE study that showed an significantly elevated RR (1.40 [1.03–1.89]) only for the highest decile of cumulative mobile phone use (≥1640 hr) (The INTERPHONE Study Group 2010). This indicates that if a causally relevant exposure occurs only in a small proportion of the population, the effect might not be detectable in population-level data.

Benign Brain Tumors

The incidence of acoustic neuroma (vestibular schwannoma) and meningioma also have been evaluated with respect to mobile phone use in case–control and cohort studies. An International Agency for Research on Cancer (IARC) review panel judged that the evidence of tumorigenicity from these studies was insufficient for meningioma but somewhat more suggestive for acoustic neuroma (Baan et al. 2011).

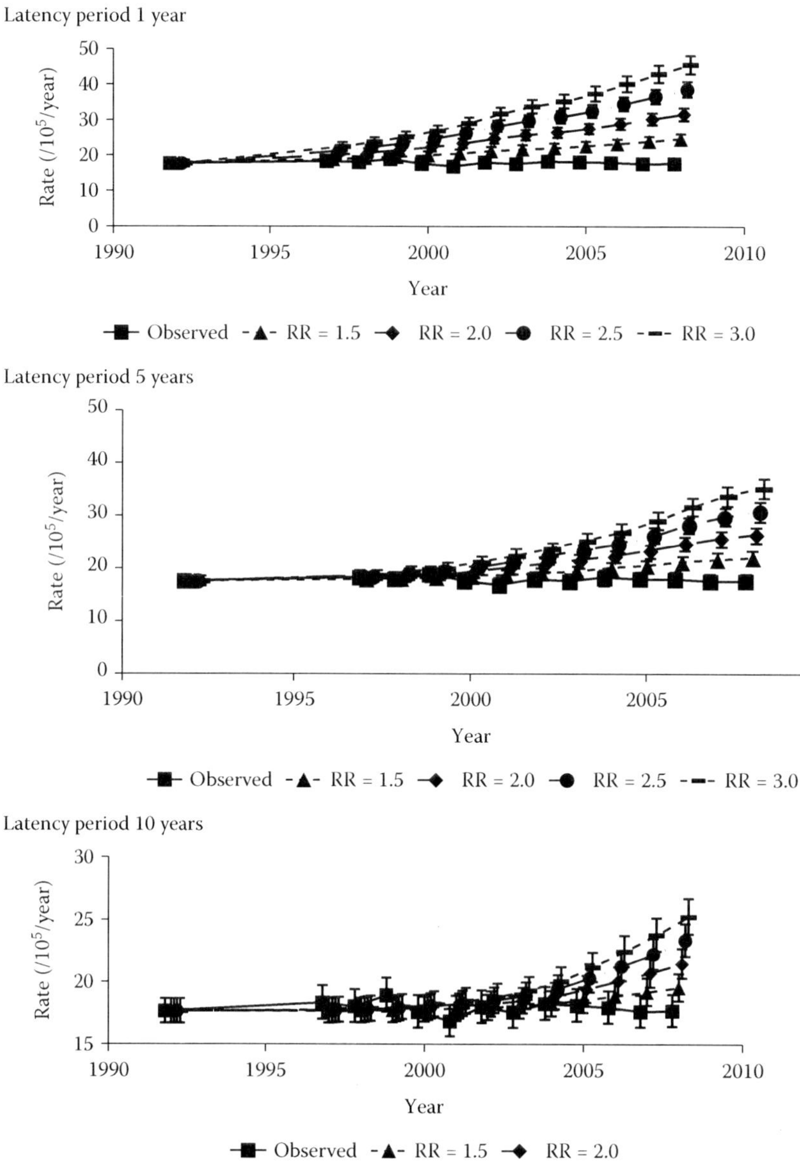

FIGURE 14.3
Observed and projected incidence rates (with 95% CIs) of malignant glioma in non-Hispanic white people, by latency period and various assumed levels of RR associated with ever using a phone. (Modified after Little M. P. et al., *BMJ* 344, e1147, 2012.) Note differences in scale for y-axis.

Tumor registry data may be less well suited to analyses of time trends for benign intracranial tumors of the nervous system, and the data about time trends is less extensive than for glioma. Acoustic neuromas and most meningiomas are histogically benign and often are slow growing (Kleihues and Cavenee 1997). They may be present for years before coming to diagnosis, so the etiologically relevant time period for exposure is unclear. In the United States, SEER did not start systematically collecting data about benign brain tumors until 2004 (Horner et al. 2009), and the data are too limited to assess time trends in incidence.

The Central Brain Tumor Registry of the United States (CBTRUS 2010) has collected incidence data for benign tumors for a longer time, but there is some question about whether the completeness of diagnosis and reporting has been stable over time; specifically, completeness of reporting might have increased over time.

Tumor registries in Nordic countries have collected incidence data for benign brain tumors for many years. Deltour et al. (2009) examined the time trend in incidence of meningioma among persons aged 20–79 years between 1974 and 2003 in Denmark, Finland, Norway, and Sweden. Over this entire interval, the incidence increased by 0.8%/year among men (0.4–1.3). Between 1990 and 2003, the incidence increased by 3.8%/year among women (3.2–4.4). This increase was largely due to an increase among women in the 60- to 79-year-old age group. The authors cautioned that interpretation of the findings is complicated by variation in the completeness of diagnosis and registration (Larjavaara et al. 2008; Deltour et al. 2009). Larjavaara et al. (2011) evaluated the incidence of acoustic neuroma in these same countries between 1987 and 2007. For all countries and both sexes combined, the incidence rate increased by 3.0%/year (2.1–3.9), with considerable heterogeneity among the four countries (Table 14.1). For both sexes, the increase was greatest among persons ≥65 years old (4.0%/year among men and 3.9%/year among women). Most of the increase occurred before the late 1990s, after which time the trend was flat or declining. The increasing trend could be related to improvements in diagnosis or registration, a true increase in risk, or some combination of factors. The pattern of changes in incidence over time does not suggest that mobile phone use was responsible.

Methodological Issues

Cancer incidence and mortality data in the countries where the studies referenced above were conducted have been collected systematically for many years, beginning well before the advent of use of mobile phones. Completeness of registration is unlikely to be related to mobile phone use. Selection bias is unlikely and recall bias is a nonissue. Nonetheless, many methodological limitations and issues should be kept in mind when interpreting time trend data.

Confounding is at least a theroretical possibility. During the period of increasing mobile phone use, some other unknown exposure may have been changing in prevalence with the effect of exerting a downward influence on brain cancer incidence rates and obscuring a real increase due to mobile phone use. Other than exposure to moderate-to-high doses of ionizing radiation, which is uncommon in the general population, the only specific nongenetic risk factor linked to brain cancer with any consistency is history of allergies. A history of allergies, most notably asthma, is associated with an approximately 40% reduced risk of glioma (Chen et al. 2011). Asthma increased in prevalence worldwide over the past 50 years, but prevalence in western countries appears to have leveled off or declined in recent decades (Ekerljung et al. 2008; Douwes et al. 2011). In Sweden, prevalence of asthma in young adults was stable from 1990 to 2008 (Bjerg et al. 2011); thus, the temporality of increases in asthma do not appear to have coincided with increases in mobile phone use. Nonetheless, this illustrates the limitation of ecological studies in not having information on exposure, outcome, and potential confounders for individuals.

Population demographics could change over time in ways related to brain cancer risk. For example, in the United States, the proportion of Hispanics among whites increased considerably during the 1990s and 2000s, and white Hispanic people have a lower incidence of brain cancer than white non-Hispanic people (Horner et al. 2009; Inskip et al. 2010).

However, SEER has collected data on Hispanic ethnicity since 1992, so an effect of population admixture can be separated out. No trend in brain cancer incidence was seen for any racial or ethnic group from 1992 to 2006 (Inskip et al. 2010; Little et al. 2012).

It would be surprising if a downward effect on brain cancer incidence due to some unknown confounding factor were to exactly cancel out an upward effect due to mobile phone use, and that such an effect would be the same in all age and sex subgroups. In the United States, after upward trends in incidence in the young and elderly during the 1970s and 1980s, most likely related to improvements in diagnosis (Smith et al. 1998; Legler et al. 1999; Inskip et al. 2010), incidence rates appeared to have stabilized in the years just before the advent of mobile phones. A stable baseline facilitates evaluation of incidence patterns in years subsequent to the introduction of mobile phones and argues against the notion of trend in some other factor obscuring an effect of mobile phones.

If risk of brain cancer is only increased among long-term users and/or after a long induction (latency) period, it may be too soon for an effect to be apparent in general population incidence rates. However, even for a long mean induction time, one would expect a distribution around this mean, and sufficient time has elapsed since use of mobile telephones began that one would expect to see cases with shorter than average induction periods (Rothman 2009; Ahlbom and Feychting 2011). Regardless, a very long induction period would also mean that no associations should have been observed in the case–control studies to date (The INTERPHONE Study Group 2010; Hardell et al. 2011a, 2011b). It is logically inconsistent to interpret positive findings from selected published epidemiological studies of mobile telephones, including of relatively short-term users, as being indicative of causality while simultaneously asserting that that it is still too soon to see an increase at the population level. If continued monitoring of incidence rates over the next few years does not produce any evidence of an increase, it will become increasingly difficult to attribute the negative results to inadequate duration of follow-up of long-term heavy users of mobile phones. The assertion that brain tumors characteristically have long latencies, averaging between 20 and 40 years after exposure to a carcinogen (Kundi 2011) is not supported by the literature. For ionizing radiation, the one well-established environmental neurocarcinogen, radiation-related gliomas are apparent within 5–9 years after therapeutic irradiation during childhood (Neglia et al. 2006). In contrast to ionizing radiation, RF radiation is nonionizing and therefore unlikely to induce DNA lesions; if anything, it is more likely to be a promoting agent than an initiating agent, which implies that RF radiation-associated latency may be even shorter than 5–9 years (Little et al. 2012).

Improvements over time in precision of specification of tumor site and histology, or changes in tumor classification over time, can yield artifactual increases or decreases in incidence. If an increase is seen in one subcategory (e.g., lobe of brain or histological subtype of glioma), it is important to determine whether there was an offsetting decrease in the incidence of tumors of poorly specified location or histology. Analyses from the United States and England described above illustrate how apparent increases in incidence of brain cancers of specified location sometimes are offset by decreases in incidence of cancers of poorly specified location. With respect to brain tumor classification systems, there have been changes over time in how cancers are classified within the broad category "glioma" but not in how tumors are classified between the categories glioma, meningioma, and acoustic neuroma (Louis et al. 2007; Davis et al. 2008).

Incidence data from cancer registries for the most recent years can be biased downward due to delays in reporting and registration of cancers. Incidence rates for these years often are revised upwards later as additional cancers are identified and registered; however, this effect is relatively small, particularly after 1–2 years (Horner et al. 2009).

Indicators of changes in mobile phone use over time typically are based on surrogate measures such as number of subscribers. This is not synonomous with number of users and does not indicate level or duration of use, even at the population level. If individuals used mobile phones less in earlier years (in terms of minutes or hours per day), it is conceivable that observations for longer latency intervals are characterized by lighter levels of use in the earlier years. However, changes over time in mobile telephone technology also should be considered (see Chapter 19). Early mobile phones were analog, whereas more recent phones are digital. Other things being equal, a given duration of use for an analog phone would involve higher exposure of the head to RF radiation. Use of hands-free devices was less in the earlier years, as was the density of cellular transmission towers; again, both factors would, *ceteris paribus*, imply greater exposure to the head in earlier years. Earlier mobile phones operated at a lower frequency (800–900 MHz) than later phones (>1800 MHz). Depth of tissue penetration of RF radiation in tissue is inversely related to frequency(National Council on Radiation Protection and Measurements 1993; World Health Organization 1993). The net effect of all of these multiple factors on exposure of the brain to RF radiation in earlier versus later years is unclear.

Not considered here is another type of ecological study, namely, international comparisons of brain cancer incidence in relation to frequency of mobile phone subscriptions (de Vocht et al. 2013). These raise a different and more complex array of methodological issues compared with temporal comparisons, particularly relating to confounding and variation in completeness of brain cancer diagnosis and registration.

Summary and Comments

Ecological studies are not well suited to detecting small risks, risks associated with uncommon exposures or uncommon subtypes of brain cancer, risks confined to small demographic subgroups, risks with very long induction periods (at least not at present), or cancer outcomes with well-established, highly prevalent known causes; however, some of these limitations pertain equally to analytic, nonexperimental epidemiological studies. For example, detection of increases in risk of the order of 20% is challenging, whatever the study design.

In the matter of mobile phone use and brain cancer risk, certain weaknesses of the ecological approach are minimized and strengths are maximized. Mobile phone use certainly does not qualify as a rare exposure, even with allowance for a reasonably long induction or latency period, and the increase in use was abrupt and dramatic rather than gradual. If mobile phone use causes brain cancer, but only after a very long latency period, the existence of ongoing, high-quality, population-based cancer registries in multiple countries provides an efficient infrastructure for ongoing surveillance. The costs of conducting such surveillance, given this infrastructure, are trivial in comparison with the costs of undertaking a new case–control or cohort study aimed solely at this question, and there is no guarantee that a large analytic study would provide more informative or less ambiguous results. (This efficency argument is diminished somewhat if a mobile phone component can be incorporated into ongoing large, multipurpose cohort studies.) There are no known environmental or behavioral causes of brain cancer that are widely distributed in the population, and cancer registries can adequately account for demographic risk indicators. This lessens concerns about possible confounding. Stable brain cancer incidence rates in Western countries in the early 1990s, after increases in the 1970s and 1980s, provided a baseline from which to assess possible subsequent increases associated with mobile phone use. Due to their frequently large population coverage, population-based cancer registries provide the opportunity to conduct studies of uncommon cancers, although temporal patterns become increasingly sensitive to changes in tumor classification systems the finer the subdivision of glioma subtypes.

The value of ecological studies may be understood not so much as an alternative to analytic epidemiological studies but as a complementary source of information. They can serve as a "reality check" in terms evaluating the consistency of findings from analytic studies with population-level incidence trends. Two of the studies described above, one from northern Europe (Deltour et al. 2012) and a second from the United States (Little et al. 2012), raised questions about the consistency of incidence trends with results from one of the two case–control studies used by IARC to evaluate brain tumor risks with respect to mobile phone use (Baan et al. 2011). We suggest that results from ongoing monitoring of brain tumor incidence rates be a central component of future health risk evaluations.

Acknowledgments

This work was supported by the Intramural Research Program of the National Cancer Institute, National Institutes of Health, and U.S. Department of Health and Human Services.

References

Ahlbom, A. and Feychting, M. 2011. Mobile telephones and brain tumours. *BMJ* 343: d6605.

Aydin, D., Feychting, M., Schüz, J., Röösli, M. and the CEFALO study team. 2012. Childhood brain tumours and use of mobile phones: comparison of a case-control study with incidence data. *Environ Health* 11: 35.

Baan, R., Grosse, Y., Lauby-Secretan, B., El Ghissassi, F., Bouvard, V., Benbrahim-Tallaa, L., et al. 2011. Carcinogenicity of radiofrequency electromagnetic fields. *Lancet Oncol* 12(7): 624–6.

Bjerg, A., Ekerljung, L., Middelveld, R., Dahlen, S. E., Forsberg, B., Franklin, K., et al. 2011. Increased prevalence of symptoms of rhinitis but not of asthma between 1990 and 2008 in Swedish adults: comparisons of the ECRHS and GA(2)LEN surveys. *PLoS One* 6(2): e16082.

Cardis, E., Deltour, I., Mann, S., Moissonnier, M., Taki, M., Varsier, N., et al. 2008. Distribution of RF energy emitted by mobile phones in anatomical structures of the brain. *Phys Med Biol* 53(11): 2771–83.

CBTRUS (Central Brain Tumor Registry of the United States). 2010. *CBTRUS statistical report: primary brain and central nervous system tumors diagnosed in the United States in 2004–2006*. Hinsdale, IL: Central Brain Tumor Registry of the United States.

Chen, C., Xu, T., Chen, J., Zhou, J., Yan, Y., Lu, Y., et al. 2011. Allergy and risk of glioma: a meta-analysis. *Eur J Neurol* 18(3): 387–95.

CTIA (Cellular Telephone Industry Association). 2009. [cited September 2009]. Available from: http://www.ctia.org

CTIA (Cellular Telephone Industry Association). 2011. *CTIA semi-annual wireless industry survey 2011*. [cited April 12, 2011]. Available from: http://files.ctia.org/pdf/CTIA_Survey_MY_2011_Graphics.pdf

Davis, F. G., Malmer, B. S., Aldape, K., Barnholtz-Sloan, J. S., Bondy, M. L., Brannstrom, T., et al. 2008. Issues of diagnostic review in brain tumor studies: from the Brain Tumor Epidemiology Consortium. *Cancer Epidemiol Biomarkers Prev* 17(3): 484–9.

Deltour, I., Auvinen, A., Feychting, M., Johansen, C., Klaeboe, L., Sankila, R., et al. 2012. Mobile phone use and incidence of glioma in the Nordic countries 1979–2008: consistency check. *Epidemiology* 23(2): 301–7.

Deltour, I., Johansen, C., Auvinen, A., Feychting, M., Klaeboe, L. and Schüz, J. 2009. Time trends in brain tumor incidence rates in Denmark, Finland, Norway, and Sweden, 1974–2003. *J Natl Cancer Inst* 101(24): 1721–4.

de Vocht, F., Burstyn, I. and Cherrie, J. W. 2011a. Time trends (1998–2007) in brain cancer incidence rates in relation to mobile phone use in England. *Bioelectromagnetics* 32(5): 334–9.

de Vocht, F., Burstyn, I. and Cherrie, J. W. 2011b. Authors' reply to Kundi's comments on de Vocht et al. 2011b. "Time trends (1998–2007) in brain cancer incidence rates in relation to mobile phone use in England." *Bioelectromagnetics* 32: 675–6.

de Vocht, F., Hannam, K. and Buchan, I. 2013. Environmental risk factors for cancers of the brain and nervous system: the use of ecological data to generate hypotheses. *Occup Environ Med* 70(5): 349–56.

Douwes, J., Brooks, C., van Dalen, C. and Pearce, N. 2011. Importance of allergy in asthma: an epidemiologic perspective. *Curr Allergy Asthma Rep* 11(5): 434–44.

Ekerljung, L., Ronmark, E., Larsson, K., Sundblad, B. M., Bjerg, A., Ahlstedt, S., et al. 2008. No further increase of incidence of asthma: incidence, remission and relapse of adult asthma in Sweden. *Respir Med* 102(12): 1730–6.

Hardell, L., Carlberg, M. and Hansson Mild, K. 2011a. Pooled analysis of case-control studies on malignant brain tumours and the use of mobile and cordless phones including living and deceased subjects. *Int J Oncol* 38(5): 1465–74.

Hardell, L., Carlberg, M. and Hansson Mild, K. 2011b. Re-analysis of risk for glioma in relation to mobile telephone use: comparison with the results of the Interphone international case-control study. *Int J Epidemiol* 40(4): 1126–8.

Horner, M. J., Ries, L. A. G., Krapcho, M., Neyman, N., Aminou, R., Howlader, N., et al. 2009. *SEER cancer statistics review, 1975–2006*. National Cancer Institute. [cited September 26, 2012]. Available from: http://seer.cancer.gov/csr/1975_2006/ (accessed on January 21, 2010).

Inskip, P. D., Hoover, R. N. and Devesa, S. S. 2010. Brain cancer incidence trends in relation to cellular telephone use in the United States. *Neuro Oncol* 12(11): 1147–51.

INTERPHONE Study Group. 2010. Brain tumour risk in relation to mobile telephone use: results of the INTERPHONE international case-control study. *Int J Epidemiol* 39(3): 675–94.

Kleihues, P. and Cavenee, W. K. 1997. *Pathology and genetics of tumours of the nervous system*. Lyon, France: International Agency for Research on Cancer.

Kohler, B.A., Ward, E., McCarthy, B.J., Schymura, M.J., Ries, L.A., Eheman, C., Jemal, A., Anderson, R.N., Ajani, U.A., Edwards, B.K. 2001. Annual report to the nation on the status of cancer, 1975-2007, featuring tumors of the brain and other nervous system. *J Natl Cancer Inst*. 103(9): 714–36.

Kundi, M. 2011. Comments on de Vocht et al. "Time trends (1998–2007) in brain cancer incidence rates in relation to mobile phone use in England." *Bioelectromagnetics* 32(8): 673–4; author reply 675–6.

Larjavaara, S., Haapasalo, H., Sankila, R., Helén, P., Auvinen, A. 2008. Is the incidence of meningiomas underestimated? A regional survey. *Br J Cancer*. 99(1): 182–4.

Larjavaara, S., Feychting, M., Sankila, R., Johansen, C., Klaeboe, L., Schüz, J., Auvinen, A. 2011. Incidence trends of vestibular schwannomas in Denmark, Finland, Norway and Sweden in 1987–2007. *Br J Cancer*. 105(7): 1069–75.

Legler, J. M., Ries, L. A., Smith, M. A., Warren, J. L., Heineman, E. F., Kaplan, R. S., et al. 1999. Cancer surveillance series [corrected]: brain and other central nervous system cancers: recent trends in incidence and mortality. *J Natl Cancer Inst* 91(16): 1382–90.

Little, M. P., Rajaraman, P., Curtis, R. E., Devesa, S. S., Inskip, P. D., Check, D. P., et al. 2012. Mobile phone use and glioma risk: comparison of epidemiological study results with incidence trends in the United States. *BMJ* 344: e1147.

Louis, D. N., Ohgaki, H., Wiestler, O. D. and Cavanee, W. K. 2007. *WHO classification of tumours of the central nervous system*. Lyon, France: International Agency for Research on Cancer, p. 8.

National Council on Radiation Protection and Measurements. 1993. *A practical guide to the determination of human exposure to radiofrequency fields*. NCRP report No. 119. Bethesda, MD: National Council on Radiation Protection and Measurements.

Neglia, J. P., Robison, L. L., Stovall, M., Liu, Y., Packer, R. J., Hammond, S., et al. 2006. New primary neoplasms of the central nervous system in survivors of childhood cancer: a report from the Childhood Cancer Survivor Study. *J Natl Cancer Inst* 98(21): 1528–37.

Röösli, M., Michel, G., Kuehni, C. E. and Spoerri, A. 2007. Cellular telephone use and time trends in brain tumour mortality in Switzerland from 1969 to 2002. *Eur J Cancer Prev* 16(1): 77–82.

Rothman, K. J. 1986. *Modern epidemiology*. Boston, MA: Little Brown.

Rothman, K. J. 2009. Health effects of mobile telephones. *Epidemiology* 20(5): 653–5.

Shephard, A. 2007. Chapter 1: Use of ICT among households and individuals. In: Avery V, Chamberlain E, Summerfeld C, Zealey L, editors. *Focus on the digital age*. Norwich, UK: Office of National Statistics, p. 68.

Smith, M. A., Freidlin, B., Ries, L. A. and Simon, R. 1998. Trends in reported incidence of primary malignant brain tumors in children in the United States. *J Natl Cancer Inst* 90(17): 1269–77.

Swerdlow, A. J., Feychting, M., Green, A. C., Leeka Kheifets, L. K. and Savitz, D. A. 2011. Mobile phones, brain tumors, and the INTERPHONE study: where are we now? *Environ Health Perspect* 119(11): 1534–8.

World Health Organization. 1993. *Electromagnetic fields (300 Hz to 300 GHz)*. Geneva: World Health Organization.

15

Synthesis of Epidemiological Studies on Mobile Phone Use and the Risk of Brain Tumors

Isabelle Deltour and Joachim Schüz

CONTENTS

Introduction

Mobile phone technology was introduced in the 1980s, but it became popular in the mid-1990s. Today, the majority of people worldwide use mobile phones. For communication, mobile phones emit radiofrequency (RF) electromagnetic fields that at high levels of exposure lead to heating of human tissue (International Commission on Non-Ionizing Radiation Protection [ICNIRP] 2009). Mobile phone technologies comply with protection guidelines to prevent biologically relevant heating, but some concerns were expressed when mobile technology entered the markets on whether there were unknown biological effects related to RF fields that could lead to adverse health effects, including cancer (Microwave News 1992). Due to the increasing popularity of the technology, many research groups around the world started investigations to study whether mobile phone use increases the risk of cancer, in particular brain tumors, because the highest RF field is absorbed by the brain when the phone is held to the head. The epidemiological research focused initially on the possibility of risk in middle-aged adults, the population that was the first to use mobile phones frequently, due to initially high costs related to their use. Today, concerns relate also to children and teenagers, who have become frequent users of mobile phones; due to their smaller head size, different physiological properties, and still-developing nervous system, the question whether they would be more vulnerable to RF fields arose (Aydin et al. 2011a). The brain tumors investigated included glioma, constituting the majority of malignant brain tumors; meningioma; and vestibular schwannoma, the latter two being predominantly benign tumors. The etiology of brain tumor is poorly known.

Many epidemiological investigations have been published, including several in the recent past. Based on a review of the scientific evidence, we propose some tentative conclusions. In addition, we put these conclusions into the context of the 2011 evaluation of RF fields within the Monograph Program of the International Agency for Research on Cancer (IARC) on the evaluation of carcinogenic risks to humans, classifying RF fields as a possible carcinogen (IARC group 2B) based mainly on epidemiological findings of an increased risk of glioma and vestibular schwannoma in heavy mobile phone users (Baan et al. 2011).

Mobile Phone Use and RF Exposure Assessment: Self-Reported, Operator Information, and Environmental Exposures

Commonly, if a true causal association exists, it is anticipated that the higher the exposure, the higher the risk, that is, intensity of exposure increases the risk; the longer the exposure, the higher the risk, that is, duration increases the risk; and even more so for the cumulative exposure, that is, the product of duration and intensity. This pattern should emerge consistently in exposed populations. There are no scientifically sound reasons why there should be a different expectation related to exposure from mobile phones when it comes to this classical model of evaluation of the carcinogenic potential of agents. However, intensity and duration of exposure to RF emitted by mobile phones are not easily measured, and proxy measures such as cumulative amount of mobile phone use, modeled RF energy emitted by mobile phones (based on cumulative amount of use modified by several technical features of the networks and devices), or even simple measures such as ever use of mobile phones or ever subscriber of a mobile phone have been used.

Most studies have investigated whether the cumulative amount of use of mobile phones is associated with the risk of brain tumors. Studies have used various ways to quantify the use, almost exclusively based on self-reports from study subjects: ever/never use or ever/never regular use of a mobile phone, time since start of first use or years of duration of use, or cumulative number of calls or cumulative duration of calls. Number of calls and call time vary widely across countries, and also between individuals within countries. Mobile phone use while using hands-free kits or car phones is generally ignored in the cumulative exposure, on the basis that the exposure to the head is minimal. For the same reason, only voice calls but not text messaging are counted in the number of calls.

Validation studies were performed to compare the self-reported mobile phone use measured by traffic records from the network operators or the handsets. All validation studies showed that it was very difficult for participants to accurately recall their past and even current use of mobile phones (Vrijheid et al. 2006a, 2009a; Schüz and Johansen 2007; Inyang et al. 2009; Aydin et al. 2011b). The comparison between objectively measured use of mobile phones and self-reported use in the Interphone questionnaire, 6–12 months after the recording of the objective measures, showed that healthy adult volunteers, on average, slightly underestimated their monthly number of calls (ratio = 0.9) but overestimated the monthly call time by 40% (ratio = 1.4). There was a very large spread of under- to overestimation: for call time, 95% of participants reported between 8 times below their actual call time up to 17 times above, with the remaining 5% doing even worse. The reporting appeared to depend on level of use with underestimation at low levels of use and overestimation at high levels of use, spuriously increasing the range of exposures (Vrijheid et al. 2006a). Moreover, a differential bias over time emerged because cases showed a stronger tendency toward overreporting than controls, based on limited data (Vrijheid et al. 2009a). Children and adolescents had poorer recall than adults when their answers to the CEFALO questionnaire were compared to lifetime operator data. Cases overestimated their duration of calls by 52% and controls by 163%, but the difference between the groups was only borderline significant (p = .07); overall, the duration of calls was overestimated by 120% (Aydin et al. 2011b). Simulation studies showed that errors of these magnitudes could seriously distort the shape of a dose–effect relationship if one existed and that differential errors between cases and controls could induce spuriously positive or negative effect estimates in absence of a true underlying association (Vrijheid et al. 2006b; Aydin et al. 2011b).

In a smaller study among adolescents, correlation between self-reported and recorded duration was very low (correlation coefficient = 0.1), with negative correlations observed for some investigated subgroups (Inyang et al. 2009). In another study of 14-year-old teenagers, laterality was also poorly reported (Kappa coefficient = 0.3) (Inyang et al. 2010). No data are currently available on the accuracy of reporting of the use of hands-free kits.

The relationship between use of the mobile phone and cumulative energy emitted by the phone has been investigated (Berg et al. 2005) in second generation GSM phones. The correlation between use, that is, number of calls and duration of calls, and total cumulative energy based on measured output power was fair: 0.50 and 0.48, respectively. When an attempt was made to model cumulative RF energy at the tumor location using the Interphone dataset, cumulative call time was also the self-reported parameter most associated to the modeled cumulative energy at the tumor location (Kappa coefficient = 0.68) (Cardis et al. 2011b), with first and second generation phones. This measure of exposure is most related to the safety protection standards for exposure to RF fields that are expressed in terms of maximum specific absorption rate (SAR) per unit mass of tissue, that is, energy per unit time and unit mass.

With first and second generation networks, the main source of exposure to RF fields for the head was the use of mobile phones by the subject, due to the proximity of the head to the RF field-emitting antenna. Environmental exposure from base stations and other sources was negligible compared to exposure received from close-to-body devices, in particular the use of mobile phones. With the development of third and fourth generation networks, the distribution of exposures in the population keeps evolving, but contributions from close-to-body exposure decreases due to modifications of the technology, increasing comparatively the relative proportion of far field sources in the total exposure (Joseph et al. 2010). In addition to the change in relevant exposure sources, the lower output power from the latest mobile phone technologies results in orders of magnitude lower exposure in users; measurement studies show 100- to 1000-fold decrease in emissions related to mobile phone use (Persson et al. 2012). Very likely, a heavy user of today's technologies has a lower exposure to the brain than the occasional mobile phone users of the early 1990s.

In summary, exposure assessment in mobile phone epidemiological studies faces great challenges. Recall errors, both systematic and random, as well as differential recall, are competing errors that potentially lead to under- as well as overestimation of risks, and they are superimposed on the classical potential sources of biases of epidemiological studies, such as confounding and selection bias. The latter has also been identified to strongly impact on the results, as in the Interphone study, a study in which mobile phone use prevalence was different between participants and nonparticipants, with higher participation rates in mobile phone users compared with nonusers (Vrijheid et al. 2009b) (Table 15.1). Modeling error scenarios in an attempt to correct them, led to high variability in the results—from underestimation to artificially creating spurious relationships depending on the assumptions made; in particular, several equally likely scenarios of weight of each bias and error led to very different interpretations (see Table 15.1).

Epidemiological Studies on Brain Tumors: Glioma and Meningioma

Studies were done using three different epidemiological study designs, namely, ecological (incidence time trend), case–control, and cohort studies (see Chapters 12–14). All approaches have advantages and disadvantages, and each complements the interpretation of the others. The main limitation of the few cohort studies conducted so far are the crude exposure indicators; hence, they had limited potential to detect small risk increases or risk increases

TABLE 15.1

Reporting Errors Identified in Epidemiological Studies of Mobile Phone Use and the Risk of Brain Tumors

Source of Error	Most Likely Effect
Proxy Measure	
Using mobile phone use as proxy for RF energy absorbed by the head (Berg et al. 2005; Cardis et al. 2011a,b)	Fair correlation leading to nondifferential random error likely to underestimate a true association, if any
Using subscriber status as proxy for mobile phone use (error due to users not subscribing for a mobile phone and subscribers not using their phone) (Schüz and Johansen 2007)	Leading to nondifferential random error likely to underestimate a true association, if any; however, if applied to early times of mobile phone use where a large proportion of the population did not use a mobile phone, the underestimation is rather modest
Reporting Errors	
Random errors in accuracy of reporting actual mobile phone use (Vrijheid et al. 2006a,b)	Large error likely to underestimate a true association, if any
Systematic nondifferential error of underestimating of mobile phone use of low users and overestimating of mobile phone use of high users (Vrijheid et al. 2006a,b)	Likely to inflate a true or spurious association, if any
Differential error by tendency of cases to overreport mobile phone use more strongly than controls (Vrijheid et al. 2009a; INTERPHONE Study Group 2010)	Could possibly create spurious association or inflate true association
Selection Bias	
Cohorts: Healthy cohort effect as early mobile phone users were of higher socioeconomic status (Schüz et al. 2006)	Representing protective effect for cancers related to unhealthy lifestyle (e.g., tobacco use), but unclear effect for brain tumors, if any
Case–control: underrepresentation of mobile phone users among controls (Vrijheid et al. 2009c)	Creating spurious protective effect

in small subgroups, such as the very heavy users of mobile phones. Case–control studies implemented a comprehensive exposure assessment, but the validation studies showed the presence of several competing errors that prevented firm conclusions from studies of this design. Ecological studies are usually not very strong designs in epidemiology, due to the possible ecological fallacy, but given the distinct exposure time trends, they have some value: mobile phone use became extremely popular within very few years. Moreover, this occurred at different points in time in different segments of the populations; consequently, incidence time trends should follow the same patterns, in particular, showing an increase first in middle-aged men, should a causal association exist. Uncommonly and importantly, incidence time trends can be used as consistency checks of the results from the case–control and cohort studies: the relative risk estimates of the analytic studies can be translated into expected excess cases and the observed incidence time trends can be compared with the predicted incidence time trends based on these excess case estimations.

Case–Control Studies Show Some Associations but Recall Errors and Biases Are Present

Hardell Studies

Among the first projects to investigate the possible association between mobile phone use and brain tumor risk were the studies performed by the Hardell group in northern Sweden (Hardell et al. 2006, 2011, 2013; Carlberg et al. 2013). Their first studies included cases diagnosed up to 2003.

The Hardell case–control studies reported markedly increased risks of malignant brain tumors in relation to any use of mobile phone (odds ratio [OR] = 1.3; 95% confidence interval [CI], 1.1–1.6) for ever using a mobile phone up to 2003, for the risk of malignant brain tumors) (Hardell et al. 2011). A recent study by the same authors involving cases diagnosed between 2007 and 2009 reported associations of the same order of magnitude, namely, OR = 1.6 (95% CI, 1.0–2.7) for any use of mobile phone, for malignant brain tumors. ORs were highest in the groups of highest exposure. No association was observed for meningioma (Hardell et al. 2006; Carlberg et al. 2013). Noteworthy, in these publications, the methods lack methodological detail; for example, it is unclear whether controls are selected at random from population lists, and whether matching of controls to cases was performed at random. Unadjusted analyses and separate analyses by sex were not reported, and descriptive data on sociodemographics of the sample were generally not provided. In these studies, the self-reported use has never been validated against objective records, leaving open questions about the magnitude of the biases and errors, and their impact on the results.

The numbers of cases recorded in the cancer registries in the Nordic countries (Deltour et al, 2012), and in the USA (Little et al, 2012) are far lower than what the reported risks by the previously mentioned studies by Hardell and colleagues would predict. The incidence rate time trends in registry data are not compatible with those relative risks. Moreover, such levels of risks have not been reproduced in any other epidemiological study on this topic (see Chapter 14). It is therefore plausible that the increased risks of malignant brain tumors observed in the Hardell studies are attributable to biases and errors.

Interphone and Other Case–Control Studies

Smaller case–control studies were conducted in the United States before Interphone; they did not show any association for either glioma or meningioma; however, the latency periods between first exposure and diagnoses of brain tumors in those studies were very short (see Chapter 13).

Interphone was a large multinational case–control study of brain tumors coordinated by IARC (Cardis et al. 2007; see Chapter13). No association was observed for meningioma (INTERPHONE Study Group 2010). For glioma, the OR for being a regular user of a mobile phone was 0.8 (95% CI, 0.7–0.9) (INTERPHONE Study Group 2010). Relative risks did not vary between the categories of duration of use, that is, mobile phone users for ≥10 years of use had no different risk from users for a shorter duration. Analyses by number of calls did not show any increased risks either. The analyses by lifetime cumulative call time showed an increased risk only among those who were the heaviest users (OR = 1.4; 95% CI, 1.0–1.9). Those heaviest users were defined as having a cumulative call time of 1640 hr or more in their lifetime, equivalent to roughly the top 10% of regular mobile phone users in the control group. Similarly elevated OR were observed in some sensitivity analyses such as when heaviest users who had a tumor in the temporal lobe were compared with their controls. However, in analyses stratifying by time since start of use and level of use, the only significantly increased risk was among the heaviest users who had started use in the most recent time period (1–4 years before the interview) (OR = 3.8; 95% CI, 1.3–11.4). Such a time pattern whereby risk for heavy and intensive users would be increased almost fourfold is unlikely, as again, an effect in the incidence time trends would be detectable by 2008, but no such effect can be observed in the data (Deltour et al. 2012). Overall, using the consistency check of the Interphone results with observed incidence rate time trends shows that the overall decrease in risk of regular users is not compatible with incidence time trends due to the lack of any decrease in rates, but that the small risk increase in very heavy users cannot be completely ruled out.

Further analyses on subsets of the Interphone data were conducted. An attempt was made to construct a gradient of cumulative exposure to RF fields, based on modeling of the energy absorbed at the tumor location. This model included a large number of parameters, and some of them were based on reliable sources of information such as the technical features of the devices (Vrijheid et al. 2009c). The main determinant of the reconstructed exposure to RF fields, however, remained the self-reported use, thereby carrying over the recall biases and errors into the estimation of energy absorbed at the tumor location. Self-reported laterality was also a major determinant of exposure, and prone to recall biases and errors (Schüz 2009). Although the reconstruction of the exposure was useful to lay out and analyze its determinants for future epidemiological studies, the calculations have not mitigated biases and errors in this dataset.

In practice, the results of these analyses on the data from five Interphone countries yielded an increased risk of glioma in the most highly exposed quintile of exposure (OR = 1.7; 95% CI, 1.0–2.7) (Cardis et al. 2011a), which is also incompatible with the stable incidence rate time trends in the glioma rates as recorded in cancer registries of the United States and the Nordic countries.

Methodological approaches to limit the impact of laterality biases were developed and applied to the data of seven other Interphone countries (Larjavaara et al. 2011b). A dichotomization of the tumors between the external and the internal tumors (>5 cm from the point of the exposure vs. <5 cm) was conducted to compare these two groups. No associations were found.

Inconsistent results were obtained in the CEFALO case–control study among children and adolescents (Aydin et al. 2011a). Overall, the OR of being regular user was slightly elevated (OR = 1.4; 95% CI, 0.9–2.0), but there were no dose–response trends observed by duration of use, amount of use or when looking at the laterality of the tumors.

Cohort Studies Do Not Show Associations but Are Limited by Crude Exposure Information

The Danish cohort of early private subscribers of mobile phones is a cohort set up retrospectively, with prospective follow-up through record linkage. All Danish mobile phone companies provided the identification information and the year of subscription of their customers for all mobile phone numbers of the country until 1995. Persons who subscribed to mobile phone services in their own names were identified in the Danish population, but subscriptions in the name of a company rather than an individual could not be allocated. Analyses over increasing time of follow-up have been published (Johansen et al. 2001; Schüz et al. 2006; Frei et al. 2011). The successive analyses adjusted for further information on the cohort members as potential confounding factors (e.g., education, income); none showed increased risks of any cancer or benign brain tumors. The strengths of this study are that it is a nationwide study, so that the entire Danish adult population is either classified as "mobile subscriber before 1995" or "later mobile subscriber/never subscriber"; the nationwide cancer registry provides the cases in these two groups, so follow-up is virtually complete and nondifferential. These early individual subscribers do not appear to have higher rates of brain tumors than the rest of the population 1, 6, or 13 years after the start of use of mobile phones; noteworthy, rates of glioma have not increased overall in Denmark either (Deltour et al. 2009). These early users had also the highest exposures, because mobile phones in the 1980s and early 1990s had much higher output power than technologies of today (see above).

The U.K. Million Women study is a prospective cohort study conducted in the United Kingdom, in which about 800,000 women aged 50–64 years answered questions on their use of mobile phone between 1999 and 2005 (The Million Women Study Collaborative Group 1999). No differences in risk of glioma or meningioma were observed by level or duration of use for these women followed for 7 years, on average (Benson et al. 2013a). Similar results were obtained when follow-up was prolonged by 2 years until end of 2011 (Benson et al. 2013b). Limitations are that mobile phone use was only assessed at two points in time with two relatively broad questions, so that for instance long-term heavy users could not be analyzed as a separate group.

Ecological Studies of Incidence Time Trends

Ecological studies of incidence time trends have not shown detectable effects of mobile phone use anywhere; these studies will answer the research question in the long run, unless the population attributable fraction is too small to be detected.

Surveillance of incidence rate time trends of the brain tumors in high-quality cancer registries is the ultimate way to monitor this question, given that the increase in prevalence of use of mobile phone has been extremely rapid in all countries worldwide, and today the majority of the global population use mobile phones. For example, use of mobile phone is >100 subscriptions per 100 inhabitants since 2005 in the Nordic countries. However, high-quality population-based cancer registries that are complete, accurate, relatively rapid, and allow use of their data for epidemiological research exist only in a few countries. Where data were available, in England until 2007, in the Nordic countries i.e., Denmark, Finland, Norway, and Sweden, until 2008, or in the United States until 2008, rates of glioma were stable or increased slowly (de Vocht et al. 2011; Deltour et al. 2012; Little et al. 2012). Meningioma rates are increasing; however, the increase started before the introduction of mobile phones and is likely to be due to a combination of better diagnosis and more complete registration (Deltour et al. 2009). Rates in children were stable in Sweden, England, and the United States, and they did not appear to be influenced by changing environmental factors (Inskip et al. 2010; Aydin et al. 2011a, 2012; de Vocht et al. 2011).

Incidence data are especially robust for glioma, of which most are rather rapidly growing glioblastoma and therefore may also have a shorter induction time than the slowly growing benign tumors.

Epidemiological Studies on Vestibular Schwannoma (Acoustic Neuroma)

The pattern of risk observed in case–control studies for acoustic neuroma is similar to that of glioma, but it is based on smaller numbers. Case–control studies conducted by the Hardell group in Sweden showed markedly increased risks of OR = 1.6 (95% CI = 1.2–2.2) for all mobile phone use increasing to OR = 4.5 (95% CI, 2.1–9.5) with latency >20 years (Hardell et al. 2013). In contrast, the Interphone case–control study showed a slightly and nonsignificantly increased risk restricted to the heaviest users (OR = 1.3; 95% CI, 0.9–2.0); notably, however, the risk in the second highest user group was significantly decreased (OR = 0.5; 95% CI, 0.3–0.8), leading to a somewhat confusing dose–response pattern (INTERPHONE Study Group 2011). In Interphone, the risk was further elevated when longer latencies were considered by disregarding exposures in the past 5 years before diagnosis (OR = 2.8 [95% CI, 1.5–5.2] for the heaviest users, and OR = 0.6 [95%, CI 0.3–1.1] for the second highest decile); such an analysis is perhaps justified for this slow growing tumor. In the Danish cohort, only the risk among men could be analyzed, due to the small number of long-term female users (Schüz et al. 2011b). Risk in subscribers of mobile phones for >11 years (15 exposed cases) was compared with risk in subscribers for <11 years and nonsubscribers, with no evidence of a difference in risk (relative risk [RR] = 0.9; 95% CI, 0.5–1.5). Moreover, although the majority of mobile phone users self-reported using their mobile phone predominantly on the right side of the head, no side predominance was observed among acoustic neuroma cases. Earlier analyses had yielded similar results (Johansen et al. 2001; Schüz et al. 2006). In the U.K. Million Women study that, by design, includes only women, when the follow-up was considered until end of 2009, an RR of 2.5 (95% CI, 1.1–5.6) was observed for users of >10 years (based on eight exposed cases) and a significant trend with duration of use (Benson et al. 2013a). However, when two extra years of follow-up were added to reach the end of 2011, none of these risks were significant any longer: the risk for those who had used mobile phone the longest was 1.2 (95% CI, 0.6–2.3), and the trend test with duration of use was no longer significant ($p = .3$) (Benson et al. 2013b). The incidence of this rare tumor has been studied in England, where the rates were stable among men and women (Benson et al. 2013); in the Nordic countries, the analysis of incidence rates showed marked differences between countries, and increases that could be attributable to improved diagnostics and detection (Larjavaara et al. 2011a).

Conclusions

Glioma

Table 15.2 shows a summary of effect estimates from the most influential epidemiological studies on mobile phone use and glioma risk, that is, those large enough to have sufficient statistical power to detect potentially modest risk increases and covering latency periods

TABLE 15.2

Overview of Most Influential Epidemiological Analytic Studies on Mobile Phone Use and the Risk of Brain Tumors

Study (Reference)	Study Population	Diagnostic Period	Result for "Ever Use" (95% CI)	Result for "10+ Years of Use"	Result for Highest Category of Cumulative Use
Glioma					
Hardell I (Hardell et al. 2011)	Cases = 1148; controls = 2438	1997–2003	1.3 (1.1–1.6)[a]	2.5 (1.8–3.3)	>2000 hr; 3.2 (2.0–5.1)
Hardell II (Hardell et al. 2013)	Cases = 593[b]; controls = 1368	2007–2009	1.6 (1.0–2.7)[a,b]	>10–15 years: 1.3 (0.8–2.2)[b]; >15–20 years: 1.5 (0.8–2.6)[b]; >20–25 years: 1.9 (1.1–3.5)[b]; >25 years: 2.9 (1.4–5.8)[b]	>2376 hr[b]; 2.8 (1.6–4.8)
Interphone (INTERPHONE Study Group 2010)	Cases = 2708; controls = 2972	2000–2004	0.8 (0.7–0.9)[a]	1.0 (0.8–1.3)	≥1640 hr; 1.4 (1.0–1.9)
Danish subscriber cohort (Frei et al. 2011)	Cases = 3664; cohort size = 3.21 million persons; 3.8 million exposed person-years	Up to end of 2007	Men: 1.1 (1.0–1.2)[c]; women 1.0 (0.7–1.4)[c]	Men: 1.0 (0.8–1.3); women: 1.0 (0.6–2.0)	NA
U.K. Million Women study (Benson et al. 2013a, 2013b)	Cases = 875; cohort size = 791, 710 persons; 5.8 million person-years[d]	Up to end of 2011	0.9 (0.8–1.1)	0.8 (0.6–1.1)	Daily use; 0.8 (0.6–1.1)[d]
Meningioma					
Hardell I (Hardell et al. 2006)	Cases = 916; controls = 2162	1997–2003	Digital phone 1.1 (0.9–1.3)[a,e]	Digital phone; 1.3 (0.5–3.2)[e]	>1000 hr; digital phone: 0.7 (0.3–1.4)
Hardell II (Carlberg et al. 2013)	Cases = 709; controls = 1368	2007–2009	1.0 (0.7–1.4)[a]	>10–15 years: 1.0 (0.7–1.4); >15–20 years: 1.0 (0.6–1.5); >20–25 years: 0.8 (0.5–1.4); >25 years: 1.2 (0.6–2.3)	>2376 hr; 1.3 (0.8–1.9)
Interphone (INTERPHONE Study Group, 2010)	Cases = 2409; controls = 2662	2000–2004	0.8 (0.7–0.9)[a]	0.8 (0.6–1.1)	≥1640 hr; 1.2 (0.8–1.6)

Continued

TABLE 15.2 (*Continued*)

Overview of Most Influential Epidemiological Analytic Studies on Mobile Phone Use and the Risk of Brain Tumors

Study (Reference)	Study Population	Diagnostic Period	Result for "Ever Use" (95% CI)	Result for "10+ Years of Use"	Result for Highest Category of Cumulative Use
Danish subscriber cohort (Frei et al. 2011)	Cases = 1757; cohort size = 3.21 million persons, 3.8 million exposed person-years	Up to end of 2007	Men: 0.8 (0.6–1.1)[c]; women: 1.0 (0.7–1.5)[c]	Men: 0.9 (0.6–1.4); women: 0.9 (0.5–1.9)	NA
U.K. Million Women study (Benson et al. 2013a, 2013)	Cases = 397; cohort size = 791, 710 persons, 5.8 million person-years[d]	Up to end of 2011	1.0 (0.8–1.3)	1.1 (0.8–1.5)	Daily use[d]; 1.1 (0.7–1.9)
Vestibular Schwannoma (Acoustic Neuroma)					
Hardell I and II (Hardell et al. 2013)	Cases = 316; controls = 3530	1997–2003 pooled with 2007–2009	1.6 (1.2–2.2)[a]	>10–15 years: 2.1 (1.3–3.5); >15–20 years: 2.1 (1.0–4.2); >20 years: 4.5 (2.1–9.5)	>1486 hr; 2.6 (1.5–4.4)
Interphone (INTERPHONE Study Group, 2011)	Cases = 1105; controls = 2145	2000–2004	0.8 (0.7–1.0)[a]	0.8 (0.5–1.1)	≥1640 hr; 1.3 (0.9–2.0)
Danish subscriber cohort (Schüz et al. 2011b)	Cases = 806; cohort size = 2.9 million persons	Up to end of 2006	NA	Men: 0.9 (0.5–1.5); women: NA	NA
U.K. Million Women study (Benson et al. 2013a, 2013)	Cases = 126; cohort size = 791, 710 persons	Up to end of 2011	1.2 (0.8–1.7)	1.2 (0.6–2.3)	Daily use[b]; 1.4 (0.6–3.1)

Note: NA, not applicable.

[a] Up to 1 year before reference date.

[b] Numbers and risks for malignant brain tumor cases; results for glioma (n = 546) not reported separately but are described as "similar."

[c] Subscribers.

[d] Numbers and risks for follow-up to end of 2009.

[e] Numbers taken from table 2 of Hardell et al. (2006); slightly different numbers can be found in table 4 of Hardell et al. (2006) for the same risks.

of at least 10 years, with more details presented in other chapters (see Chapters 12–14). Figure 15.1 contrasts the observed glioma incidence rate among 40- to 59-year-old males in the Nordic countries with the incidence rate that is predicted if there was a true risk increase with mobile phone use of 100%, 50%, and 20%. Altogether, these results show little evidence of an association between mobile phone use and glioma risk. In 2011, the IARC expert group concluded that there was limited evidence in humans based on the two case–control studies reporting an increased risk in heavy users (Baan et al. 2011). This judgment was however done before several recent publications became available: more recent incidence time trends (Deltour et al. 2012; Little et al. 2012), an update of the Danish cohort

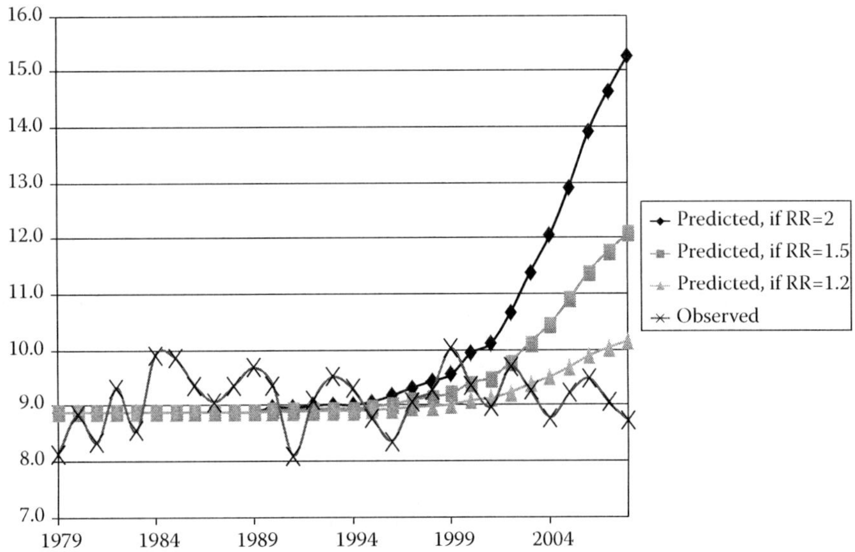

FIGURE 15.1
Glioma incidence rate observed in the Nordic countries among 40- to 59-year-old men; mobile phone-related RR increases of 1.2, 1.5, and 2, respectively, are highly implausible. (From Deltour I et al., *Epidemiology* 23, 301–307, 2012.)

(Frei et al. 2011), a new study from northern Sweden (Carlberg et al. 2013; Hardell et al. 2013), and the results of the U.K. Million Women study (Benson et al. 2013, 2013b). These more recent results appear to attenuate the earlier observations. In addition, the strong contrast between the effect estimates from the Hardell studies and the lack of any change in incidence rates from cancer registries calls into question the usefulness of the former studies for risk assessment. Based on the time trend incidence data to date, the small risk increase observed in Interphone is still possible, but unlikely (Deltour et al., 2012) and may indeed be a result of recall bias. Alternatively, the apparent inconsistency between the small risk increase in Interphone heaviest users and all the other studies not detecting increased risks could conceivably be explained by the strong decrease in output power of the latest mobile phone technologies in recent years and more optimized networks with respect to power adaptation, so that today's heaviest users do not accumulate the same magnitude of RF energy as the early heavy users. According to that argument, even if there was once a true risk for the heaviest users up to 2004, the risk would not have persisted later and it would have become undetectable after that time and in future studies. Whether that would be consistent with the available cohort data remains to be evaluated.

Another important question remains whether it is likely that an association could occur in the future as the studies reviewed here are inadequate if the lag between exposure and malignancy is longer than 15–20 years. Several observations can be put forward against this hypothesis. From the statistical perspective, incidence rates have followed their secular time trends in all age groups, sex, and countries where this has been studied that are all independent confirmations of each other, for a tumor that is very well recorded in high-quality cancer registries. Cohort studies have not shown increased risks even in the categories of longest use either. One may also hypothesize that should an association exist, the differences between individuals would be expected to lead to earlier effects in some persons than in others, with the induction time being the average of the individuals time to effect, so that even if the peak effect would occur later, some early indication should become detectable already now. From the exposure point of view, however, epidemiological studies available today reflect the potential effect of the exposure of the studied populations in the past. No large-scale epidemiological study has produced results on third and fourth generation mobile phones and networks; Cosmos, a large-scale prospective cohort study in Europe (Schüz et al. 2011a) and the U.K. Million Women study may help answer this question (Benson et al. 2013b).

Data (Aydin et al. 2011a; de Vocht et al, 2011) about glioma risk in children or young adults do not demonstrate increased risks, but data are scarce. Monitoring of time trends does not show any changes in incidence time trends. Studies should also investigate start of using mobile phones as a child with possible effects during adulthood that—due to the short time since introduction of mobile phone technology—could not be investigated in studies conducted until now.

Meningioma

To date, the epidemiological evidence on RF emitted by mobile phones shows no evidence of an overall effect on the risk of meningioma (Table 15.2). Incidence data from cancer registries are less reliable for meningioma than for glioma, because the registration is possibly incomplete and changes in the registration processes are likely to have an influence on the meningioma rates. Because this tumor is very slow growing and the currently available studies were not be able to detect effects after latency times of ≥20 years, further monitoring is needed, especially monitoring time trends in high-quality registries. This is, however, a challenge due to the underregistration mentioned above (Larjavaara et al. 2008) and to underdiagnosis of otherwise asymptomatic tumors because it makes the ascertainment of cases in any study dependent on the level of diagnostic activity (Morris et al. 2009). Little is known on the determinants of diagnostic activity, aside from availability of improved and affordable diagnostic tools (computed tomography [CT] scans, then later magnetic resonance imaging [MRI] scans), and awareness and attitude of doctors and patients toward mild disease symptoms, such as migraine.

Vestibular Schwannomas

For vestibular schwannomas, also called acoustic neuromas, the evidence is also gradually becoming more consistent (Table 15.2), but due to the very low incidence of this tumor, estimates of risks have been more variable across studies. Again, the IARC monograph concluded that there was limited evidence in humans mainly influenced by the findings of increased risks in heavy users in the two case–control studies, but the more recent results from the cohorts and from incidence time trends appear to attenuate these findings. Incidence rates time trend studies for vestibular schwannoma are not very reliable due to underreporting of this tumor even in high-quality cancer registries (Larjavaara et al. 2011a). The strongest finding for overall mobile phone use comes from the two cohorts, the Danish subscriber study (Schüz et al. 2011) and the U.K. Million Women study, (Benson et al. 2013b) not showing any increased risks but again with limited power to address specifically the very heavy users. The Interphone study is therefore the only remaining study to indicate increased risks, if one assumes that the analyses of the Hardell studies on vestibular schwannoma have the same limitations as their glioma results. Given that an early symptom of vestibular schwannoma is hearing difficulties, it is also conceivable that mobile phones act almost as diagnostic tools, used by the individuals themselves to detect hearing problems earlier than nonusers. This, mixed with random fluctuations due to small numbers, could explain the differences in some of the results. Due to the expected slow growth of vestibular schwannoma, further follow-up is needed, ideally using long-term prospective studies (Schüz et al. 2010).

Perspectives

In conclusion, data accrued until now suggest that common use of mobile phones is not related to an increased risk of any type of brain tumor. Some uncertainty remains with regard to very heavy use of mobile phones, in particular phones of the first and second generation with high output power, because study results are conflicting; due to lack of increase in incidence time trends and the findings from the cohort studies, even small risks are unlikely, but they cannot be ruled out. Two open questions remain: first, in particular for the slowly growing tumors, the latency period of studies to date may have been too short to detect any effect, assuming that such an effect would only become detectable after >20 years since first exposure; and second, long-term effects of having started to use a mobile phone as a child, and possibly being more vulnerable, have not been addressed by studies to date. However, some of these questions can be addressed by smart monitoring of incidence time trends of high-quality cancer registries. Because the exposure is so common, even small risks or risks in subpopulations should not be missed, meriting further research on possible adverse effects related to mobile phone use.

References

Aydin, D., Feychting, M., Schüz J., et al. (2011a). Mobile phone use and brain tumors in children and adolescents: a multicenter case-control study. *J Natl Cancer Inst*. 103(16): 1264–1276.

Aydin, D., Feychting, M., Schüz J., et al. (2011b). Impact of random and systematic recall errors and selection bias in case-control studies on mobile phone use and brain tumors in adolescents (CEFALO study). *Bioelectromagnetics*. 32(5): 396–407.

Aydin, D., Feychting, M., Schüz, J., et al. (2012). Childhood brain tumours and use of mobile phones: comparison of a case-control study with incidence data. *Environ Health*. 11: 35.

Baan, R., Grosse, Y., Lauby-Secretan, B., et al. (2011). Carcinogenicity of radiofrequency electromagnetic fields. *Lancet Oncol*. 12(7): 624–626.

Benson, V. S., Pirie, K., Schüz, J., et al. (2013a). Mobile phone use and risk of brain neoplasms and other cancers: prospective study. *Int J Epidemiol*. 42(3): 792–802.

Benson, V. S., Pirie, K., Schüz, J., et al. (2013b). Authors' response to: the case of acoustic neuroma: commenton mobile phone use and risk of brain neoplasms and other cancers. *Int J Epidemiol*. epub 10.1093.

Berg, G., Schüz, J., Samkange-Zeeb, F., et al. (2005). Assessment of radiofrequency exposure from cellular telephone daily use in an epidemiological study: German validation study of the international case-control study of cancers of the brain—INTERPHONE-study. *J Expo Anal Environ Epidemiol*. 15(3): 217–224.

Cardis, E., Armstrong, B. K., Bowman J. D., et al. (2011a). Risk of brain tumours in relation to estimated RF dose from mobile phones: results from five Interphone countries. *Occup Environ Med*. 68(9): 631–640.

Cardis, E., Richardson, L., Deltour, I., et al. (2007). The INTERPHONE study: design, epidemiological methods, and description of the study population. *Eur J Epidemiol*. 22(9): 647–664.

Cardis, E., Varsier, N., Bowman, J. D., et al. (2011b). Estimation of RF energy absorbed in the brain from mobile phones in the Interphone study. *Occup Environ Med*. 68(9): 686–693.

Carlberg, M., Söderqvist, F., Hansson Mild, K., et al. (2013). Meningioma patients diagnosed 2007—2009 and the association with use of mobile and cordless phones: a case—control study. *Environ Health*. 12(1): 60.

Deltour, I., Auvinen, A., Feychting, M., et al. (2012). Mobile phone use and incidence of glioma in the Nordic countries 1979–2008: consistency check. *Epidemiology*. 23(2): 301–307.

Deltour, I., Johansen, C., Auvinen, A., et al. (2009). Time trends in brain tumor incidence rates in Denmark, Finland, Norway, and Sweden, 1974–2003. *J Natl Cancer Inst*. 101(24): 1721–1724.

De Vocht, F., Burstyn, I. and Cherrie, J. W. (2011). Time trends (1998–2007) in brain cancer incidence rates in relation to mobile phone use in England. *Bioelectromagnetics*. 32(5): 334–339.

Frei, P., Poulsen, A. H., Johansen, C., et al. (2011). Use of mobile phones and risk of brain tumours: update of Danish cohort study. *BMJ*. 343: d6387.

Hardell, L., Carlberg, M. and Hansson Mild, K. (2006). Pooled analysis of two case-control studies on the use of cellular and cordless telephones and the risk of benign brain tumours diagnosed during 1997–2003. *Int J Oncol*. 28(2): 509–518.

Hardell, L., Carlberg, M. and Hansson Mild, K. (2011). Pooled analysis of case-control studies on malignant brain tumours and the use of mobile and cordless phones including living and deceased subjects. *Int J Oncol*. 38(5): 1465–1474.

Hardell, L., Carlberg, M., Söderqvist, F., et al. (2013). Case-control study of the association between malignant brain tumours diagnosed between 2007 and 2009 and mobile and cordless phone use. *Int J Oncol*. 43(6): 1833–1845.

Inskip, P. D., Hoover, R. N. and Devesa, S. S. (2010). Brain cancer incidence trends in relation to cellular telephone use in the United States. *Neuro Oncol*. 12(11): 1147–1151.

International Commission on Non-Ionizing Radiation Protection. (2009). ICNIRP statement on the guidelines for limiting exposure to time-varying electric, magnetic, and electromagnetic fields (up to 300 GHz). *Health Phys*. 97(3): 257–258.

INTERPHONE Study Group. (2010). Brain tumour risk in relation to mobile telephone use: results of the INTERPHONE international case-control study. *Int J Epidemiol*. 39(3): 675–694.

INTERPHONE Study Group. (2011). Acoustic neuroma risk in relation to mobile telephone use: results of the INTERPHONE international case-control study. *Cancer Epidemiol*. 35(5): 453–464.

Inyang, I., Benke, G., McKenzie, R., et al. (2010). A new method to determine laterality of mobile telephone use in adolescents. *Occup Environ Med*. 67(8): 507–512.

Inyang, I., Benke, G., Morrissey, J., et al. (2009). How well do adolescents recall use of mobile telephones? Results of a validation study. *BMC Med Res Methodol*. 9: 36.

Johansen, C., Boice, J. D., Jr., McLaughlin, J. K., et al. (2001). Cellular telephones and cancer—a nationwide cohort study in Denmark. *J Natl Cancer Inst*. 93(3): 203–207.

Joseph, W., Frei, P., Röösli, M., et al. (2010). Comparison of personal radio frequency electromagnetic field exposure in different urban areas across Europe. *Environ Res*. 110(7): 658–663.

Larjavaara, S., Feychting, M., Sankila, R., et al. (2011a). Incidence trends of vestibular schwannomas in Denmark, Finland, Norway and Sweden in 1987–2007. *Br J Cancer*. 105(7): 1069–1075.

Larjavaara, S., Haapasalo, H., Sankila, R., et al. (2008). Is the incidence of meningiomas underestimated? A regional survey. *Br J Cancer*. 99(1): 182–184.

Larjavaara, S., Schuz, J., Swerdlow, A., et al. (2011b). Location of gliomas in relation to mobile telephone use: a case-case and case-specular analysis. *Am J Epidemiol*. 174(1): 2–11.

Little, M. P., Rajaraman, P., Curtis, R. E., et al. (2012). Mobile phone use and glioma risk: comparison of epidemiological study results with incidence trends in the United States. *BMJ*. 344: e1147.

Microwave News. (1992). *Cellular Telephone Radiation Blamed for Brain Tumor*. XII(3): 1.

Million Women Study Collaborative Group. (1999). The Million Women Study: design and characteristics of the study population [peerreviewed research]. Available from: http://breast-cancer-research.com/vol1no1/19aug99/research/1

Morris, Z., Whiteley, W. N., Longstreth, W. T., Jr., et al. (2009). Incidental findings on brain magnetic resonance imaging: systematic review and meta-analysis. *BMJ*. 339: b3016.

Persson, T., Törnevik, C., Larsson, L. E., et al. (2012). Output power distributions of terminals in a 3G mobile communication network. *Bioelectromagnetics*. 33(4): 320–325.

Schüz, J. (2009). Lost in laterality: interpreting "preferred side of the head during mobile phone use and risk of brain tumour" associations. *Scand J Public Health*. 37(6): 664–667.

Schüz, J., Elliott, P., Auvinen. A., et al. (2011a). An international prospective cohort study of mobile phone users and health (Cosmos): design considerations and enrolment. *Cancer Epidemiol*. 35(1): 37–43.

Schüz, J., Jacobsen, R., Olsen, J. H., et al. (2006). Cellular telephone use and cancer risk: update of a nationwide Danish cohort. *J Natl Cancer Inst*. 98(23): 1707–1713.

Schüz, J. and Johansen, C. (2007). A comparison of self-reported cellular telephone use with subscriber data: agreement between the two methods and implications for risk estimation. *Bioelectromagnetics*. 28(2): 130–136.

Schüz, J., Steding-Jessen, M., Hansen, S., et al. (2011b). Long-term mobile phone use and the risk of vestibular schwannoma: a Danish nationwide cohort study. *Am J Epidemiol*. 174(4): 416–422.

Vrijheid, M., Armstrong, B. K., Bédard, D., et al. (2009a). Recall bias in the assessment of exposure to mobile phones. *J Expo Sci Environ Epidemiol*. 19(4): 369–381.

Vrijheid, M., Cardis, E., Armstrong, B. K., et al. (2006a). Validation of short term recall of mobile phone use for the Interphone study. *Occup Environ Med*. 63(4): 237–243.

Vrijheid, M., Deltour, I., Krewski, D., et al. (2006b). The effects of recall errors and of selection bias in epidemiologic studies of mobile phone use and cancer risk. *J Expo Sci Environ Epidemiol*. 16(4): 371–384.

Vrijheid, M., Mann, S., Vecchia, P., et al. (2009b). Determinants of mobile phone output power in a multinational study: implications for exposure assessment. *Occup Environ Med*. 66(10): 664–671.

Vrijheid, M., Richardson, L., Armstrong, B. K., et al. (2009c). Quantifying the impact of selection bias caused by nonparticipation in a case-control study of mobile phone use. *Ann Epidemiol*. 19(1): 33–41.

16

Electromagnetic Fields, Symptoms and Idiopathic Environmental Intolerance

Christos Baliatsas and G. James Rubin

CONTENTS

Introduction

Various terms have been given to physical symptoms that lack a clear pathological basis (Engel et al. 2003), including nonspecific physical symptoms (NSPS), medically unexplained (physical) symptoms (MUPS/MUS), functional somatic syndromes, and psychosomatic symptoms (Wessely et al. 1999; Kirmayer et al. 2004; Van den Berg et al. 2005). Such symptoms are common in the general population and among users of primary and secondary health-care services (Kroenke et al. 1990; Reid et al. 2001; Kroenke 2003). In the absence of a medical explanation, sufferers of NSPS sometimes attribute their condition to low-level exposure to environmental factors, adopting labels for their condition that specify a putative cause such as sick building syndrome (SBS), multiple chemical sensitivity (MCS), or Gulf War syndrome (Barsky and Borus 1999; Das-Munshi et al. 2006; Iversen et al. 2007). Although environmental pollutants often can and do cause symptoms, as in the case of carbon monoxide poisoning (Prockop and Chichkova 2007), these particular attributions are highly controversial.

Electromagnetic fields (EMFs) are another environmental exposure suspected by some of causing NSPS (Genuis and Lipp 2011). As with other attributions, this, too, is controversial. Currently, there is no generally accepted bioelectromagnetic mechanism to explain how exposure to EMF levels below regulatory limits might cause symptoms (AGNIR 2012). Thus, although the attribution of NSPS to low-level EMF exposure is often referred to in the lay press and by sufferers as "electromagnetic hypersensitivity" (EHS) and similar terms (Table 16.1), the more etiologically neutral term "idiopathic environmental intolerance attributed to EMF" (IEI-EMF) has been recommended by the World Health Organization (WHO) (Hillert et al. 2006) and has been gaining broader acceptance among the scientific community. IEI-EMF is a complex condition. Although attribution of NSPS to EMF exposure is one of the primary criteria for the identification of people with IEI-EMF (Baliatsas et al. 2012b), this criterion is not

TABLE 16.1
Synonymous Terms for IEI-EMF in the Peer-Reviewed Literature

Hypersensitivity (HS) to EMFs or electricity/electromagnetic hypersensitivity (EHS) or electrohypersensitivity
(Subjective/self-reported/perceived) Sensitivity to EMFs or electricity/electromagnetic sensitivity or electrosensitivity (ES)
Environmental annoyance attributed to EMFs
Idiopathic environmental intolerance attributed to EMFs
Electrical hypersensitivity syndrome
Electromagnetic distress syndrome
Electromagnetic intolerance
Environmental illness

necessarily sufficient; several additional or alternative characteristics have also been observed, such as experience of symptoms soon after the individual is exposed to EMF or believe themselves to have been exposed, self-report of being (hyper)sensitive to EMF, and the absence of a medical diagnosis that could explain the reported symptoms (Baliatsas et al. 2012b).

The symptoms associated with IEI-EMF are heterogenous; no consistent syndrome has been identified (Hillert et al. 2002; Eltiti et al. 2007; Rubin et al. 2010; Augner et al. 2012). The symptoms most commonly attributed to EMF exposure are predominantly neurovegetative and dermatological, such as headache, fatigue, dizziness, sleep disturbances, lack of concentration, skin redness, and tingling and burning sensations (Hillert et al. 1999, 2002; WHO 2005). The estimated prevalence of IEI-EMF also varies and is partly dependent on how strict the case definition is (Baliatsas et al. 2012b). For example, in epidemiological studies, when symptom attribution is the only criterion used to identify people with IEI-EMF, prevalence can reach 18% (Mohler et al. 2010), whereas estimates of between 1.5% and 5% have been given when more demanding criteria are applied (Hillert et al. 2002; Levallois et al. 2002; Schreier et al. 2006; Schröttner and Leitgeb 2008). Regardless of the criteria used, IEI-EMF seems to be more common in women and people >40 years of age (Baliatsas et al. 2012b).

IEI-EMF is often associated with social, occupational, and mental impairment (Röösli et al. 2004; Carlsson et al. 2005; Tseng et al. 2011); in the most extreme cases, sufferers have been known to retreat almost entirely from modern society to escape from the EMFs that seem to harm them (Boyd et al. 2012).

Resolving the controversial issues of the existence of a causal association between nonionizing EMF and NSPS and the etiology of IEI-EMF should lay the foundation for the development of evidence-based interventions for people suffering from the condition and for risk communication strategies with the general public.

In this chapter, we give an overview and critical evaluation of the relevant literature on experimental and observational studies in this area, addressing methodological concerns along with implications for future research.

Current Evidence from Studies of Experimental and Observational Design

Experimental Studies

Human laboratory studies involving randomization and double-blinding can produce high-quality evidence about the causal associations between an exposure and a health outcome (Atkins et al. 2004). Many experimental studies have therefore been conducted to test whether exposure to EMF can trigger or exacerbate NSPS. Participants in these studies, mostly people with IEI-EMF, are typically exposed to different (low) levels of radiofrequency (RF) and/or extremely low-frequency (ELF) electromagnetic fields (Rubin et al. 2005, 2010). These studies have primarily examined outcomes such as self-reported symptoms; physiological and cognitive changes; and the individual ability to discriminate active from inactive (sham) provocations, an ability sometimes described as "electrosensibility" (Leitgeb and Schröttner 2003; Kwon et al. 2008).

Early "provocation studies" focused on the effect of video display units (VDUs) due to increased health concerns and symptom attribution among office workers (Berg et al. 1992). Thirteen studies were performed between 1982 and 2000; all included participants with symptoms attributed to VDU use (Rubin et al. 2005). Only two trials reported a weak significant association that was either not replicated in follow-up (Oftedal et al. 1995, 1999) or attributable to problems with conducting multiple comparisons (Sjöberg and Hamnerius 1995).

More recent experiments have investigated the report of symptoms in relation to near-field exposure from mobile phone handsets, mainly of the Global System for Mobile Communications (GSM) type, producing a peak specific absorption rate (SAR) in the head, up to 2 W/kg (Christ and Kuster 2005; Röösli et al. 2010a; Röösli and Hug 2011). Summarized evidence does not suggest an effect of exposure in people with IEI-EMF (Rubin et al. 2005, 2010; Röösli and Hug 2011; Augner et al. 2012; Kwon et al. 2012). In a few studies, significant effects on NSPS have been sporadically observed (Hillert et al. 2008; Kim et al. 2008), but chance and comparability issues between cases and controls were considered as plausible explanations for these effects (Rubin et al. 2005, 2010; Röösli and Hug 2011). For example, in a study in a sample of sensitive and nonsensitive regular users of Terrestrial Trunked Radio (TETRA) handsets, the only effect of exposure was an unexpected reduction in skin sensations after exposure to continuous wave signals, and this effect was only observed among participants who felt that they were sensitive to TETRA signals (Nieto-Hernandez et al. 2011).

Provocation studies using far-field exposure from mobile phone base stations have predominantly focused on the GSM 900 and/or GSM 1800 and/or Universal Mobile Telecommunication System (UMTS) frequency range, with exposure levels ranging between 1 and 10 V/m (Röösli et al. 2010a). In several investigations, there was no symptom increase during active exposure (Regel et al. 2006; Furubayashi et al. 2009; Wallace et al. 2010). Three randomized double-blind trials (Zwamborn et al. 2003; Eltiti et al. 2007; Riddervold et al. 2008) showed significant associations between exposure and various symptom scores, although again these associations are suspected to be caused by methodological issues. In the Zwamborn et al. (2003) study, results may have been affected by demographic differences between cases and controls; in the Eltiti et al. (2007) study, the order in which sham and active exposures were presented to the participants affected the outcome variables (Rubin et al. 2010; Röösli et al. 2010a); and in the Riddervold et al. (2008) study, additional analyses demonstrated that symptom increase during the experiment occurred due to baseline differences in symptom scores.

In total, 29 experiments have investigated whether there is an association between exposure to various EMF triggers and changes in objectively measured physiological reactions among people with IEI-EMF (Rubin et al. 2011). Only five of them reported statistically significant results, suggesting an effect of exposure on pupillary reflex (Rea et al. 1991), visual attention and perception (Trimmel and Schweiger 1998), heart rate and blood pressure (Hietanen et al. 2002), and indicators of sleep quality (Mueller and Schierz 2004; Arnetz et al. 2007). Again, however, these findings were isolated, unreplicable, and occasionally contradictory (Rubin et al. 2005, 2010), often suffering from shortcomings such as unbalanced ordering of exposure conditions (Hietanen et al. 2002).

According to most of the studies investigating ELF-EMF in relation to symptoms and/or physiological reactions, active exposure did not seem to cause negative acute effects (Wennberg et al. 1994; Lyskov et al. 2001; Wenzel et al. 2005). Only one study showed significant improvements in sleep quality during shielding of exposure, but this was attributed to several participants successfully circumventing the study's blinding (Leitgeb et al. 2008).

Although most of these provocation studies have taken place in the laboratory, several field interventions have been conducted, combining the advantages of a naturalistic setting with controlled exposure conditions, achieved by activating or deactivating an exposure equipment or by using shielding curtains to protect participants from pre-existing EMFs (Oftedal et al. 1995, 1999; Flodin et al. 2000; Heinrich et al. 2007; Danker-Hopfe et al. 2008; Leitgeb et al. 2008; Augner et al. 2009). None of these studies showed an increase of NSPS under active exposure conditions that did not exceed 0.43 V/m (Röösli et al. 2010a); unexpectedly, a significant effect on "calmness" was observed in one of the studies (Augner et al. 2009). Finally, evidence supported by meta-analytic evaluations indicates that individuals with self-reported IEI-EMF are unlikely to detect the presence or absence of active exposure (Röösli 2008; Röösli et al. 2010a).

Despite these findings, it would be a mistake to assume that participants simply fail to react during laboratory testing. In fact, quite the opposite is true. Numerous provocation studies have observed participants with IEI-EMF experiencing symptoms that can be quite severe, but these symptoms tend to occur when participants believe they are being exposed regardless of whether these beliefs are accurate (Wenzel et al. 2005; Wilén et al. 2006; Rubin et al. 2006; Eltiti et al. 2007; Oftedal et al. 2007; Hillert et al. 2008; Leitgeb et al. 2008; Furubayashi et al. 2009; Szemerszky et al. 2010; Wallace et al. 2010). This provides strong evidence that psychological factors may be sufficient to trigger the symptoms that typify IEI-EMF.

In summary, findings from a considerable number of studies do not support an effect of low-level RF- or ELF-EMFs on acute NSPS, and relevant cognitive and physiological reactions in people with IEI-EMF or asymptomatic controls. In the few studies where such an effect was observed, results were inconsistent or contradictory, and methodological shortcomings such as unblinding, differences in baseline demographics between cases and controls, errors due to multiple testing, and poor counterbalancing of the order in which exposure was presented seemed to be plausible explanations (Rubin et al. 2005; Röösli 2008; Rubin et al. 2010, 2011; Röösli et al. 2010a; Röösli and Hug 2011).

Yet despite this experimental literature, some questions remain unanswered. In particular, although these experiments provide good evidence that short-term exposure to EMF is not responsible for triggering NSPS, they are unable to test whether longer-term exposure, lasting weeks, months or years, might cause symptoms. Observational studies provide the only pragmatic way of answering this question.

Observational Studies

A growing body of epidemiological evidence has been published in the last decade on the possible association between residential exposure to RF-EMFs and report of NSPS in the general population (Röösli et al. 2010a; Baliatsas et al. 2012a). Most of these studies are of cross-sectional design, although there are several recent studies that have begun to report the findings from prospective cohorts (Frei et al. 2012; Mohler et al. 2012).

Main EMF sources of investigation were primarily mobile phones and mobile phone base stations. In contrast to experimental literature, only one observational study has investigated exposure–outcome associations in people with self-reported IEI-EMF (Röösli et al. 2010b), whereas research on the potential effects of ELF-EMF is limited and partly outdated (Li et al. 2002).

In all epidemiological studies, outcome assessment has been based on self-report instruments, typically assessing the presence of headache, sleep disturbances, dizziness, fatigue, concentration difficulties, and skin problems (Röösli et al. 2010a; Baliatsas et al. 2012a). In terms of exposure characterization, a major distinction between studies has been considered (Röösli 2008; Baliatsas et al. 2012a): some studies are assessing exposure levels based on self-reported measures (Sandström et al. 2001; Santini et al. 2003) that rely exclusively on the participants' subjective estimations of the magnitude of exposure to an EMF source; but other studies, usually referred to as actual exposure studies, are using objective methods such as distance between geocoded address and the nearest base station (Blettner et al. 2009), field strength spot measurements (Berg-Beckhoff et al. 2009), personal dosimeters (Thomas et al. 2008; Heinrich et al. 2011), and exposure prediction modeling (Mohler et al. 2010). Although the measured exposure levels differ across these studies, they are always considerably lower than the safety limits established by the International Commission of Non-Ionizing Radiation Protection (ICNIRP 1998, 2009). It is noteworthy that in most of the studies the exposure cut-off point for the highest exposure groups did not exceed 0.5 V/m (Röösli et al. 2010a).

Overall, the majority of existing studies do not suggest an effect of actual exposure on NSPS (Thomas et al. 2008; Berg-Beckhoff et al. 2009; Mohler et al. 2010, 2012; Heinrich et al. 2011; Frei et al. 2012). Relevant systematic reviews and meta-analyses consistently conclude that there is insufficient evidence or no evidence in favor of an actual exposure effect on acute or chronic symptoms in the everyday environment (Röösli et al. 2010a; Baliatsas et al. 2012a).

Despite the lack of statistical significance, it is striking that persons classified as "highly exposed" tend to report NSPS more frequently or severely than those in the "unexposed"/"low exposed" category (Baliatsas et al. 2012a). The lack of sufficient exposure contrast and/or the small prevalence of people with IEI-EMF could be potential explanations for missing a genuine exposure effect if one existed; however, it is also possible that spurious associations might be observed due to selection bias that cause an effect overestimation, positive-outcome bias in peer review literature and exposure misclassification (Röösli et al. 2010a; Baliatsas et al. 2012a).

One particularly clear finding from this literature is the tendency for studies that rely on self-reported exposure to find stronger symptomatic effects than the better quality studies that use objective measures of actual exposure, especially for neurological symptoms (Baliatsas et al. 2012a). Because people are not able to accurately self-estimate the magnitude of their exposure to EMF sources (Inyang et al. 2008; Vrijheid et al. 2009; Frei et al. 2010; Shum et al. 2011; Hutter et al. 2012), the association between symptoms and perceived exposure may partly be due to the existence of a nocebo effect, in which perceived exposure produces a self-fulfilling expectation of symptoms (Rubin et al. 2006; Röösli 2008). Other explanations, including recall and selection biases, may also contribute to this effect. Methodological quality therefore is an important determinant of the existence and strength of significant associations. Studies with a higher risk of bias, mainly regarding exposure assessment, sample selection, and adjustment for confounders, tend to report more significant symptomatic effects (Röösli et al. 2010a; Röösli and Hug 2011; Baliatsas et al. 2012a).

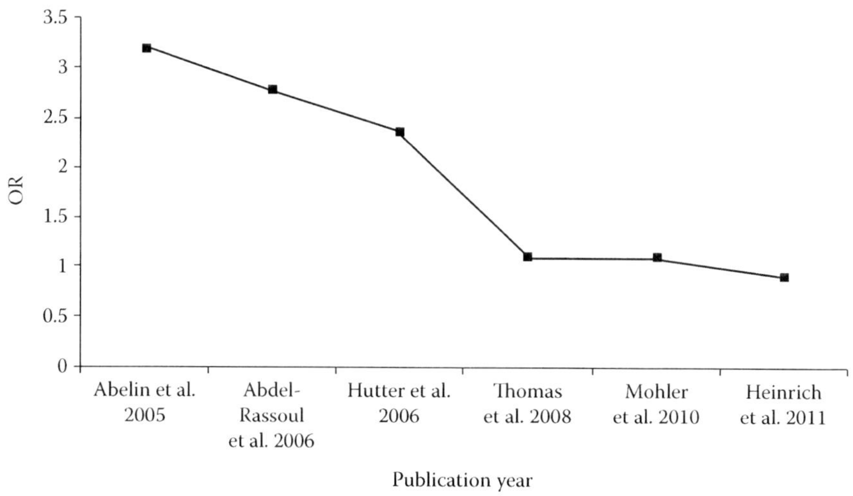

FIGURE 16.1
Odds ratios (when available) of actual residential EMF exposure on sleep problems versus year of publication of observational studies. Based on the included studies of the review of Baliatsas et al. (2012a).

More recent studies, with a lower probability of bias and that used more advanced exposure characterization methods such as personal dosimeters and exposure prediction modeling, have been less likely to demonstrate significant results (Thomas et al. 2008; Heinrich et al. 2010; Mohler et al. 2010, 2012; Frei et al. 2012). A representative example is illustrated in Figure 16.1, in relation to sleep problems as a health outcome.

The principal advantage of observational studies over provocation experiments lies in the fact that they assess long-term exposure and outcomes in large population samples. Considering the results of recent well-designed investigations that used more systematic and validated approaches in exposure characterization (Heinrich et al. 2011; Frei et al. 2012; Mohler et al. 2012), the evidence for absence of an effect of RF-EMFs on NSPS becomes more solid. However, besides the cross-sectional nature of most of these investigations, minimization of bias related to study design and use of valid exposure proxies constitute a major challenge.

Methodological Considerations for Future Research

Experimental and observational studies have different merits and limitations (Table 16.2); combining their findings is essential if we are to draw conclusions about the possibility of an exposure effect from EMFs on NSPS. Based on the current evidence from laboratory trials, observational studies, and field interventions, an association between low-level EMFs and NSPS and IEI-EMF cannot be supported. This conclusion has now been reached by numerous reviewers who have considered this literature (Table 16.3). For both types of study, methodological quality and the belief of being exposed seem to be key determinants of the reported associations. Nevertheless, before ruling out an

TABLE 16.2

Basic Methodological Merits and Limitations of the Two Main Study Designs in the Research Field of EMF and NSPS

Study Design	Advantages	Disadvantages
Experimental/provocation	Controlled exposure conditions	Investigation of only short-term exposure and effects, small sample size
Epidemiological/observational	Daily life/long-term exposure and effects	Increased risk for exposure misclassification and selection, information and confounding bias

TABLE 16.3

Reviews Published in the Last Decade on the Association between Nonionizing EMF Exposure and Nonspecific Symptoms and/or Relative Physiological Effects, in the General Population and/or EMF-Sensitive Subgroups

Reference	Type of Review and No. of Articles Included	Design and Population of the Reviewed Studies	Overall Conclusion
Levallois 2002	Narrative, 17	Observational and experimental/provocation studies, general population and EMF-sensitive samples	Insufficient evidence
Rubin et al. 2005	Systematic and meta-analysis, 31	Provocation studies, EMF-sensitive samples	No association
Seitz et al. 2005	Systematic, 13	Observational and experimental/provocation studies, general population and EMF-sensitive samples	Invalid and contradictory evidence
Röösli 2008	Systematic and meta-analysis, 16	Observational and experimental/provocation studies, general population and EMF-sensitive samples	Insufficient and contradictory evidence; most observational studies report an association, whereas most experimental studies do not
Kundi and Hutter 2009	Narrative, 10	Observational and experimental/provocation studies, general population and EMF-sensitive samples	Insufficient and contradictory evidence; several observational studies report an association that is weakly supported by experimental studies
Rubin et al. 2010	Systematic, 15	Provocation studies, EMF-sensitive samples	No association
Röösli et al. 2010a	Systematic and meta-analysis, 17	Observational and experimental/provocation studies, general population and EMF-sensitive samples	Insufficient evidence/no association
Rubin et al. 2011	Systematic, 29	Provocation studies, EMF-sensitive samples	No association
Röösli and Hug 2011	Narrative, 23	Observational and experimental/provocation studies, general population and EMF-sensitive samples	No association
Augner et al. 2012	Systematic and meta-analysis, 17	Provocation studies, general population and EMF-sensitive samples	No association
Baliatsas et al. 2012a	Systematic and meta-analysis, 20	Observational studies, general population	Insufficient evidence/no association

effect of actual exposure, several considerations should be taken into account in future research.

In terms of experimental studies, it is still possible that EMFs trigger acute symptoms in a small part of the population, but the existing studies failed to recruit patients with genuine sensitivity to the exposure and were therefore unable to demonstrate findings that could support a bioelectromagnetic mechanism (Rubin et al. 2011). The development of validated case-definition criteria for IEI-EMF could facilitate the identification of homogeneous groups possibly susceptible to EMFs (Baliatsas et al. 2012a). The limited number of people who participate in provocation studies is a related issue, especially if we consider the fact that the average number of participants is generally lower in studies that do not show any significant effect of exposure (Rubin et al. 2011). In addition, details regarding the performance of power calculations are usually missing in the literature (Rubin et al. 2010, 2011).

Laboratory experiments should be randomized and double blinded, preferably including an asymptomatic control group, that could be an important component not only for testing the exposure effects but also the adequacy of the blinding (Flodin et al. 2000). Properly designed provocation studies should control for background environmental exposures, other than EMF, such as noise, that could influence the assessed outcomes. Because it has been reported that the symptoms attributed to EMF can last for days after exposure (Röösli et al. 2004), controlling also for the so-called carry-over or hangover effects by assessing baseline symptom severity and using intervals between provocations is of importance to prevent symptom overreporting between the sessions. In addition, when studies investigate cognitive and/or physiological responses, participants' stress due to lack of familiarity with the experimental process should be minimized, with the use of a habituation session, before the main study (Rubin et al. 2011). Finally, it is remarkable that most researchers in this area typically fail to publish the protocols for their studies in advance, a practice that has been required in medical research for several years to prevent publication bias or the selective reporting of findings (De Angelis et al. 2004). It is high time that this practice is adopted more generally by colleagues working in this field.

Regarding observational studies, the identification of clear exposure contrasts is challenging and improvements in characterization of personal exposure should be a priority (Bolte and Eikelboom 2012). Given that even the most advanced assessment methods could misclassify exposure (Bolte et al. 2011), the combination of modeled exposure of base stations and high-voltage overhead power lines and exposimeter measurements in a prediction model seems to be a sound approach for epidemiological studies (Frei et al. 2009, 2010). Another challenge is to correct potential bias similar to the so-called healthy worker effect (see Chapter 3).

Characterization of ELF-EMF in relation to NSPS should be a main priority for future epidemiological studies because research on this association is, surprisingly, almost negligible. Assessment of mobile/wireless phone use based on self-reports should be gradually replaced by more objective measurements, such as operator data and billing records, because its validity as an actual exposure proxy is limited (Inyang et al. 2008; Shum et al. 2011; Hutter et al. 2012). In addition, more effort should be made to reduce sources of information/awareness and sample selection bias; these are strongly related to how exposure and outcome measurements are described to study participants (e.g., awareness beforehand of study objectives and exposure status) and stratification (e.g., whether more people with increased concerns related to EMF tend to participate, important differences between responders and nonresponders).

In all epidemiological studies on EMF to date, NSPS have been measured exclusively by self-reported questionnaires that vary considerably in terms of frequency and severity assessment (Baliatsas et al. 2012a). Because only a clinical examination can exclude medical disorders and determine whether a symptom has an organic explanation, it is still unknown whether the symptoms reported in epidemiological studies can appropriately be labeled as unexplained/nonspecific. Another limitation of using solely self-reported outcomes is the possibility of spurious exposure–outcome associations, induced by information and selection bias (Röösli et al. 2010a). Combining registry-based symptoms from general practices with EMF exposure data would constitute a useful alternative approach.

Moreover, there is limited epidemiological evidence on the association of actual exposure to EMFs with NSPS in children and adolescents as well as people with IEI-EMF (Heinrich et al. 2010, 2011; Röösli et al. 2010b); further research would contribute to a more comprehensive picture for the potential exposure effects in such population groups. Additional cohort studies and field experiments would add extra value to the existing state of the art due to their relative design advantages.

Finally, accounting for societal and individual factors is crucial for the comprehension of the impact of environmental factors (Page et al. 2006). Some studies have shown an association between psychological variables and NSPS in relation to EMFs (Rubin et al. 2006, 2008; Österberg et al. 2007; Landgrebe et al. 2008; Johansson et al. 2010; Szemerszky et al. 2010; Baliatsas et al. 2011); however, evidence on their mediating or moderating role is lacking. It is promising that more recent studies, in contrast with those before 2006, consider a sufficient number of confounders in the analyses and tend to investigate the effect of perceived exposure as well, together with some psychological components. Perceived exposure should be assessed as a separate conceptual entity, indicating the general belief of the magnitude of being exposed, and not as a proxy of actual exposure.

Conclusions

There is no convincing evidence to date to support a causal association between EMFs and nonspecific physical symptoms; however, there is still scope for better research, particularly in observational studies. Modern developments in exposure and outcome measurement and further investigation of the role of potentially explanatory factors will provide answers to some of the final remaining questions in this area. This is not only important for the epidemiological assessment of the electromagnetic exposure effects on public health but also for the management, care, and support of the people involved.

Acknowledgments

C.B. contributed this chapter during his PhD program and would like to thank Dr. Irene van Kamp and Prof. Erik Lebret for their support.

References

Abdel-Rassoul, G., El-Fateh, O.A., Salem, M.A., et al. 2006. Neurobehavioral effects among inhabitants around mobile phone base stations. *Neurotoxicology* 28: 434–40.

Abelin, T., Altpeter, E. and Röösli, M. 2005. Sleep disturbances in the vicinity of the short-wave broadcast transmitter Schwarzenburg. *Somnologie* 9: 203–9.

AGNIR. 2012. *Health Effects from Radiofrequency Electromagnetic Fields. Report of the Advisory Group on Non-ionising Radiation*. Doc HPA, London: Health Protection Agency, RCE-20.

Arnetz, B.B., Akerstedt, T., Hillert, L., et al. 2007. The effects of 884 MHz GSM wireless communication signals on self-reported symptoms and sleep (EEG). An experimental provocation study. *PIERS Online* 3: 1148–50.

Atkins, D., Best, D., Briss P.A., et al. 2004. Grading quality of evidence and strength of recommendations. *BMJ* 328: 1490.

Augner, C., Florian, M., Pauser, G., et al. 2009. GSM base stations: short-term effects on well-being. *Bioelectromagnetics* 30: 73–80.

Augner, C., Gnambs, T., Winker, R., et al. 2012. Acute effects of electromagnetic fields emitted by GSM mobile phones on subjective well-being and physiological reactions: a meta-analysis. *Sci Total Environ* 424: 11–15.

Baliatsas, C., Van Kamp, I., Bolte, J., et al. 2012a. Non-specific physical symptoms and electromagnetic field exposure in the general population: can we get more specific? A systematic review. *Environ Int* 41: 15–28.

Baliatsas, C., Van Kamp, I., Kelfkens, G., et al. 2011. Non-specific physical symptoms in relation to actual and perceived proximity to mobile phone base stations and powerlines. *BMC Public Health* 11: 421.

Baliatsas, C., Van Kamp, I., Lebret, E., et al. 2012b. Idiopathic environmental intolerance attributed to electromagnetic fields (IEI-EMF): a systematic review of identifying criteria. *BMC Public Health* 12: 643.

Barsky, A.J. and Borus J.F. 1999. Functional somatic syndromes. *Ann Intern Med* 130: 910–21.

Berg, M., Arnetz, B.B., Lidén, S., et al. 1992. Techno-stress. A psychophysiological study of employees with VDU-associated skin complaints. *J Occup Environ Med* 34: 698–701.

Berg-Beckhoff, G., Blettner, M., Kowall, B., et al. 2009. Mobile phone base stations and adverse health effects: phase 2 of a cross-sectional study with measured radio frequency electromagnetic fields. *Occup Environ Med* 66: 124–30.

Blettner, M., Schlehofer, B., Breckenkamp, J., et al. 2009. Mobile phone base stations and adverse health effects: phase 1 of a population-based, cross-sectional study in Germany. *Occup Environ Med* 66: 118–23.

Bolte, J., Van der Zande, G. and Kamer, J. 2011. Calibration and uncertainties in personal exposure measurements of radiofrequency electromagnetic fields. *Bioelectromagnetics* 32: 652–63.

Bolte, J.F.B. and Eikelboom, T. 2012. Personal radiofrequency electromagnetic field measurements in the Netherlands: exposure level and variability for everyday activities, times of day and types of area. *Environ Int* 48: 133–42.

Boyd, I., Rubin, G.J. and Wessely, S. 2012. Taking refuge from modernity: 21st century hermits. *J R Soc Med* 105: 523–5.

Carlsson, F., Karlson, B., Orbaek, P., et al. 2005. Prevalence of annoyance attributed to electrical equipment and smells in a Swedish population, and relationship with subjective health and daily functioning. *Public Health* 119: 568–77.

Christ, A. and Kuster, N. 2005. Differences in RF energy absorption in the heads of adults and children. *Bioelectromagnetics* 26(Suppl 7): S31–44.

Danker-Hopfe, H., Dorn, H., Sauter, C., et al. 2008. *Untersuchung der Schlafqualität bei Anwohnern einer Basisstation. Experimentelle Studie zur Objektivierung möglicher psychologischer und physiologischer Effekte unter häuslichen Bedingungen* [Study of Sleep Quality among People Living Near a Mobile Phone Base Station: Experimental Study of Possible Psychological and Physiological Effects in the Everyday Environment]. Abschlussbericht erstellt im Auftrag des Bundesamtes für Strahlenschutz [Final report commissioned by the Federal Office for Radiation Protection]. Berlin, Germany: Deutsches Mobilfunk Forschungsprogramm, p. 252. In German.

Das-Munshi, J., Rubin, G.J. and Wessely, S. 2006. Multiple chemical sensitivities: a systematic review of provocation studies. *J Allergy Clin Immunol* 118: 1257–64.

De Angelis, C., Drazen, J.M., Frizelle, F.A., et al. 2004. Clinical trial registration: a statement from the International Committee of Medical Journal Editors. *Ann Intern Med* 141: 477–8.

Eltiti, S., Wallace, D., Ridgewell, A., et al. 2007. Does short-term exposure to mobile phone base station signals increase symptoms in individuals who report sensitivity to electromagnetic fields? A double- blind randomized provocation study. *Environ Health Perspect* 115: 1603–8.

Engel, C.C. 2003. Somatization and multiple idiopathic physical symptoms: relationship to traumatic events and posttraumatic stress disorder. In: Schnurr, P.P. and Green, B.L., eds. *Trauma and Health: Physical Health Consequences of Exposure to Extreme Stress*. Washington, DC: American Psychological Association.

Flodin, U., Seneby, A. and Tegenfeldt, C. 2000. Provocation of electric hypersensitivity under everyday conditions. *Scand J Work Environ Health* 26: 93–8.

Frei, P., Mohler, E., Braun-Fahrländer, C., et al. 2012. Cohort study on the effects of everyday life radio frequency electromagnetic field exposure on non-specific symptoms and tinnitus. *Environ Int* 38: 29–36.

Frei, P., Mohler, E., Bürgi, A., et al. 2009. A prediction model for personal radio frequency electromagnetic field exposure. *Sci Total Environ* 408: 102–8.

Frei, P., Mohler, E., Bürgi, A., et al. 2010. Classification of personal exposure to radio frequency electromagnetic fields (RF-EMF) for epidemiological research: evaluation of different exposure assessment methods. *Environ Int* 36: 714–20.

Furubayashi, T., Ushiyama, A., Terao, Y., et al. 2009. Effects of short-term W-CDMA mobile phone base station exposure on women with or without mobile phone related symptoms. *Bioelectromagnetics* 30: 100–13.

Genuis, S.J. and Lipp, C.T. 2011. Electromagnetic hypersensitivity: fact or fiction? *Sci Total Environ* 414: 103–12.

Heinrich, S., Kühnlein, A., Thomas, S., et al. 2010. Association between exposure to radiofrequency electromagnetic field exposure and acute well-being in children and adolescents. *Environ Health* 9: 75.

Heinrich, S., Ossig, A., Schlittmeier, S., et al. 2007. Elektromagnetische Felder einer UMTS-Mobilfunkbasisstation und mögliche Auswirkungen auf die Befindlichkeit—eine experimentelle Felduntersuchung [Electromagnetic fields from a UMTS base station and possible effects on well-being—an experimental field study]. *Umweltmed Forsch Prax* 12: 171–80. In German.

Heinrich, S., Thomas, S., Praml, G., et al. 2011. The impact of exposure to radio frequency electromagnetic fields on chronic wellbeing in young people—a cross sectional study based on personal dosimetry. *Environ Int* 37: 26–30.

Hietanen, M., Hamalainen, A.M. and Husman, T. 2002. Hypersensitivity symptoms associated with exposure to cellular telephones: no causal link. *Bioelectromagnetics* 23: 264–70.

Hillert, L., Akerstedt, T., Lowden, A., et al. 2008. The effects of 884 MHz GSM wireless communication signals on headache and other symptoms: an experimental provocation study. *Bioelectromagnetics* 29: 185–96.

Hillert, L., Berglind, N., Arnetz, B.B., et al. 2002. Prevalence of self-reported hypersensitivity to electric or magnetic fields in a population-based questionnaire survey. *Scand J Work Environ Health* 28: 33–41.

Hillert, L., Hedman, B.K., Söderman, E., et al. 1999. Hypersensitivity to electricity: working definition and additional characterization of the syndrome. *J Psychosom Res* 47: 429–38.

Hillert, L., Leitgeb, N. and Meara, J. 2006. Working group report. In: Mild, K.H., Repacholi, M., Van Deventer, E., et al., eds. *Electromagnetic Hypersensitivity*. Proceedings: International Workshop on EMF Hypersensitivity. Prague: WHO, pp. 15–26.

Hutter, H.P., Ehrenhöfer, L., Freuis, E., et al. 2012. Poor-to-moderate agreement between self and proxy interviews of mobile phone use. *Bioelectromagnetics* 33: 561–7.

Hutter, H.P., Moshammer, H., Wallner, P., et al. 2006. Subjective symptoms, sleeping problems, and cognitive performance in subjects living near mobile phone base stations. *Occup Environ Med* 63: 307–13.

International Commission on Non-Ionizing Radiation Protection. 1998. Guidelines for limiting exposure to time-varying electric, magnetic, and electromagnetic fields (up to 300 GHz). *Health Phys* 74: 494–522.

International Commission on Non-Ionizing Radiation Protection. 2009. ICNIRP statement on the guidelines for limiting exposure to time-varying electric, magnetic, and electromagnetic fields (up to 300 GHz). *Health Phys* 97: 257–8.

Inyang, I., Benke, G., McKenzie, R., et al. 2008. Comparison of measuring instruments for radiofrequency radiation from mobile telephones in epidemiological studies: implications for exposure assessment. *J Expo Sci Environ Epidemiol* 18: 134–41.

Iversen, A., Chalder, T. and Wessely, S. 2007. Gulf war illness: lessons from medically unexplained symptoms. *Clin Psychol Rev* 27: 842–54.

Johansson, A., Nordin, S., Heiden, M., et al. 2010. Symptoms, personality traits, and stress in people with mobile phone-related symptoms and electromagnetic hypersensitivity. *J Psychosom Res* 68: 37–45.

Kim, D.W., Lee, J.H., Ji, H.C., et al. 2008. Physiological effects of RF exposure on hypersensitive people by a cell phone. *30th Annual International IEEE EMBS Conference* 20–25 August, New Jersey, pp. 2322–5.

Kirmayer, L.J., Groleau, D., Looper, K.J., Dominicé, M. 2004. Explaining medically unexplained symptoms. *Can J Psychiatr* 49: 663–72.

Kroenke, K. 2003. Patients presenting with somatic complaints: epidemiology, psychiatric comorbidity and management. *Int J Methods Psychiatr Res* 12: 34–43.

Kroenke, K., Arrington, M.E. and Mangelsdorff, A.D. 1990. The prevalence of symptoms in medical outpatients and the adequacy of therapy. *Arch Intern Med* 150: 1685–89.

Kundi, M., and Hutter, H.P. 2009. Mobile phone base stations—effects on wellbeing and health. *Pathophysiology* 16: 123–35.

Kwon, M.K., Choi, J.Y., Kim, S.K., et al. 2012. Effects of radiation emitted by WCDMA mobile phones on electromagnetic hypersensitive subjects. *Environ Health* 11: 69.

Kwon, M.S., Koivisto, M., Laine, M., et al. 2008. Perception of the electromagnetic field emitted by a mobile phone. *Bioelectromagnetics* 29: 154–9.

Landgrebe, M., Frick, U., Hauser, S., et al. 2008. Cognitive and neurobiological alterations in electromagnetic hypersensitive patients: results of a case-control study. *Psychol Med* 38: 1781–91.

Leitgeb, N. and Schröttner, J. 2003. Electrosensibility and electromagnetic hypersensitivity. *Bioelectromagnetics* 24: 387–94.

Leitgeb, N., Schröttner, J., Cech, R., et al. 2008. EMF-protection sleep study near mobile phone base stations. *Somnologie* 12: 234–43.

Levallois, P. 2002. Hypersensitivity of human subjects to environmental electric and magnetic field exposure: a review of the literature. *Environ Health Perspect* 110: 613–18.

Levallois, P., Neutra, R., Lee, G., et al. 2002. Study of self-reported hypersensitivity to electromagnetic fields in California. *Environ Health Perspect* 110: 619–23.

Li, C.Y., Chen, P.C., Sung, F.C., et al. 2002. Residential exposure to power frequency magnetic field and sleep disorders among women in an urban community of Northern Taiwan. *Sleep* 25: 428–32.

Lyskov, E., Sandstrom, M. and Mild, K.H. 2001. Provocation study of persons with perceived electrical hypersensitivity and controls using magnetic field exposure and recording of electrophysiological characteristics. *Bioelectromagnetics* 22: 457–62.

Mohler, E., Frei, P., Fahrländer, C.B., et al. 2010. Effects of everyday radiofrequency electromagnetic-field exposure on sleep quality: a cross-sectional study. *Radiat Res* 174: 347–56.

Mohler, E., Frei, P., Fröhlich, J., et al. 2012. Exposure to radiofrequency electromagnetic fields and sleep quality: a prospective cohort study. *PLoS One* 7: e37455.

Mueller, C.H. and Schierz, C. 2004. Project NEMESIS: double-blind study on effects of 50Hz EMF on sleep quality, physiological parameters and field perception in people suffering from electrical hypersensitivity. In: Mild, K.H., Repacholi, M., Van Deventer, E., et al., eds. *Electromagnetic Hypersensitivity*. Proceedings of the International Workshop on EMF Hypersensitivity. Prague: WHO, pp. 107–21.

Nieto-Hernandez, R., Williams, J., Cleare, A.J., et al. 2011. Can exposure to a terrestrial trunked radio (TETRA)-like signal cause symptoms? A randomized double-blind provocation study. *Occup Environ Med* 68: 339–44.

Oftedal, G., Nyvang, A. and Moen, B.E. 1999. Long-term effects on symptoms by reducing electric fields from visual display units. *Scand J Work Environ Health* 25: 415–21.

Oftedal, G., Straume, A., Johnsson, A., et al. 2007. Mobile phone headache: a double-blind, sham-controlled provocation study. *Cephalalgia* 27: 447–55.

Oftedal, G., Vistnes, A.I. and Rygge, K. 1995. Skin symptoms after the reduction of electric-fields from visual-display units. *Scand J Work Environ Health* 21: 335–44.

Österberg, K., Persson, R., Karlson, B., et al. 2007. Personality, mental distress, and subjective health complaints among persons with environmental annoyance. *Hum Exp Toxicol* 26: 231–41.

Page, L.A., Petrie, K.J. and Wessely, S.C. 2006. Page Psychological responses to environmental incidents: a review and a proposed typology. *J Psychosom Res* 60: 413–22.

Prockop, L.D. and Chichkova, R.I. 2007. Carbon monoxide intoxication: an updated review. *J Neurol Sci* 262: 122–30.

Rea, W.J., Pan, Y., Fenyves, E.J., et al. 1991. Electromagnetic field sensitivity. *J Bioelectricity* 10: 241–56.

Regel, S.J., Negovetic, S., Röösli, M., et al. 2006. UMTS base stationlike exposure, well-being, and cognitive performance. *Environ Health Perspect* 114: 1270–5.

Reid, S., Wessely, S., Crayford, T., et al. 2001. Medically unexplained symptoms in frequent attenders of secondary health care: retrospective cohort study. *BMJ* 322: 767.

Riddervold, I.S., Pedersen, G.F., Andersen, N.T., et al. 2008. Cognitive function and symptoms in adults and adolescents in relation to rf radiation from UMTS base stations. *Bioelectromagnetics* 29: 257–67.

Röösli, M. 2008. Radiofrequency electromagnetic field exposure and non-specific symptoms of ill health: a systematic review. *Environ Res* 107: 277–87.

Röösli, M., Frei, P., Mohler, E., et al. 2010a. Systematic review on the health effects of exposure to radiofrequency electromagnetic fields from mobile phone base stations. *Bull World Health Organ* 88: 887–96.

Röösli, M. and Hug, K. 2011. Wireless communication fields and non-specific symptoms of ill health: a literature review. *Wien Med Wochenschr* 9–10: 240–50.

Röösli, M., Mohler, E. and Frei, P. 2010b. Sense and sensibility in the context of radiofrequency electromagnetic field exposure. *Comptes Rendus Physique* 11: 576–84.

Röösli, M., Moser, M., Baldinini, Y., et al. 2004. Symptoms of ill health ascribed to electromagnetic field exposure—a questionnaire survey. *Int J Hyg Environ Health* 207: 141–50.

Rubin, G.J., Cleare, A.J. and Wessely, S. 2008. Psychological factors associated with self-reported sensitivity to mobile phones. *J Psychosom Res* 64: 1–9.

Rubin, G.J., Das Munshi, J. and Wessely, S. 2005. Electromagnetic hypersensitivity: a systematic review of provocation studies. *Psychosom Med* 67: 224–32.

Rubin, G.J., Das Munshi, J. and Wessely, S. 2006. A systematic review of treatments for electromagnetic hypersensitivity. *Psychother Psychosom* 75: 12–18.

Rubin, G.J., Hahn, G., Everitt, B., et al. 2006. Are some people sensitive to mobile phone signals? A within participants, double-blind, randomised provocation study. *Br Med J* 332: 886–9.

Rubin, G.J., Hillert, L., Nieto-Hernandez, R., et al. 2011. Do people with idiopathic environmental intolerance attributed to electromagnetic fields display physiological effects when exposed to electromagnetic fields? A systematic review of provocation studies. *Bioelectromagnetics* 32: 593–609.

Rubin, G.J., Nieto-Hernandez, R. and Wessely, S. 2010. Idiopathic environmental intolerance attributed to electromagnetic fields (formerly 'electromagnetic hypersensitivity'): an updated systematic review of provocation studies. *Bioelectromagnetics* 31: 1–11.

Sandström, M., Wilen, J., Oftedal, G., et al. 2001. Mobile phone use and subjective symptoms. Comparison of symptoms experienced by users of analogue and digital mobile phones. *Occup Med* 51: 25–35.

Santini, R., Santini, P., Danze, J.M., et al. 2003. [Symptoms experienced by people in vicinity of base stations: II/Incidences of age, duration of exposure, location of subjects in relation to the antennas and other electromagnetic factors]. *Pathol Biol* 51: 412–15.

Schreier, N., Huss, A. and Röösli, M. 2006. The prevalence of symptoms attributed to electromagnetic field exposure: a cross-sectional representative survey in Switzerland. *Soz Praventivmed* 51: 202–9.

Schröttner, J. and Leitgeb, N. 2008. Sensitivity to electricity—temporal changes in Austria. *BMC Public Health* 8: 310.

Seitz, H., Stinner, D., Eikmann, T., et al. 2005. Electromagnetic hypersensitivity (EHS) and subjective health complaints associated with electromagnetic fields of mobile phone communication—a literature review published between 2000 and 2004. *Sci Total Environ* 349: 45–55.

Shum, M., Kelsh, M.A., Sheppard, A.R., et al. 2011. An evaluation of self-reported mobile phone use compared to billing records among a group of engineers and scientists. *Bioelectromagnetics* 32: 37–48.

Sjöberg, P. and Hamnerius, Y. 1995. Study of provoked hypersensitivity reactions from a VDU. In: Katjalainen, K.B., ed. *Electromagnetic Hypersensitivity*. 2nd Copenhagen Conference. Denmark: Danish Association for Electromagnetic Hypersensitivity, Randers, pp.101–10.

Szemerszky, R., Köteles, F., Lihi, R., et al. 2010. Polluted places of polluted minds? An experimental sham-exposure study on background psychological factors of symptom formation in "idiopathic environmental intolerance attributed to electromagnetic fields." *Int J Hyg Environ Health* 213: 387–94.

Thomas, S., Kühnlein, A., Heinrich, S., et al. 2008. Personal exposure to mobile phone frequencies and well-being in adults—across- sectional study based on dosimetry. *Bioelectromagnetics* 29: 463–70.

Trimmel, M. and Schweiger, E. 1998. Effects of an ELF (50 Hz, 1 mT) electromagnetic field (EMF) on concentration in visual attention, perception and memory including effects of EMF sensitivity. *Toxicol Lett* 96–97: 377–82.

Tseng, M.M., Lin, Y.P. and Cheng, T.J. 2011. Prevalence and psychiatric comorbidity of self-reported electromagnetic field sensitivity in Taiwan: a population-based study. *J Formos Med Assoc* 110: 634–41.

Van Den Berg, B., Grievink, L., Yzermans, J., et al. 2005. Medically unexplained physical symptoms in the aftermath of disasters. *Epidemiol Rev* 27: 92–106.

Vrijheid, M., Armstrong, B.K., Bédard, D., et al. 2009. Recall bias in the assessment of exposure to mobile phones. *J Expo Sci Environ Epidemiol* 19: 369–81.

Wallace, D., Eltiti, C., Ridgewell, A., et al. 2010. Do TETRA (Airwave) base station signals have a short-term impact on health and well-being? A randomized double-blind provocation study. *Environ Health Perspect* 118: 735–41.

Wennberg, A., Franzén, O. and Paulsson, L.E. 1994. Electromagnetic field provocations of subjects with 'electric hypersensitivity'. In: Simunic, D., ed. *COST Meeting on Electromagnetic Hypersensitivity*. Graz, Austria: COST, pp. 133–9.

Wenzel, F., Reissenweber, J. and David, E. 2005. Cutaneous microcirculation is not altered by a weak 50 Hz magnetic field. *Biomed Tech* 50: 14–18.

Wessely, S., Nimnuan, C. and Sharpe, M. 1999. Functional somatic syndromes: one or many? *Lancet* 354: 936–9.

Wilén, J., Johansson, A., Kalezic, N., et al. 2006. Psychophysiological tests and provocation of subjects with mobile phone related symptoms. *Bioelectromagnetics* 27: 204–14.

WHO. 2005. Electromagnetic fields and public health. Fact Sheet No. 296. World Health Organization. Available from: http://www.emfandhealth.com/WHO_EMSensitivity.pdf

Zwamborn, A.P.M., Vossen, S.H.J.A., Van Leersum, B.J.A.M., et al. 2003. *Effects of Global Communication System Radio-Frequency Felds of Well Being and Cognitive Functions of Human Subjects with and Without Subjective Symptoms*. The Hague, Netherlands: Netherlands Organisation for Applied Scientific Research (TNO).

Part III

The Broader Perspective

17

How Can Future Epidemiology and Laboratory Studies Complement Each Other?

Frank S. Barnes and Ben Greenebaum

CONTENTS

Introduction

There is substantial controversy about whether low levels of intermittent exposures to both electric and magnetic fields, particularly those at power frequencies (50 and 60 Hz) and radio frequencies (RF) from, for example, cell phones and base stations, can lead to health effects such as child leukemia and brain tumors. As discussed in other chapters in this book, there have been a large number of epidemiological studies of the possibility that these fields can lead to these or other health effects. Some of these studies indicate a correlation between exposures and various health effects that is just above the statistically significant level, but others do not.

Although epidemiology studies are primarily aimed at illuminating the possible link between field exposure and human health, the primary issues of public concern, laboratory studies often have the additional goals of finding ways to promote health through medical diagnosis and treatment and of increasing basic biological knowledge. There are three distinct questions to ask concerning laboratory–epidemiology interactions, to further our knowledge:

1. What information in the literature from lab studies could be useful for shaping future epidemiological research?
2. What information from lab studies would, if known, make future epidemiological research more effective?
3. What information from current or future epidemiological research would be useful for shaping future lab research?

The objective of this chapter is to suggest several possible ways in which future epidemiological studies might be conducted so that the conditions that lead to possible health effects are better understood. In addition, we would hope that these future studies will help define experiments that may be conducted in the lab or with animals that will help us understand how to go from the physics of field interactions through the chemistry and biological effects to health effects. Throughout this chapter, we use "lab studies" as a generic term to mean any type of nonepidemiological research, including engineering work, theoretical studies, and computer simulations.

As basic background, we assume that biological systems are complex and that the application of particular electric, magnetic, or RF fields may lead to different effects on different people or test organisms or on the same individual at different times. Although unidentified differences in exposure or other experimental conditions undoubtedly explain part of the apparently inconsistent lab results in the literature, differences in the status of the exposed biological system explain others. For example, Nindl et al. (1997) find that previous exposure of the medium and whether cells are in log growth affects thymidine uptake. Second, we assume that biological systems contain a large number of feedback and repair systems so that if a perturbation causes changes in such things as growth rates, free radical concentrations, or DNA functions, these changes may be compensated for or corrected, but may also be amplified (Barnes 2007). Thus, observed changes in the biology many times will not lead to an observable health effect. However, there may well be times that these fields coupled with other stresses or random fluctuations may lead to changes that are amplified enough so as to lead to observable health effects. Short-term exposures to cell phone radiation have been shown to lead to changes in the electroencephalogram (EEG) (Croft et al. 2008), but it is unlikely that these changes in the EEG lead to long-term health effects. Another example is that blood pressure and heart rate go up during exercise and can lead to positive health effects. However, extended high blood pressure and rapid heart rates may lead to undesirable health effects.

We look at low-level fields as possible signaling mechanisms that can be amplified by processes in the body that lead that changes where the necessary energy comes from metabolic processes. Such signals can lead to either positive, negative, or no health effects, depending on the state of the individual (e.g., Nindl et al. 1997). The initial signaling events due to the electric or magnetic fields are likely to occur rapidly while the fields are being applied, resulting in changes in, for example, molecular configurations, ion concentrations in a local region, or energy states that may be more lasting. These may then be coupled to changes in chemical reactions that may be amplified and modify biological processes (Barnes 2007). For example, the fields can affect the formation of an enzymatic catalyst that initiates a process leading to cell growth or the modification of other molecules or proteins (Paunesku and Woloschak 2007). Furthermore, 10 MHz magnetic fields of 50–250 μT with a static background field of 45 μT can lead to changes in free radical concentrations of hydrogen peroxide (H_2O_2) and to both positive and negative changes in the growth HD1080 cell (C. F. Martino, personal communication 2013). These changes, taking place in a dynamic environment that includes other naturally occurring changes, may affect feedback and feed-forward loops. Only a small fraction of these changes might lead to health effects, some of which may not show up for years.

In this chapter, we outline how information derived from lab and animal studies might be used to help define parameters for the exposed populations and controls that will lead to a better understanding of which conditions lead to health effects and which do not. We outline several possible reasons why this might be so, and we make some suggestions on ways that future epidemiological studies and future laboratory, engineering, theoretical, and computer simulation studies might complement each other to be more effective in identifying conditions that lead to biologically significant changes. Three themes emerge prominently from our discussions:

1. Cell phones, in addition to being a source of field exposures that are to be studied, should be considered as tools that, with the help of apps yet to be developed and possibly also with external sensors, may be able to record and store exposure parameters, call data, and some aspects of subjects' physiological states.
2. In addition to field intensity and frequency, important exposure parameters may include pulsing patterns, timing of intermittent exposures, and exposure to additional fields, particularly exposure to a steady magnetic field from the earth or other sources.
3. The subjects' physiological states, including not only stress from other external factors but also their health and any psychological stress, may also be important.

Physical Models for the Effects of Electric and Magnetic Fields on Biological Systems

A key issue is how weak electric and magnetic fields can affect a biological system in the presence of large thermal and other environmental fluctuations. Many models have been proposed that predict exposure conditions that can be expected to be sensed by biological systems. For example, it has been shown that the lifetime for conversion between the singlet and triplet states of radial pairs are a function of the steady-state and low-frequency background magnetic fields (Brocklehurst and McLaughlan 1996; Woodward et al. 2002; Timmel and Henbest 2004). This change in turn may lead to changes in free radical concentration and to changes in the growth rates of some cancers and other cells (Martino et al. 2010). It currently appears that this mechanism may well be behind many other observed field effects from static or slowly varying fields to high frequencies, because the time scale for free radical conversion is very much shorter than even the period of RF oscillations. Direct epidemiological tests of the free radical hypothesis are not likely, but if their characteristic downstream products could be identified, they could possibly be used as end points in epidemiological or lab studies. Radical oxygen species (ROS) are often discussed in this context; although they are very short lived, their effects or reaction products sometimes are not. For example, H_2O_2 concentrations could be measured. One active area of research is the role of cryptochromes in animal navigation and other processes. These light-sensitive dyes appear to have significant radical pair activity and have been implicated in navigation mechanisms in birds and possibly other organisms (Phillips et al. 2010; Maeda et al. 2012). They also play important photoactive roles in plants; Maeda et al. (2012) have shown that they are sensitive to very weak magnetic fields in *Arabidopsis*.

The earth's static magnetic field (direct current [DC] or SMF) seems to have an important influence on biological systems of different complexity. Exposure to reduced field conditions (SMFs <23 μT) for times as short as 20 min have been shown to affect the kinetics of the conformation of chromatin in human fibroblasts (VH-10) and lymphocytes (Belyaev et al. 1997). Also, this kind of exposure has been shown to affect growth rates of cancer (fibrosarcoma HT1080, colorectal HCT116) (Martino et al. 2010; Martino 2011) and primary human umbilical vein endothelial cell (HUVEC; Martino 2011) and bovine pulmonary artery endothelial cell (PAEC) cell lines (Martino et al. 2010; Martino 2011) for exposures as short as 24 hr. Prolonged exposure to reduced SMF environments has been shown to affect regeneration of planarians (Novikov et al. 2007, 2008) and to induce developmental abnormalities in the Japanese newt (Asashima et al. 1991). A large body of experimental data demonstrate in most cases suppression of growth processes, cell division and differentiation, as well as significant changes at the cellular and subcellular level and altered ionic balance, enzyme activities, and several metabolic processes (Belyavskaya 2001; Volpe 2003; Belyavskaya 2004; Galland and Pazur 2005). Extensive effects have also been observed in warm-blooded animals exposed to these environments. In chicks, it was found that when incubated in a reduced SMF environment, significantly more exogenous noradrenalin was needed for memory consolidation than in the control (Xiao et al. 2009). Similarly, day-old chicks hatching from an altered SMF environment displayed long-term memory impairment in a one-trial passive avoidance task (Wang et al. 2003).

A reduced static field can lower nociception in mice on a single application (Choleris et al. 2002) and increase it after multiple exposures (Prato et al. 2005) and can cause marked disturbances of blood and lymph circulation in the myocardium (Nepomnyashchikh et al. 1997), as well as induction of early embryogenesis in vitro and effects on their reproduction capacity in vivo (Fesenko et al. 2010). In combination with weak alternating (AC) fields, this reduced static field effect seems to depend on the product of field strength and frequency (Prato et al. 2011). Tipping et al. (1999) found that *Drosophila melanogaster* larvae reared in weak static fields expressed several genes differently, as well as changing expression on exposure to weak extremely low-frequency (ELF) fields. This insect has been shown to become completely amnesiac after 10 generations in a reduced background magnetic field environment (Zhang et al. 2004); and in similar conditions, it has been shown in our laboratory how such exposures affect susceptibility to ionizing radiation on the same organism (Portelli et al. 2012).

Interestingly, this same exposure has been shown to modulate H_2O_2 concentrations in human pancreatic cancer cells (AsPC-1) and HT1080 cells (Martino and Castello 2011). Because free radical concentrations are thought to be significant in cancer growth, a hypothesis can be proposed that magnetic fields outside the range of the earth's magnetic field of 23–65 μT may either increase or decrease the incidence of a given type of cancer or its growth rate. Epidemiologists could consider whether a sufficient group of subjects exists to compare those routinely subject to magnetic fields outside the normal 23-65 μT range with subjects only exposed to the usual fields.

RF fields in the 200 kHz region are being used to treat cancer, with exposures that are in the range of a few hundred volts per meter (Fonkem and Wong 2012). In addition, transitions between the singlet and triplet states of free radicals are expected to be modified by exposures to magnetic fields in the range of 1–10 MHz for static magnetic fields of 50 μT. Ashraf et al. (2008) have data at field strengths of volts per meter at 900 MHz on the change in direction and speed of neutrophils migrating up a concentration gradient of C-amp. Similar effects have been seen at 1800 MHz by Tiwari and Singh (2012).

A wide range of data from a variety of laboratories seem to indicate a resonance-like interaction between alternating and SMFs, with certain combinations of SMF intensity, alternating field intensities, and frequencies being especially effective (for a review, see Liboff 2007). Several theories have tried to explain this resonance-like behavior in the data, but none so far are without significant flaws (Binhi 2002). Nevertheless, many of the experiments appear sound, and the phenomena would have to be explained (or explained away) by any eventual mechanism. Any epidemiological test of these combinations would require exposure data beyond current capabilities.

Another known mechanism of interaction is the direct force exerted by the earth's magnetic field on internal strings of magnetite grains in magnetotactic bacteria that use the direction of the field to migrate into a preferred environment (Frankel and Blakemore 1989). Magnetic particles are known to be present in other organisms, such as fish, but their function and mechanism of interaction are not clear (Eder et al. 2012). It has long been known that both ambient light and the earth's magnetic field are used by birds to navigate in migration, but the mechanism is complicated and not fully understood. Wiltschko et al. (2010) review the magnetic field's apparent role in each of two mechanisms: one mechanism apparently interacting with light-sensing mechanisms in the visual system and the other mechanism possibly with high magnetic permeability particles in the beak. The first mechanism is thought to be related to the radical pair hypothesis mentioned above; and the second, to more direct magnetic interactions.

Basis for Biological Responses to Exposures

Growth processes are slow, and the body has many repair mechanisms that usually correct undesirable perturbations. Data by Litovitz et al. (1994, 1997) show that a coherent signal for >1 s and repetitive pulses for minutes or hours are required for some biological systems to modify the growth of chick embryos. Here, noise and random signals are effectively ignored by these systems. This makes sense in that we would not want a single perturbation such as exposure to fields from lightning at long distance to cause health problems. Thus, it is likely that repetitive exposures will be needed to modify cancer growth processes or to trigger a condition that does not get repaired. A new epidemiology study might include acquiring information on exposure conditions that meet some of these conditions: repeated exposures to periodical fields applied for extended periods in addition to or instead of cell phones, before the identification of the cancer. For example, someone might be working in the vicinity of a radio or TV transmitter, an RF antitheft device, or an RF sealing machine. These results could be compared with, for example, the average exposure value or wire line configurations to see whether they provide larger odds ratios for the incidence of cancers with these exposure conditions than the other metrics. For example, the type of signal transmission, such as Global System for Mobile Communications (GSM) or code division multiple access (CDMA), can be recorded along with the length of the call and the field strength or power transmitted by a smart phone as discussed below. The periodicity for pulses in a GSM signal might be checked to see whether it has an integer relation to the cycle time of Ca^{2+} in neutrophils or heart rates. Phone call rates might be checked to see whether they might have periodicity that could be related to a cell growth cycle. In addition, for example, the number of calls per day or per week can be recorded and compared with various biological processes.

Some data indicate that 50 Hz magnetic fields at low levels and electromagnetic fields at cell phone frequencies can induce biological responses similar to other low-level stress (Litovitz et al. 1994; Han et al. 1998; Korneva et al. 1999; Di Carlo et al. 2000; Shallom et al. 2002). These stress responses, such as activation of the immune system or heat-shock proteins (HSPs), can lead to either positive or negative health effects. For example, HSPs can protect DNA and other molecules from damage by elevated temperatures; however, they can also protect cancer-generating DNA from thermal destruction during therapy. The possibility exists that infrequent exposure to cell phone frequencies might stimulate the body's defense system and position the immune system to attack conditions leading to formation of cancer cells, whereas overstimulation of the immune system could leave insufficient reserves to respond to new challenges. Thus, a future epidemiology study might include questions about other sources of stress that might occur in time periods associated with heavy cell phone use. For example, some cell phone users are also exposed to high temperatures, heavy exercise, or difficult personal problems such as financial or relationship problems. These subjects might be compared with people who use cell phones but experience relatively lower levels of other sources of stress. At least in principle, measurements of stress-related blood factors might be correlated with the epidemiological data, although research has only found elevated levels in lab animals for intense RF exposures (Black and Heynick 2003). Tokalov and Gutzeit (2004) have shown increases in HL60 HSPs from acute myeloid leukemia cells for 50 Hz fields, with a maximum increase for magnetic fields in the range of 40–80 μT. Robertson et al. (2007) have reviewed results for the role of magnetic fields in cytoprotection and repair. Zmyslony et al. (2004) have shown that for lymphocytes exposed to 40 μT, 50 Hz magnetic fields directed along the earth's static magnetic field that there was a decrease of fluorescence in relation to nonexposed samples, indicating a reduction in reactive oxygen.

Related to the forgoing possibilities would be epidemiological studies that divide subjects according to the presence or absence of pre-existing health conditions that might weaken their bodies' defense systems. Elderly people are known to recover from injuries more slowly than younger people. People who have diseases that tax or are a result of problems with the body's immune system, including arthritis and other autoimmune diseases or severe allergies, might be compared with healthy control groups for differences in the incidence of cancer or for changes in their immune system conditions. Fields have inconsistently been found to affect various aspects of the immune system, although not necessarily at environmental intensities (Tremblay et al. 1996; Black and Heynick 2003). Hence seeing whether less-healthy subgroups of epidemiological study populations are more sensitive to field exposure could be useful, at least in principle, although accurately classifying subjects and getting sufficient statistical precision could be difficult.

Another example related to the way the state of a biological system responds to an electromagnetic stimulus comes from the work of Blackman et al. (1993). His neuroblastoma cells required stimulation with nerve growth factor in vitro to produce neurite outgrowth and were most responsive to electromagnetic fields when stimulated with the amount of nerve growth factor (NGF) that gave approximately half of the maximum response to the growth factor. Fields no longer influenced the level of outgrowth with an excess of growth factor. Further exploration of what he has called the "hairy edge" phenomenon (C. F. Blackman, personal communication, 1992) might lead to explanations of sometimes conflicting laboratory results and could allow more intelligent grouping of epidemiological samples in analysis. This thinking may be further supported by the clinical success with using pulsed electromagnetic fields to begin the healing process in nonunion bone fractures, but evidence for benefit with fresh fractures is less convincing (Pilla 2007).

Because the body is an electrochemical system, electric fields can also perturb the system. Environmental electric fields are hard to measure because conducting objects in the room distort the fields. Because the body is a relatively good conductor, changes in position, including sitting or standing, change the field distribution both in the room and also inside the body. At 50 or 60 Hz, electric fields do not penetrate much below the surface of the skin. However, the body has sensory systems located close to the surface, and it is not clear whether repetitive stimulations at field strengths <5 kV/m are unimportant. Typical electric fields in residential and business environments range from 1 to 100 V/m;

under power lines, they range up to approximately 10 kV/m. However, as noted, these field strengths are attenuated extremely strongly at the surface of the body by a factor of as much as 10^5–10^6 (Kaune and Gillis 1981). Ambient 50/60 Hz magnetic fields induce very weak electric fields and currents inside the body, <200 mV/m at 50 Hz when exposed to a uniform 1 μT field (Dawson et al. 1997; Dawson and Stuchly 1998; Kavet et al. 2001). Both alternating and static electric fields of a few tenths of a volt per meter have been shown to affect cell growth patterns in vitro (Pullar 2011). In addition, both directly applied and magnetically induced electric fields, applied for as little as 20 min a day, have been shown to induce the repair of broken bones for exposures at typical field strengths considerably higher than the environmental fields (Pilla 2007). Because in epidemiological studies both the nonexposed and the exposed populations are likely to be exposed to static electric fields on the order of 100 V/m, the study might look to see whether there is a difference between populations at this level and those that are regularly exposed to fields >1 kV/m for periods in excess of 5 min. Most ELF epidemiology studies have been concerned with the magnetic field exposure at 50 or 60 Hz. Adding information concerning electric field exposure to the analysis might prove interesting.

It is well known that the human body is an adaptive system and that our response to the same stimulations at different times leads to different responses. Allergic reactions are classic cases of repeated exposures leading to either stronger or weaker reactions than a single exposure. Experiments on various electromagnetic field effects have also shown both increases and decreases in similar measurements, as noted above. Nonelectromagnetic environmental conditions such as light can also influence the response to fields (e.g., Prato et al. 2000). The work of Litovitz and others at Catholic University indicates the coherence of the applied fields over periods of several minutes is important in how much HSP was generated in response to the fields (Di Carlo et al. 2000, and references therein). This group also found that superimposing noise on an ELF or RF field decreased the HSP response (Litovitz et al. 1994, 1997, and references therein). Han et al. (1998) found HSP response to ELF fields in cultured breast cells was stronger after several short exposures, compared with a single longer exposure. Therefore, the timing, repetition rate, and accompanying coherent or incoherent field exposures could be factors to be considered in epidemiological studies with sufficient exposure data. It could be interesting to see whether exposure over various periods of time would show temporal patterns in biological or health effects. Other evidence cited above indicates that reduced static magnetic field also can have effects. Thus, a future epidemiological study on the effects of 50 or 60 Hz exposures might include measurement of the static magnetic field as well as estimating the fraction of time that the subject was exposed to magnetic fields that were >65 μT and <23 μT for periods of >5 min.

Among other problems that have been discussed in this book and that would assist future epidemiology is the need to find biological changes that are definitively linked to both electromagnetic field exposure and, hence perhaps, also to a human health effect. Laboratory studies are the only practical way to provide this information. The wide body of sometimes conflicting data concerning effects on DNA, RNA, and proteomics, as well as in cell cultures and whole animals, has given some weight but far from full support to the public concern that has fueled epidemiological searches for cancers (for reviews, see McCormick 2007; Michaelson et al. 2007; Paunesku and Woloschak 2007). Animal behavior, although harder to measure than biochemical or cell-level changes, could be more sensitive indicators. For example, Prato et al. (2011) have used nociception in mice; and Rogers and coworkers (Greenebaum 1995) have done an extensive set of experiments using various observed behaviors in baboons to show changes due to fields. D'Andrea et al. (2003)

have reviewed animal behavioral studies at microwave frequencies; and more recently, Sienkiewicz et al. (2005) examined human behavioral studies at low and high frequencies. All show mixed results. If it does develop that an easily measured human physiological or behavioral end point is found, for instance, an increase in a blood chemical or a distinctive change in a behavioral test, an epidemiological study could use the biological change as an end point with greater statistical precision compared with using a full-blown disease. Other research that could be useful could be additional study into correlations between exposure parameters and their surrogates and confounders, such as correlating locations near power lines with their electromagnetic fields and traffic corridors with their airborne pollutants. Further studies of the accuracy of epidemiological questionnaire recall data could also be candidates.

Biological processes have a wide variety of oscillating systems with varying cycle times. For example, hearts beat periodically with variations that respond to the body's need for oxygen. We sleep typically once every 24 hr. Growing cells divide periodically. These systems have different sensitivities to perturbations at different parts of the cycle. Cells in the nervous system, including the brain, fire at periodic rates in many different patterns. It is a general property of oscillating system that they can be injection locked to external signals that are periodic at frequencies that are near their natural oscillating frequency. Electromagnetic fields of similar frequency and modulated RF fields have been shown to influence the firing rate of neurons (Barnes 1992, 2007; Bawin et al. 1996; Beckett et al. 2003; Sundaram et al. 2009). Croft et al. (2008) have shown that exposure to fields from cell phones can modify the EEG in humans. Subjects' EEGs during sleep were mildly affected by RF exposure, but in a way that varied between subjects (Loughran et al. 2012). Epileptic-like pulses in vitro can be desynchronized with single, short pulses (e.g., Durand and Warman 1984), and recent neural network simulations have shown similar results (Monteforte and Wolf 2012). Slower neurological and other physiological cycles such as NAD(P)H oscillations can also be suppressed or amplified with properly synchronized fields, sometimes at very low intensities (Chan and Nicholson 1986; Rosenspire et al. 2000, 2001). Thus, lab experiments could be of interest that investigate whether modulation or frame repetition rates in the cell phone could change responses to natural signals that are required for the proper functioning of the body. If so, this difference could then be tested in an epidemiological study, possibly a study integrated with physiological sensors linked to cell phones, as discussed below.

Characterizing and Measuring Exposures

The two most obvious missing pieces of information in understanding how fields affect health are knowledge of the mechanism(s) through which electromagnetic fields affect biological systems; and, barring that, knowledge of the specific exposure parameter(s) that are relevant. We have already commented on several studies that could be pointing to candidate mechanisms or clues to the type or timing of fields that could important. They will not be elaborated upon here. Epidemiological studies of those ideas require the acquisition or reanalysis of very large amounts of data. Any such epidemiological study would require high-quality data on the specific exposure parameters for each of the proposed mechanisms including measurements of the SMF for testing a free radical hypothesis. The design and testing of appropriate instruments would be necessary.

Most epidemiological work to date studying effects of mobile phones has not classified analyzed subjects by exposure characteristics other than number and length of mobile telephone calls and the number of years of use, although other characteristics such as the type of phone system used (frequency bands and pulsing pattern, if any, are sometimes collected; see Cardis et al. 2011; Little et al. 2012). Therefore, a "natural experiment" related to different RF pulsing patterns may already be available in existing data bases: in the United States, two mobile phone systems have long been in simultaneous use, the continuous wave, frequency-modulated CDMA system and the GSM system that uses pulsed RF. The two systems have also competed elsewhere, but the relative numbers of subscribers to each has historically not been as equal as in the United States. Both systems have approximately the same average power, but the GSM pulses are instantaneously stronger, occur at 217 Hz, and are accompanied by 217 Hz magnetic pulses in the handset, when transmitting (Heger and Mohr 1995; Cardis et al. 2011). An epidemiological study could not tease out which of these was the crucial factor if there are differences between the systems; that would be a question best suited for lab studies. To date, lab comparisons of the two systems, whether in cells or whole organisms, have not been done to our knowledge, but lab findings of differences would be strong motivation for gathering epidemiological data to compare users of the two systems. However, because datasets may already exist that could be reanalyzed to compare the two systems, the effort should be considered, even in the absence of lab comparisons. As in any epidemiological study, numbers of subjects and confounders would be an issue. First-generation CW phone technology probably should not be considered because of the greater power levels and longer time since use, but how to handle those who used both systems is clearly a relevant issue. Obviously, studies would have to be done quite carefully, because any non-negative findings would have implications for existing commercial operators.

To our knowledge, the intensity of exposure to a large number of subjects, either average or instantaneous, or other detailed characteristics of exposure have not been directly measured. All phones only expose the user to the most intense power occasionally during a call, mostly while the user is speaking; and phones transmit at lower than their maximum rated power whenever that is sufficient for the nearest base station. Instrumentation to measure, record, and summarize the various parameters characterizing this personal exposure does not exist to our knowledge and would require considerable engineering to design and test. A possible way of acquiring some of these data required to better characterize exposures (e.g., personal exposure, monitoring aspects of subjects' physiological status, or having the subject keep a reliable log of an activity) could be through an app for smart phones, adding a sensor attachment if necessary. Such an app may currently be under development (M. Röösli, personal communication, 2013). Such an app could, for example, record the strength of the cell phone power transmission, length of the call, time of day, and possibly the base station signal intensity and the position where the phone is being held relative to the body. This app could also record parameters such as the current consumed by the device at the time of the call so as to determine the background magnetic fields generated by the functioning phone. Alternatively, the phone could at least prompt the subject to enter time and other information and store it for later processing. There is also sufficient computational power in smart phones that some of these data could be processed and stored so as to reduce the amount of data that would need to be processed centrally by the investigator. Because of the large number of possible parameters of the cell phone electric and magnetic fields that might be relevant in terms of exposure, the large variability in the potential sensitivity of the person using the phone, and the very large amounts of data that would be accumulated, it is likely that pattern recognition programs will need to be used to look for the numerous possible correlations between the exposure characteristics and the possible health effects of interest.

Surveys have been published of the general environmental level of high-frequency fields that, although very much weaker than those of a transmitting mobile phone, have caused concern among some members of the public because of their omnipresence. These one-time surveys have used laboratory grade or portable personal dosimetry instruments and have measured separately the field strengths for various phone and broadcast bands, and some have used these measurements to estimate specific absorption rate (SAR) in a person. The results have varied widely according to type of location and country (e.g., Joseph et al. 2012).

ELF studies of exposure parameters have been discussed in other chapters of this book. Some studies, particularly those of workers in an occupational environment but also some household or general environmental surveys, have measured parameters other than average intensity, such as peak intensities, length of time at various field intensities, or number of transients. Equipment to measure some of these parameters in the range of 50–60 Hz has been developed, especially the EMDEX, EMDEX Lite, and PROTON devices; and their use has given some indication of occupational and environmental exposure conditions (e.g., Kaune 1993). But the reliability of one device, PROTON, has been questioned (Leeper and Wertheimer 2002). Such devices have only been used in producing surveys of short-term exposures of small samples of individuals rather than large populations. They cannot, of course, produce retrospective information. As with high-frequency exposures, new methods of capturing long-term, individual low-frequency field exposures would be needed for epidemiological studies that examined more exposure parameters in detail, again possibly using cell phone apps.

Although measuring and subdividing a pool of subjects according to the multitude exposure parameters would seem to be an impossible task, one means of simplifying the analysis would be the concept of "effects functions" (Morgan and Nair 1992). Although first discussed in terms of 50–60 Hz effects, the general concept applies to RF research as well. Certain measured combinations of amplitude, pulsing, and exposure time are consistent with some and inconsistent with other models of what field parameters are effective. For instance, dividing subjects between high and low exposure, with few disease cases at low or no exposure, could be consistent with several models, but not with others. The disease could be occurring after exposure above a threshold that is over the low/high dividing point, after exposure within a "window" that is above that threshold but below a higher level at which few subjects are exposed, after a high enough exposure for a certain period of time, or with an incidence that rises with intensity. In this example, the data would not be consistent with a threshold below the dividing point, exposure that required weak but pulsed fields, or exposure that was within well-defined frequency windows. The magnetite hypothesis is consistent with thresholds or rising incidence with intensity, but not with windows; the free radical hypothesis is very dependent on a narrow intensity window but not on frequency. Zhang et al. (1997) give an example of a set of parameters at which low-frequency household background fields would dominate appliance exposure and another set where appliance exposures dominate. Because multiple field combinations are most easily tested in the lab or using computer simulation, this information would help shape future epidemiological studies.

Should one or more combinations of exposure parameters show a larger probability of producing biological effects, whether because of a hypothetical mechanism, biological experimentation, or measured intensities, epidemiological datasets may exist that could be reanalyzed with these exposure parameters in mind. It is also possible that "natural experiments" may exist within existing datasets that could be exploited with relatively little effort.

Conclusions

We have discussed several different results from the laboratory about possible reasons biological or health effects may be occurring as a result of exposures to electric and magnetic fields. Because the lines between the questions we posed in the Introduction section are not easy to define, we have addressed many individual topics separately in this chapter. However, it should be clear that each topic raises some substantive issues and may suggest future research that falls under one or more of these three lines of questioning. We summarize each in turn below:

1. Information in the literature from lab studies that could be useful for shaping future epidemiological research: Several bodies of lab data with specific combinations of fields and frequencies or temporal structure, as well as the free radical model for interactions, could be used to shape dosimetry in future studies. The wide body of somewhat conflicting data concerning effects on DNA, RNA, and proteomics has given some weight to the public concern that has fueled epidemiological search for cancers, but it also could be examined in the future to influence decisions to include or exclude other diseases in epidemiological studies. Surveys of environmental and occupational exposures in the literature have been highly important in structuring past and, presumably, future epidemiological work. Some hypotheses based on lab data might be tested using reanalysis of existing datasets, such as comparing users of GSM and CDMA phone systems.
2. Information from lab studies that would, if known, make future epidemiological research more effective: Engineering research to improve dosimetry, particularly long-term, widespread personal dosimeters, to incorporate more detailed information about exposure, could reduce a significant amount of epidemiological uncertainty. Both surveys and instrumentation, including perhaps mobile phone–based techniques using an app for recording the exposure parameters, are likely to be needed. Behavioral studies concerning subjects' recall of historical use also have potential to reduce uncertainty. Lab studies of all types that can reduce current conflicts in the literature or test hypotheses, such as those involving free radicals, could narrow the focus of future epidemiology.
3. Information from current or future epidemiological research that could be useful for shaping future lab research: Epidemiological results might motivate deeper research into the cellular level progenitors of leukemia, in the case of ELF fields, or certain head tumors, in the case of RF, although in neither case is the epidemiological evidence of an association strong. Although many types of cancer have not been examined epidemiologically, the results of the other types that have been studied would discourage their use in lab tests except when generic cell processes are studied.

Epidemiological studies suggested by the questions that frame this chapter will, in general, not be easy and may not be currently possible for several reasons. First, enough study subjects may not be available to produce sufficient statistical precision for the low-incidence diseases that are of concern. Some systems seem affected in the laboratory, sometimes increasing and sometimes decreasing a measured end point; and the extent to which these results can be used in epidemiological studies with sufficient statistical power should be considered. In the same vein, some of the more plausible proposed theories for mechanisms by which low levels of electric and magnetic fields might affect biological processes could help define epidemiological studies' biological end points or divisions in exposure conditions by which subjects should be classified. Numbers become an increasingly difficult situation when the pool of cases is subdivided according to an increasing number of exposure characteristics or the subjects' physiological or other status.

In addition, some studies might require information that is not in an existing data base or is difficult for a new study to get reliably, such as each subject's historical or concurrent exposure, behavior, or physiological measurement data. Finally, choosing the best way to classify subjects by exposure and other relevant data and removing effects of confounders is quite difficult, as has been discussed in other chapters. Designs of future studies should recognize that there are amplifier and feedback systems in the body so that exposures at different times can lead to different effects. Conditions that are likely to be confounders, such as stress and other health conditions, that can either increase or suppress the sensitivity to electric and magnetic field exposures should be considered. It should be noted that biological effects are not necessarily health effects, and some laboratory experiments resulting in detectable changes, especially in cells instead of whole organisms, may not result in changes in things epidemiology can consider, because there are many corrective feedback loops and the sensitivity may change as a function other conditions.

However, we suggest that scientists doing both lab and epidemiology studies should consider the present state of both types of work when planning new studies. We particularly suggest attention to developing ways for cell phones to be used as individual monitors, to exposure to steady magnetic fields and the time sequences of all field exposures, and to changes in the physiological status of the various subjects, including their exposures to other possible stressors.

References

Asashima M, Shimada K, Pfeiffer CJ. 1991. Magnetic shielding induces early developmental abnormalities in the newt, Cynops pyrrhogaster. *Bioelectromagnetics* 12: 215–224.

Ashraf A, Cheema M, Tambawala M, Laterza R, Zhou E, Rathnabharathi K, et al. 2008. Effects of 900-MHz radio frequencies on chemotaxis of human neutrophils in vitro. *IEEE Trans Biomed Eng* 55(2): 795–797.

Barnes FS. 1992. Some engineering models for interactions of electric and magnetic fields with biological systems. *Bioelectromagnetics* 13(Suppl 1): 67–85.

Barnes FS. 2007. Interaction of DC and ELF electric fields with biological materials and systems. In: Barnes FS, Greenebaum B, eds. *Bioengineering and Biophysical Aspects of Electromagnetic Fields*. Boca Raton, FL: CRC Press, pp. 115–156.

Bawin SM, Satmary WM, Jones RA, Adey WR, Zimmerman G. 1996. Extremely-low-frequency magnetic fields disrupt rhythmic slow activity in rat hippocampal slices. *Bioelectromagnetics* 17: 388–395.

Beckett EA, McGeough CA, Sanders KM, Ward SM. 2003. Pacing of interstitial cells of Cajal in the murine gastric antrum: neurally mediated and direct stimulation. *J Physiol* 553: 545–559.

Belyaev IY, Alipov YD, Harms-Ringdahl M. 1997. Effects of zero magnetic field on the conformation of chromatin in human cells. *Biochim Biophys Acta* 1336: 465–473.

Belyavskaya NA. 2001. Ultrastructure and calcium balance in meristem cells of pea roots exposed to extremely low magnetic fields. *Adv Space Res* 28: 645–650.

Belyavskaya NA. 2004. Biological effects due to weak magnetic field on plants. *Adv Space Res* 34: 1566–1574.

Binhi V. 2002. *Magnetobiology: Underlying Physical Problems*. London: Academic Press, 473 pp.

Black DR, Heynick LN. 2003. Radiofrequency (RF) effects of blood cells, cardiac, endocrine, and immunological functions. *Bioelectromagnetics* 24(Suppl 6): S187–S195.

Blackman CF, Benane SG, House DE. 1993. Evidence for direct effect of magnetic fields on neurite outgrowth. *FASEB J* 7: 801–806.

Brocklehurst B, McLauchlan K. 1996. Free radical mechanism for the effects of environmental electromagnetic fields on biological systems. *Int J Radiat Bio* 69: 3–24.

Cardis E, Varsier N, Bowman JD, Deltour I, Figuerola J, Mann S, et al. 2011. Estimation of RF energy absorbed in the brain from mobile phones in the Interphone study. *Occup Environ Med* 68: 686–693.

Chan CY, Nicholson C. 1986. Modulation by applied electric fields of purkinje and stellate cell activity in the isolated turtle cerebellum. *J Physiol* 371: 89–114.

Choleris E, Del Seppia C, Thomas AW, Luschi P, Ghione G, Moran GR, et al. 2002. Shielding, but not zeroing of the ambient magnetic field reduces stress-induced analgesia in mice. *Proc Biol Sci* 269: 193–201.

Croft RJ, Hamblin DL, Spong J, Wood AW, McKenzie RJ, Stough C. 2008. The effect of mobile phone electromagnetic fields on the alpha rhythm of human electroencephalogram. *Bioelectromagnetics* 29: 1–10.

D'Andrea JA, Adair ER, de Lorge JO. 2003. Behavioral and congnitive effects of microwave exposure. *Bioelectromagnetics* 24(Suppl 6): S39–S62.

Dawson TW, Caputa K, Stuchly MA. 1997. A comparison of 60 Hz uniform magnetic and electric induction in the human body. *Phys Med Biol* 42: 2319–2329.

Dawson TW, Stuchly MA. 1998. Effects of skeletal muscle anisotropy on human organ dosimetry under 60 Hz uniform magnetic field exposure. *Phys Med Biol* 43: 1059–1074.

Di Carlo AL, Mullins JM, Litovitz TA. 2000. Electromagnetic field-induced protection of chick embryos against hypoxia exhibits characteristics of temporal sensing. *Bioelectrochemistry* 52: 17–21.

Durand DM, Warman EN. 1984. Desynchronization of epileptiform activity by extracellular current pulses in rat hippocampal slices. *J Physiol* 480: 527–537.

Eder SH, Cadiou H, Muhamad A, McNaughton PA, Kirschvink JL, Winklhofer M. 2012. Magnetic characterization of isolated candidate vertebrate magnetoreceptor cells. *Proc Natl Acad Sci U S A* 109: 12022–12027.

Fesenko EE, Mezhevikina LM, Osipenko MA, Gordon RY, Khutzian SS. 2010. Effect of the "zero" magnetic field on early embryogenesis in mice. *Electromagn Biol Med* 29(1-2):1–8.

Fonkem E, Wong ET. 2012. NovoTTF-100A: a new treatment modality for recurrent glioblastoma. *Expert Rev Neurother* 12: 895–899.

Frankel RB, Blakemore RP. 1989. Magnetite and magnetotaxis in microorganisms. *Bioelectromagnetics* 10: 223–237.

Galland P, Pazur A. 2005. Magnetoreception in plants. *J Plant Res* 118: 371–389.

Greenebaum B, ed. 1995. Experiments on exposure of nonhuman primates to 60 Hz electric and magnetic fields. *Bioelectromagnetics* 16(Suppl 3): 122.

Han L, Lin H, Head M, Jin M, Blank M, Goodman R. 1998. Application of magnetic field-induced heat shock protein 70 for presurgical cytoprotection. *J Cell Biochem* 71: 577–583.

Heger W, Mohr W. 1995. Pro's and con's of TDMA and CDMA for digital cellular networks. *SBMO/IEEE MTT-S Proceedings of Microwave and Optoelectronics Conference*, vol. 2., pp. 810–815.

Joseph W, Frei P, Röösli M, Vermeeren G, Bolte J, Thuróczy G, et al. 2012. Between-country comparison of whole-body SAR from personal exposure data in urban areas. *Bioelectromagnetics* 33: 682–694.

Kaune WT. 1993. Assessing human exposure to power-frequency electric and magnetic fields. *Environ Health Perspect* 101(Suppl 4): 121–133.

Kaune WT, Gillis MF. 1981. General properties of the interaction between animals and ELF electric fields. *Bioelectromagnetics* 2: 1–11.

Korneva HA, Grigoriev VA, Isaeva EN, Kaloshina SM, Barnes FS. 1999. Effects of low-level 50 Hz magnetic fields on the level of host defense and on spleen colony formation. *Bioelectromagnetics* 20: 57–62.

Leeper E, Wertheimer N. 2002. Potential motion related bias in the worn dosimeter measurements of two childhood leukemia studies. *Bioelectromagnetics* 23: 390–397.

Liboff AR. 2007. The ion cyclotron resonance hypothesis. In: Barnes FS, Greenebaum B, eds. *Bioengineering and Biophysical Aspects of Electromagnetic Fields*. Boca Raton, FL: CRC Press, pp. 261–292.

Litovitz TA, Krause D, Montrose CJ, Mullins JM. 1994. Temporally incoherent magnetic fields mitigate the response of biological systems to temporally coherent magnetic fields. *Bioelectromagnetics* 15: 399–409.

Litovitz TA, Penafiel LM, Farrel JM, Krause D, Meister R, Mullins JM. 1997. Bioeffects induced by exposure to microwaves are mitigated by superposition of ELF noise. *Bioelectromagnetics* 18: 422–430.

Little MP, Rajaraman P, Curtis RE, Devesa SS, Inskip PD, Check DP, et al. 2012. Mobile phone use and glioma risk: comparison of epidemiological study results with incidence trends in the United States. *BMJ* 344: e1147.

Loughran SP, McKenzie RJ, Jackson ML, Howard ME, Croft RJ. 2012. Individual differences in the effects of mobile phone exposure on human sleep: rethinking the problem. *Bioelectromagnetics* 33: 86–93.

Maeda K, Robinson AJ, Henbest KB, Hogben HJ, Biskup T, Ahmad M, et al. 2012. Magnetically sensitive light-induced reactions in cryptochrome are consistent with its proposed role as a magnetoreceptor. *Proc Natl Acad Sci U S A* 109: 4774–4779.

Martino CF. 2011. Static magnetic field sensitivity of endothelial cells. *Bioelectromagnetics* 32: 506–508.

Martino CF, Castello PR. 2011. Modulation of hydrogen peroxide production in cellular systems by low level magnetic fields. *PLoS One* 6: e22753.

Martino CF, Portelli L, McCabe K, Hernandez M, Barnes F. 2010. Reduction of the earth's magnetic field inhibits growth rates of model cancer cell lines. *Bioelectromagnetics* 31: 649–655.

McCormick DL. 2007. Evaluation of the toxicity and potential oncogenicity of extremely low frequency magnetic fields in experimental animal model systems. In: Barnes FS, Greenebaum B, eds. *Biological and Medical Aspects of Electromagnetic Fields*. Boca Raton, FL: CRC Press, pp. 21–52.

Michaelson SM, Elson EC, Anderson LE. 2007. Interaction of nonmodulated and pulse-modulated radio frequency fields with living matter: experimental results. In: Barnes FS, Greenebaum B, eds. *Biological and Medical Aspects of Electromagnetic Fields*. Boca Raton, FL: CRC Press, pp. 53–148.

Monteforte M, Wolf F. 2012. Dynamic flux tubes form reservoirs of stability in neuronal circuits. *Phys Rev X* 2: 041007.

Morgan MG, Nair I. 1992. Alternative functional relationships between ELF field exposure and possible health effects: report on an expert workshop. *Bioelectromagnetics* 13: 335–350.

Nepomnyashchikh, LM, Klinnikova, MG, Lushnikova, EL, Molodykh, OP, Ashcheulova, NV. 1997. Effect of hypogeomagnetic field on tissue and intracellular reorganization of mouse myocardium. *Bull Exp Biol Med* 124: 1021–1024.

Nindl G, Swez JA, Miller JM, Balcavage WX. 1997. Growth stage dependent effects of electromagnetic fields on DNA synthesis of Jurkat cells. *FEBS Lett* 414: 501–506.

Novikov VV, Sheiman IM, Fesenko EE. 2007. Multimodal effects of nearly complete geomagnetic field deprivation on fission of the planarian Dugesia tigrina. *Biophysics* 52: 498.

Novikov VV, Sheiman IM, Fesenko EE. 2008. Effect of weak static and low-frequency alternating magnetic fields on the fission and regeneration of the planarian Dugesia (Girardia) tigrina. *Bioelectromagnetics* 29: 387–393.

Paunesku T, Woloschak GE. 2007. Effects of radiofrequency and extremely low-frequency electromagnetic field radiation on cells of the immune system. In: Barnes FS, Greenebaum B, eds. *Biological and Medical Aspects of Electromagnetic Fields*. Boca Raton, FL: CRC Press, pp. 1–20.

Phillips JB, Jorge PE, Muheim R. 2010. Light-dependent magnetic compass orientation in amphibians and insects: candidate receptors and candidate molecular mechanisms. *J R Soc Interface* 7(Suppl 2): S241–256.

Pilla AA. 2007. Mechanisms and therapeutic applications of time-varying and static magnetic fields. In: Barnes FS, Greenebaum B, eds. *Biological and Medical Aspects of Electromagnetic Fields*. Boca Raton, FL: CRC Press, pp. 351–411.

Portelli L, Madapatha D, Martino C, Hernandez M, Barnes F. 2012. Reduction of the ambient magnetic field inhibits Drosophila melanogaster ability to survive ionizing radiation. *Bioelectromagnetics* 33: 706–709.

Prato FS, Desjardins-Holmes D, Keenliside LD, DeMoor JM, Robertson JA, Stodilka RZ, et al. 2011. The detection threshold for extremely low frequency magnetic fields may be below 1000 nT-Hz in mice. *Bioelectromagnetics* 32: 561–569.

Prato FS, Kavaliers M, Thomas AW. 2000. Extremely low frequency magnetic fields can either increase or decrease analgaesia in the land snail depending on field and light conditions. *Bioelectromagnetics* 21: 287–301.

Prato FS, Robertson JA, Desjardins D, Hensel J, Thomas AW. 2005. Daily repeated magnetic field shielding induces analgesia in CD-1 mice. *Bioelectromagnetics* 26: 109–117.

Pullar CE, ed. 2011. *The Physiology of Bioelectricity in Development, Tissue Regeneration, and Cancer*. Boca Raton, FL: CRC Press, 304 pp.

Robertson JA, Thomas AW, Bureau Y, Prato FS. 2007. The influence of extremely low frequency magnetic fields on cytoprotection and repair. *Bioelectromagnetics* 28: 16–30.

Rosenspire AJ, Kindzelskii AL, Petty HR. 2000. Biophysical interferon-g and sinusoidal electric fields signal by modulating NAD(P)H oscillations in polarized neutrophils. *Biophys J* 79: 3001–3008.

Rosenspire AJ, Kindzelskii AL, Petty HR. 2001. Pulsed DC electric fields couple to natural NAD(P)H oscillations in HT-1080 fibrosarcoma cells. *J Cell Sci* 114: 1515–1520.

Shallom JM, Di Carlo AL, Ko D, Penafiel LM, Nakai A, Litovitz TA. 2002. Microwave exposure induces Hsp70 and confers protection against hypoxia in chick embryos. *J Cell Biochem* 86: 490–496.

Sienkiewicz Z, Jones N, Bottomley A. 2005. Neurobehavioural effects of electromagnetic fields. *Bioelectromagnetics* 26(Suppl 7): S116–S126.

Sunderam S, Chernyy N, Peixoto N, Mason JP, Weinstein SL, Schiff SJ, et al. 2009. Seizure entrainment with polarizing low-frequency electric fields in a chronic animal epilepsy model. *J Neural Eng* 6: 046009.

Timmel CR, Henbest KB. 2004. A study of spin chemistry in weak magnetic fields. *Philos Trans A Math Phys Eng Sci* 362: 2573–2589.

Tipping DR, Chapman KE, Birley AJ, Anderson M. 1999. Observations on the effects of low frequency electromagnetic fields on cellular transcription in Drosophila larvae reared in field-free conditions. *Bioelectromagnetics* 20: 129–131.

Tiwari R, Singh R. 2012. Impact on cell mobility of human beings due to exposure to emf radiation. UACEE. *Int J Adv Electron Eng* 2(3). Available at www.seekdl.org/nm.php?id=490 (accessed online July 11, 2013).

Tokalov V, Gutzeit HO. 2004. Weak electromagnetic fields (50 Hz) elicit a stress response in human cells. *Environ Res* 94: 145–151.

Tremblay L, Houde M, Mercier G, Gagnon J, Mandeville R. 1996. Differential modulation of natural and adaptive immunity in Fischer rats exposed for 6 weeks to 60 Hz linear sinusoidal continuous-wave magnetic fields. *Bioelectromagnetics* 17: 373–383.

Volpe P. 2003. Interactions of zero-frequency and oscillating magnetic fields with biostructures and biosystems. *Photochem Photobiol Sci* 2: 637–648.

Wang X, Xu M, Li B, Li D, Jiang J. 2003. Long-term memory was impaired in one-trial passive avoidance task of day-old chicks hatching from hypomagnetic field space. *Chin Sci Bull* 48: 2454–2457.

Wiltschko R, Gehring D, Denzau S, Güntürkün O, Wiltschko W.2010. Interaction of magnetite-based receptors in the beak with the visual system underlying 'fixed direction' responses in birds. *Front Zool* 13:24.

Woodward JR, Timmel CR, Hore PJ, McLauchlan KA. 2002. Oscillating magnetic field effects on chemical reaction yields. *Riken Rev* 44: 79–81.
Xiao Y, Wang Q, Xu M-L, Jiang J-C, Li B. 2009. Chicks incubated in hypomagnetic field need more exogenous noradrenaline for memory consolidation. *Adv Space Res* 44: 226–232.
Zhang J, Nair I, Morgan MG. 1997. Effects function simulation of residential appliance field exposures. *Bioelectromagnetics* 18: 116–124.
Zmyslony M, Rajkowska E, Mamrot P, Politanski P, Jajte J. 2004. The effect of weak 50 Hz magnetic fields on the number of free oxygen radicals in rat lymphocytes in vitro. *Bioelectromagnetics* 25: 607–612.

18

What If? The Public Health Perspective

Michael Kundi and Hans-Peter Hutter

CONTENTS

Introduction

The International Agency for Research on Cancer (IARC) has classified power frequency magnetic fields (MFs) (extremely-low frequency [ELF]-MFs) in 2001 (IARC 2002) and radiofrequency (RF) electromagnetic fields (EMFs) in 2011 (IARC 2013) as possible human carcinogens (group 2B) mainly based on limited evidence from epidemiological studies. Additional studies on ELF-MFs published since 2002 increased the credibility of an association with childhood leukemia. For RF-EMFs, the classification was based on large epidemiological studies of an association between mobile phone use and brain tumors (Hardell et al. 2006a, 2006b; INTERPHONE Study Group 2010). Although the studies of MFs in the power frequency range allow for an overall estimate of the risk from increased levels of MFs, the duration of mobile phone use is too short for the time being to arrive at a final estimate of the risk and hence of the public health impact.

From the public health point of view, there are two main problems entangled with the issue of a relationship between EMFs and long-term health effects: (1) what is the burden of disease possibly related to the exposure within the population and what are the consequences for the health care system, and (2) which measures are appropriate to protect public health.

Although the first problem can be dealt with solely by scientific methods and knowledge about the structure of the health care systems, the second problem is related to risk management and thus cannot be treated without value judgments and considerations about costs and benefits.

There remain a lot of misunderstandings and deceptive opinions about the role of science in the development of measures to protect public health. As Popper (1945) states, it is logically impossible to derive a proposal for a policy from scientific statements about facts. This applies to the derivation of exposure standards as well. The level of protection such standards should provide as well as the level of proof for health-relevant effects that should be considered in these derivations cannot be established by science but must be introduced from public discourse.

In this chapter, the possible effects of mobile phone use on the trends in brain tumor incidence. If there is an association, an increased risk should be detectable by now, however only in those countries with an early onset of mobile phone use in a significant proportion of the population. In other countries this increase may only occur in the next few decades.

Second, we also explore the attributable numbers of childhood leukemia cases estimated by applying different scenarios. This is of interest given the possibility that a considerable proportion, especially of early childhood cases of acute lymphoblastic leukemia (ALL), is due to residential exposure to MFs.

And third, a possible relationship between well-being and exposure to RFs in the low-dose range from mobile phone base stations and other transmitters is discussed, and the possible consequences for public health delineated.

Impact of Mobile Phone Use on Brain Tumor Incidence Trends. What if There Is a Relationship?

Incidence trends of brain tumors have been reported for several regions: for Nordic countries (Lönn et al. 2004; Deltour et al. 2009, 2012), England (de Vocht et al. 2011a), Australia (Dobes et al. 2011), and the United States (Little et al. 2012) (see Chapter 14).

Although in these studies overall trends showed little indication of increasing incidences, there are several intriguing findings concerning incidence trends in subgroups of brain tumors. Little et al. (2012) report significant annual percent changes (APCs) for high-grade glioma (APC, 0.64%; 95% confidence interval [CI], 0.33–0.95) and for temporal lobe glioma (APC, 0.73%; 0.23–1.23). In England, malignant brain tumors show increasing incidences in the temporal lobe for men (APC, 2.7%; 2.3–3.0) and women (APC, 2.0%; 1.6–2.4%). The increase was stronger after 2000 (De Vocht et al. 2011b). In Australia (Dobes et al. 2011), no overall increase in brain tumor incidences was found, whereas an increase was noted for malignant brain tumors over the study period 2000–2008 (APC, 3.9%; 2.4–5.4).

The above-mentioned studies explored whether the incidence trends could be linked to the secular increase of mobile phone use. Therefore, some general remarks about the chance to detect an effect from mobile phone use in overall incidence trends are necessary.

For glioma, studies in atomic bomb survivors and patients receiving therapeutic X-rays indicate latencies of about 20–30 years in adults. Also, for other brain tumors, latencies of at least 20 years have been estimated (Kranzinger et al. 2001; Mohyuddin et al. 2003; Sadetzki et al. 2005; Umansky et al. 2005; Salvati et al. 2008). Given the long latency period, if use of a mobile phone would carry a risk in the sense of initiating brain tumors, this would be expected in individuals having used mobile phones before 1988. Prevalence of use, however, was insignificant in the late 1980s, with a penetration rate of <1%, on average, in Organisation for Economic Co-operation and Development (OECD) countries. Contribution of this small segment of the population to the brain tumor incidence would therefore also be insignificant unless there is a very high risk. The relative increase is given by the formula $1 + 0.01 \times (\psi - 1)$ if the fraction of users is 1% and ψ is the relative risk. With an annual variation of brain tumor incidence typically of about ±12%, the relative risk must significantly exceed 13 for an effect to be noticed at the population level.

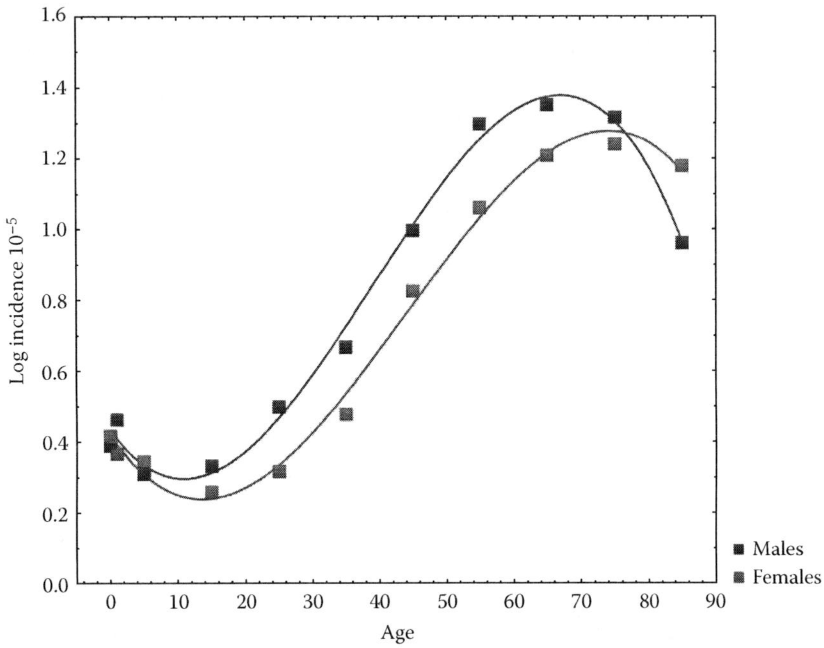

FIGURE 18.1
Age–log incidence function for malignant brain tumors (International Classification of Diseases 10: C71; data from England).

We can thus conclude that an effect on incidence cannot, for the time being, be detected, but we still need to consider a possible influence on tumor development. Assuming mobile phone use acts on already initiated brain tumors, the effect in a population of mobile phone users will be noticed by a shift of the age–incidence function. Hence, the possible increase of the relative risk is limited by the slope of the age-incidence function (actually the age-log incidence function due to the peculiar relationship between age and incidence for brain tumors). Figure 18.1 shows typical age–incidence functions in males and females.

With small variations between countries, there is an approximately linear relationship between age and log brain tumor incidence, with a slope of 0.05–0.06. This is of great importance because, as mentioned, we are concerned with an agent that might affect pre-existing tumors by increasing growth rate. If there is an effect on tumor growth rate, the effect on incidence by mobile phone use is given by the following formula (see Kundi 2010):

$$\exp(\beta + \gamma A)\left[\pi_o + \pi_c + \sum_s \pi_s \exp(\gamma \cdot s \cdot (f-1))\right] \tag{18.1}$$

where A is the age in years, β is the intercept, γ is the slope of the age–log incidence function (within the linear age range), π_s is the rate of ipsilateral mobile phone users of s years in this age group (only use at the same side of the head where the tumor is growing is counted and assumed to be half the rate of overall users of s years), π_o is the rate of non users and π_c the rate of contralateral mobile phone users, and f is the factor by which the tumor growth rate is increased due to mobile phone use.

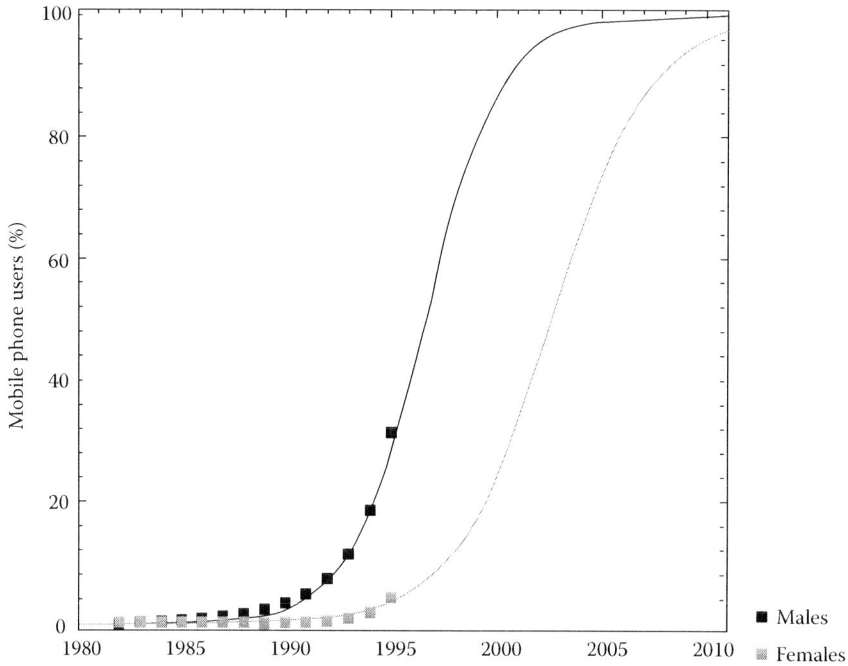

FIGURE 18.2
Estimates of mobile phone users from 1982 to 2010 in Denmark based on Johansen et al. (2001), adjusted by reports on mobile phone penetration rates and taking multiple subscriber identification modules (SIMs) per user into consideration.

There are comprehensive data about the age and sex distribution of mobile phone users, but these data cannot publicly be accessed because they are proprietary material from mobile phone providers. However, the Danish cohort study (Johansen et al. 2001) has some data about early mobile phone use and distribution of age at first use that can be extrapolated to more recent years. Based on this report and data on mobile phone penetration rates, estimates of mobile phone users between 1982 and 2010 in Denmark were obtained (Figure 18.2).

From the annual Danish population data base (Statistics Denmark; http://www.dst.dk/en), the number of individuals from each birth cohort still living in each year between 1987 and 2010 was obtained (mobile phone use before 1987 was disregarded as handheld phones were only available from this date onward).

The fraction of users by number of years of use (from 1 year to a maximum of 24 years) was calculated by applying the age- and sex-specific estimate of mobile phone users for each birth cohort and calendar year. The number of surviving users was calculated by applying the overall Danish age- and sex-specific mortality for every calendar year and weighting mortalities to consider a "healthy user" effect with 0.67 in the first year increasing to 1 in the fifth year after onset of use.

For each duration of mobile phone use, half the estimated fraction of mobile phone users was considered to have ipsilateral exposure. By applying Equation 18.1 with a slope of 0.06 and a growth rate factor of 1.5 for each calendar year of interest, the relative incidence increase (the term in square brackets in Equation 18.1) was calculated. These relative figures were multiplied by the observed incidences for 1995. Figure 18.3 shows the observed incidences of brain cancer for males and females by age groups for selected years from

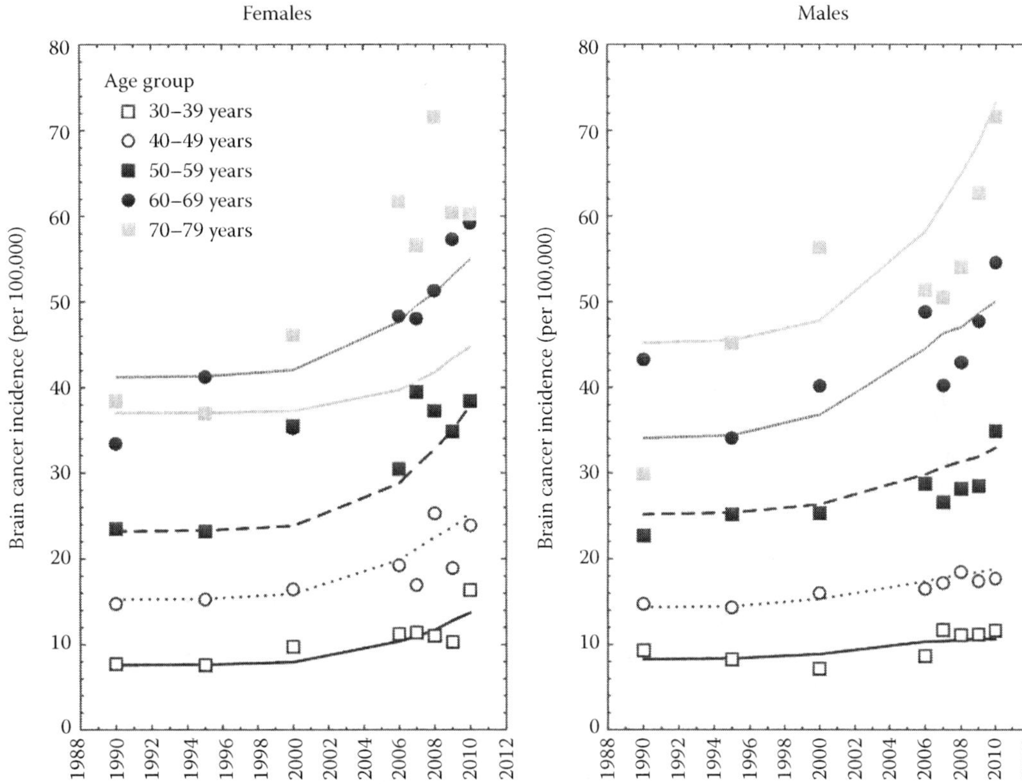

FIGURE 18.3
Observed incidence of brain cancer in Denmark for males and females (square points) and predicted increase (solid lines) of incidences based on the assumption of an increased growth rate of tumors from mobile phone use.

1990 to 2010 as well as the estimated incidences based on the assumption of a 50% increase in the growth rate of brain tumors from mobile phone use. The duration of use was calculated for each 1-year age cohort separately for each calendar year and separately for each sex. The incidences were then combined into the 10-year age groups.

As can be seen from Figure 18.3, the observed increase for the age groups 30–39 to 60–69 is quite in line with the assumption of an influence of mobile phone use. For the age group 70–79, the estimates are close for males but deviate in females. It has to be noted that the assumption of a linear age–log incidence function does not hold for this age group and that a more complex relationship, including higher order terms, exists between duration of use and incidence increase.

Obviously, the incidence trends reported from 2006 to 2010 are in good agreement with the assumption of an influence of mobile phone use on tumor growth rate. On a population level, the increase will become noticeable in other countries with lower rates of long-term users only during the next 10 years. In Nordic countries such as Denmark, because of the very early start of mobile phone use, the increase of brain tumor incidence is estimated to become apparent very soon (or is already observed), whereas in other countries still a few more years of observation are needed to detect any increase in incidences. It is also possible that the effect depends on the type of mobile phones being used. Although there is little evidence for such differential effects from the studies of Hardell et al. (2006b), it is still possible that early use of analog phones contributed more to the effect on incidence. To date, nothing can be inferred from available epidemiological evidence about the effects of 3G phones. The calculations reported above indicate that recent use contributes very little to the overall incidence, and there is no hint from these trend analyses regarding whether and how strongly these new technologies may contribute to the risk of brain cancer.

In Denmark, brain cancer incidence has increased from an absolute number of 828 cases in 1995 to 1372 cases in 2010 (i.e., a >65% increase). A great proportion of this increase can be attributed to the secular trend in mobile phone use. But even if one would not like to draw such a firm conclusion, it is evident that we cannot dismiss that the observed increase could be due to mobile phone use. It has been speculated that the increase might at least in part be due to screening programs, and this could in fact be an explanation for the much larger increase in older women as predicted; it is, however, unlikely that this holds for the young age groups as well. To put this incidence trend into perspective, it has to be stressed that the individual life-time risk to develop brain cancer is low at only 5–15 per 1000. Should this risk double, it is still only 10–30 per 1000. In relation to the life-time breast cancer risk (110–130 per 1000 females) or life-time lung cancer risk (65–85 per 1000 males), it will be still low.

From a public health perspective, a doubling of the individual risk will, in the long run, translate into a doubling of the incidence because of the almost 100% exposed individuals. A doubling of the number of brain cancer cases would put considerable strain on the health care system. Not only would a significant increase of beds need to be assigned to neurosurgeries but also a greater number of, for example, neurosurgeons, neuropathologists, neuroradiologists, and care personnel must be educated in time to respond to the increasing trend.

At present, there is suggestive evidence for an increase in malignant brain tumors; however, if benign brain tumors, accounting for about 50% of all brain tumors (Wöhrer et al. 2009), also increase, this would lead to severe problems in the neurosurgery sector. Despite this potential forthcoming crisis in the neurosurgery sector, there, as yet, is no discussion about the measures necessary to cope with the pending problem. The increase in the number of malignant brain tumor patients in Denmark, for example, is already a reality. Fortunately, oncological patients are a minority among neurosurgery patients, but the trend could lead to a situation where they account for >10% of neurosurgery patients.

General precautionary measures that are intended to reduce exposure of the head to microwaves from mobile phone use are also warranted. Such precautionary measures include information to the public by the mobile and cordless phone manufacturers about how to use these phones prudently and how to avoid high levels of exposure. Although 3G mobile phones have lower average levels of absorbed microwave energy, the substantial increase in the duration of use might outweigh the reduced exposure levels. Therefore, manufacturers should strive to develop phones designed to minimize exposure of sensitive tissues (see Chapter 19). Changed use patterns with various new applications of mobile phones today that are associated with exposure to other parts of the body, such as the head, need also be considered and the design and antenna configuration adapted to these types of use.

Attributable Number of Childhood Leukemia Cases. What Measures Are Appropriate?

Chapter 9 is dedicated to a comprehensive discussion of the evidence for an association between exposure to residential MFs and childhood leukemia. Here, only a few remarks are provided. As a consequence of two comprehensive pooled analyses (Ahlbom et al. 2000;

Greenland et al. 2000), the IARC (2002) classified power frequency MFs as possible human carcinogens based on the assessment that, although bias and confounding cannot be ruled out, the epidemiological evidence allows a causal interpretation. Another pooled analysis of six studies published after 2000 (Kheifets et al. 2010) supported the relationship found in the earlier analyses and gave almost exactly the same estimates of relative risk, if a study from Brazil with a different methodology, was left out.

Earlier criticism about possible confounders responsible for the increased risk have been refuted by evidence demonstrating that such confounders would have to be very strongly related to childhood leukemia, such that it is very unlikely they had not been already detected (Langholz 2001).

Due to the measurements conducted after diagnosis, it is quite likely that exposure misclassification occurred. However, because measurements have been carried out without knowledge of case status, such bias is likely not differential and therefore would result in a bias toward the null hypothesis of no association.

There are very few households with average exposure exceeding 0.3 or 0.4 μT. Consequently there were only few participants in this highest exposure category. Because of different participation rates for cases and controls, a shift in the latter group toward lower exposure could bias the risk estimates toward higher risks. Considering, however, that registry-only studies that relied on wire configuration or distance to power lines found effects of the same magnitude, it is unlikely that selection is a source of distorting bias. Overall, the evidence speaks in favor of an increased risk, and so far no valid counter argument has been put forward. In contrast, there is only weak support from animal experiments and mechanistic studies. But this is to be expected if the difficulties in determining effects at low exposure levels are taken into consideration. Typically, such experiments fail also for other exposures that are close to levels encountered in human environments.

Because of the ubiquitous presence of power frequency MFs in our environment, there is no truly unexposed control group. Therefore, an arbitrary reference level has to be chosen and has become standard practice to choose 0.1 μT as the upper limit for the reference category. Measurements and calculations intend to estimate the average exposure level during the etiologically relevant period, but there is no guarantee that these measurements or calculations come close to the target or that the average MF exposure during the etiologically relevant period is the correct metric at all. Assume that there is another metric, except arithmetic or geometric mean MF exposure, that correlates with these averages but that is a much stronger correlation to childhood leukemia risk, then there must be differential exposure misclassification with respect to this "correct" metric. Although this is only speculative, it demonstrates that it is too early to dismiss the possibility that there is a much higher number of cases attributable to exposure.

Given the distribution of MF exposure in controls, the population fraction of attributable cases (PAF) of childhood leukemia is given by Equation 18.2:

$$\mathrm{PAF} = \frac{\sum_i p_i(\mathrm{RR}_i - 1)}{1 + \sum_i p_i(\mathrm{RR}_i - 1)} \tag{18.2}$$

In Equation 18.2, p_i denotes the fraction of the population allocated in exposure category i, and RR_i is the relative risk for this category. The summation extends over all exposure categories except the reference category. The population-attributable number is obtained by multiplying PAF with the total number of observed cases.

Taking the odds ratios (ORs) and the exposure fractions from ELF-MF measurements reported in the pooled analysis of Ahlbom et al. (2000), the PAF of childhood leukemia is 1.6%. If exposure is dichotomized at 0.1 μT, the crude OR for exposure ≥0.1 μT is 1.56 (Table 18.1) and the PAF is 5.8%.

Calculation of the PAF

The proportion p of exposed is calculated from the controls as p = 395/3547 = 0.11, inserting into Equation 18.2: $\frac{0.11 \cdot (1.56 - 1)}{1 + 0.11 \cdot (1.56 - 1)} = 0.058.$

In developed countries, the incidence of leukemia from birth to age 15 years is about 70 per 100,000. Hence, about one to four children in 100,000 may develop leukemia before age 15 due to exposure from power frequency MFs. Given that there are around 60 million children younger than 15 years in the United States and around 73 million in the European Union (EU), and with the attributable fraction estimated above, we would expect 40–160 cases of leukemia in the United States and 49–195 cases in the EU per year due to exposure to ELF-MFs.

There is a high ecological correlation between total electric power consumption and incidence of childhood leukemia, ranging from about 1.1 per 100,000 in Africa with the lowest power consumption to about 4.4 per 100,000 in North America with the highest power consumption. This intriguing relationship has been one aspect that led to the hypothesis that the occurrence of the childhood leukemia peak characterized by a large proportion of ALL cases before age 5 years is mainly due to electrification (Milham and Ossiander 2001). Of course, electrification is only one indicator of the industrial and technological development and prosperity of a region. Therefore, other factors inevitably entangled with electric power consumption could be responsible. Unless we have a well-established mechanistic model of how exposure to MFs causes leukemia, the controversy will continue and the uncertainty about a safe exposure level will prevail.

Concerning the relationship between MFs and childhood leukemia, and in particular ALL, there is little support from animal studies and in vitro assays. It must, however, be emphasized that except one experiment using the WKAH/Hkm rat model (Bernard et al. 2008), none can be considered suitable for clarifying the issue. Even this single experiment is questionable because it used chemical induction (butylnitrosourea) producing a variety of different forms of leukemia. Aside from this problem of finding a model suitable for the study of human disease, there are more fundamental problems of experimental design which are unlikely to be solved in the near future. Lack of support from experimental evidence should therefore not prevent society from taking precautionary measures.

This uncertainty concerning a causal interpretation of the MF childhood leukemia association, however, has been used as an argument against far-reaching precautionary procedures (World Health Organization [WHO] 2007), highlighted in the two following quotes: "given the small estimated effect resulting from such a risk, the rarity of childhood leukaemia, the rarity of average exposures higher than 0.4 μT and the uncertainty in determining the relevant exposure metric ... it is unlikely that the implementation of an exposure limit based on the childhood leukaemia data and aimed at reducing average exposure to ELF magnetic fields to below 0.4 μT, would be of overall benefit to society." (WHO 2007, p. 362) and "Given the weakness of the evidence for a link between exposure to ELF-MFs and childhood leukaemia and the limited potential impact on public health,

TABLE 18.1

Estimates of True ORs and 95% CI and PAF under the Assumption of Differential and Nondifferential Sensitivities and Specificities of Exposure Measurement Using a Cutoff of 0.1 μT as the Starting Point with Data from Ahlbom et al. (2000)

Cases		Controls		Sensitivity		Specificity				
+	−	+	−	Cases	Controls	Cases	Controls	OR	95% CI	PAF (%)
Nondifferential										
359	1842	395	3152	1	1	1	1	1.56	1.33–1.81	5.8
174	2027	50	3497	0.9	0.9	0.9	0.9	5.94	4.32–8.17	6.6
232	1970	67	3480	0.7	0.7	0.9	0.9	6.09	4.62–8.04	8.8
347	1854	101	3446	0.5	0.5	0.9	0.9	6.41	5.10–8.06	13.3
695	1507	202	3346	0.3	0.3	0.9	0.9	7.65	6.47–9.06	27.4
1389	812	403	3144	0.2	0.2	0.9	0.9	13.35	11.66–15.28	58.4
Differential										
174	2027	265	3282	0.9	0.5	0.9	0.92	1.06	0.87–1.30	0.5
174	2027	214	3333	0.9	0.6	0.9	0.92	1.33	1.08–1.64	2.0
174	2027	185	3362	0.9	0.5	0.9	0.91	1.56	1.26–1.93	2.8
174	2027	179	3368	0.9	0.7	0.9	0.92	1.61	1.30–1.99	3.0
174	2027	155	3393	0.9	0.8	0.9	0.92	1.88	1.50–2.35	3.7
174	2027	149	3398	0.9	0.6	0.9	0.91	1.96	1.56–2.46	3.9
174	2027	124	3423	0.9	0.7	0.9	0.91	2.36	1.86–2.99	4.5
174	2027	107	3440	0.9	0.8	0.9	0.91	2.76	2.16–3.54	5.0
399	1802	395	3152	0.9	1	1	1	1.77	1.52–2.05	7.9
232	1970	50	3497	0.7	0.9	0.9	0.9	8.16	5.99–11.12	9.2
359	1842	202	3346	1	0.3	1	0.9	3.24	2.70–3.88	11.3
513	1688	395	3152	0.7	1	1	1	2.42	2.10–2.80	13.7
359	1842	101	3446	1	0.5	1	0.9	6.67	5.31–8.37	13.9
347	1854	67	3480	0.5	0.7	0.9	0.9	9.71	7.43–12.68	14.2
359	1842	67	3480	1	0.7	1	0.9	10.10	7.73–13.18	14.7
359	1842	50	3497	1	0.9	1	0.9	13.53	10.02–18.27	15.1
695	1507	395	3152	0.3	1	0.9	1	3.68	3.20–4.22	23.0
718	1483	395	3152	0.5	1	1	1	3.86	3.37–4.43	24.2
695	1507	101	3446	0.3	0.5	0.9	0.9	15.77	12.69–19.60	29.6
695	1507	67	3480	0.3	0.7	0.9	0.9	23.88	18.46–30.90	30.2
695	1507	50	3497	0.3	0.9	0.9	0.9	32.00	23.89–42.86	30.6
1197	1004	395	3152	0.3	1	1	1	9.51	8.31–10.87	48.7
1389	812	395	3152	0.2	1	0.9	1	13.65	11.92–15.64	58.5
1389	812	202	3346	0.2	0.3	0.9	0.9	28.40	24.05–33.54	60.9

Continued

TABLE 18.1 (*Continued*)

Estimates of True ORs and 95% CI and PAF under the Assumption of Differential and Nondifferential Sensitivities and Specificities of Exposure Measurement Using a Cutoff of 0.1 μT as the Starting Point with Data from Ahlbom et al. (2000)

Cases		Controls		Sensitivity		Specificity				
+	−	+	−	Cases	Controls	Cases	Controls	OR	95% CI	PAF (%)
1389	812	101	3446	0.2	0.5	0.9	0.9	58.51	47.14–72.63	62.0
1389	812	67	3480	0.2	0.7	0.9	0.9	88.62	68.57–114.54	62.4
1389	812	50	3497	0.2	0.9	0.9	0.9	118.74	88.73–158.89	62.6
1795	406	395	3152	0.2	1	1	1	35.28	30.36–41.00	79.2

the benefits of exposure reduction on health are unclear and thus the cost of reducing exposure should be very low." (WHO 2007, p. 372)

As shown above, the PAF, given the average measured flux densities, is very low. This is a consequence of the low proportion of "exposed" children in the population. Results of measurements in controls give an overall proportion of 11% and calculated values of only 2% of children exposed to average flux densities above 0.1 μT. Exposure misclassification may be due to use of measurements conducted after diagnosis to estimate exposure during pregnancy or early childhood, but more importantly it may be due to the wrong exposure metric.

In the presence of exposure misclassification, the relationship between the unbiased (π_I, i = 0,1,…,k) and biased (π_i^*, i = 0,1,…,k) exposure probabilities is given by Equation 18.3, with **M** the misclassification matrix with elements m_{ij} (i,j = 0,1,…,k; probability that a subject in category j is classified into category i).

$$\pi^* = \mathbf{M}\cdot\pi \tag{18.3}$$

If the matrix **M** is invertible, $\mathbf{M}^{-1}\pi^*$ gives the unbiased category probabilities. Using an additional index (0,1) to denote controls and cases, respectively, differential exposure misclassification is defined by $\mathbf{M}_0 \neq \mathbf{M}_1$.

For dichotomous exposure, Equation 18.3 can easily be solved to yield

$$\pi_{1y} = 1 - \pi_{0y}\frac{\pi^*_{1y} + SP_y - 1}{SE_y + SP_y - 1} \qquad y = 0,1 \tag{18.4}$$

where y = 0,1 denotes controls and cases, respectively; SP_y, the specificity (m_{00y}); and SE_y, the sensitivity (m_{11y}) in controls or cases (see Chapter 4). From Equation 18.4, the unbiased OR (ψ) can be calculated:

$$\psi = \frac{\pi^*_{11} + SP_1 - 1}{\pi^*_{01} + SE_0 - 1}\,\frac{\pi^*_{00} + SE_0 - 1}{\pi^*_{10} + SP_0 - 1} \tag{18.5}$$

Table 18.1 shows calculations of the "true" OR and PAF based on the assumption of differential or nondifferential exposure misclassification due to measurements of average MF using the data reported in Ahlbom et al. (2000).

To illustrate this procedure, let us assume the sensitivity in cases and controls is 0.5 and 0.7, respectively, whereas the specificity is equal in cases and controls and amounts to 0.9. The under-this-scenario–biased figures for the exposed cases and controls are shown in the table.

Group	Exposed	Unexposed	Total
Cases	359	1842	2201
Controls	395	3152	3547

To compute (under the assumptions above) the true number of exposed cases, Equation 18.4 is applied:

$$2201\frac{359/2201+0.9-1}{0.5+0.9-1}=347.$$

This is done also for controls and finally the following values are obtained:

Group	Exposed	Unexposed	Total
Cases	347	1854	2201
Controls	67	3480	3547

From this table, the corrected OR is calculated as (347 × 3480)/(1854 × 67) = 9.7.

The PAF is calculated as $\frac{16/3547\cdot(9\cdot 7-1)}{1+16/3547\cdot(9\cdot 7-1)}=0.14.$

Sensitivity and specificity can vary only within the limits defined by Equation 18.4 because the numerator must be larger than zero. Of course, the OR can take on very large values as the sensitivity in cases approaches its lower limit. Also, the PAF can be large, for example, reaching 80% if sensitivity is differentially low in cases. Similarly large PAFs are reached for nondifferential low sensitivity.

It might be argued that such low sensitivities are unlikely; however, if most of the exposed cases actually are contained in the population segment with average levels <0.1 μT, which would be the case if the correct exposure metric correlates with average flux density but with a conditional variance decreasing with increasing average exposure level, such low sensitivities are the consequence. Therefore, a much larger attributable fraction is possible.

Because the public health impact could be substantial, but the exposure component associated with leukemia risk is unknown, any measure intended to reduce the risk must target all exposure aspects (e.g., underground cable instead of overhead power lines). In contrast, the large benefit from electrification must not be compromised. Because today's equipment operates to a large extent with direct current (DC), most households have alternating current (AC)-DC transformers. This results in high costs and waste of energy. Therefore, in the past 10 years, the feasibility of direct transport of DC into households and a complete redesign of electricity supply and distribution have been investigated. The additional advantage of a substantial reduction of MF exposure should enter the public discourse about the use of electricity in our society.

Mobile Phone Base Stations. Exposure and Well-Being. What if There Is a Relationship?

Since the early 1990s, tens of thousands of base stations have been erected in countries where digital networks were introduced. Although older systems with their low number of base stations have hardly received public attention, the vast increase in base stations required for more recent systems has led to public concerns worldwide. Anecdotal reports about various effects on well-being and health have also led to an increased awareness of physicians (Leitgeb et al. 2005; Huss and Röösli 2006). Despite professional and public concerns, the WHO International EMF Project has discouraged research into effects of base stations, because it deemed research into effects of mobile phones themselves to be of higher priority (WHO 2006). As a consequence, there are few studies dealing with exposure from mobile phone base stations (see Chapter 16). Some studies used experimental short-term exposure and a few applied personal dosimetry; however, the majority were studies of chronic exposure at home or at workplaces (Table 18.2; for reviews, see Röösli 2008; Kundi and Hutter 2009).

Several experimental studies were triggered by an investigation in the Netherlands by Zwamborn et al. (2003) that found an increase in self-reported symptoms after 45 min of exposure to a base station antenna resulting in about 2.7 mW/m^2 exposure at the place where the participants, electrohypersensitive (EHS) individuals and adults without known sensitivity to EMFs, were seated. Other studies however found little evidence of a short-term response to base station signals (Regel et al. 2006; Eltiti et al. 2007, 2009; Riddervold et al. 2008; Furubayashi et al. 2009; Wallace et al. 2010, 2012). Although subjects with a self-attribution of hypersensitivity to EMFs tend to report about symptoms they experience when in the vicinity of sources of EMF exposure, it has been shown in several of these studies that upon blind exposure, these individuals are not able to discriminate between "on" and "off" conditions (see Chapter 16). However, the difficulty in testing their ability to discriminate exposure conditions in a suitable experimental setting has to be appreciated. It might be that this test situation in a laboratory environment causes too much background arousal that the EMF exposure can no longer be discriminated. Therefore, field experiments, such as Leitgeb et al. (2008), are of particular value in that they test the subjects at home in their familiar environment by using shielding devices. This experiment found no differences (concerning sleep) between shielding and sham shielding conditions. Unfortunately, no data about exposure intensity during the field-experimental conditions were reported, making it impossible to assess this outcome.

A few other field intervention studies combined the benefits of a rigorous experimental design with the advantages of observing participants in their familiar environment. Heinrich et al. (2007) studied effects of exposure to a Universal Mobile Telecommunications System (UMTS) base station that was randomly turned on and off in a double-blind procedure at workplaces in an office building. There was a tendency for decreased subjective well-being at the end of on-days but evaluation was done without considering actual exposure that varied considerably between workplaces. The same deficiency was present in the field intervention trial of Danker-Hopfe et al. (2010) where a Global System for Mobile (GMS) communications base station antenna was erected in 10 villages with no other antenna in its vicinity. The antenna was turned on and off during 10 nights in a random sequence. Sleep was monitored by electroencephalogram (EEG) and electro-oculogram (EOG) recordings in the homes of 397 participants living below 50–500 m from the experimental base station. Subjective sleep quality was assessed by morning and evening protocols. Comparison of sham and real exposure nights revealed no significant difference overall, and after subdivision into individuals with and without concerns about adverse effects of the base station. Actual exposure, which must have varied considerably, was not assessed.

Several cross-sectional studies based on personal dosimetry were conducted in Bavaria (Thomas et al. 2008, 2010; Heinrich et al. 2010, 2011). Adults as well as adolescents and children were provided with a Maschek ESM 140 dosimeter for 24 hr. Exposure to mobile phone frequencies (up- and down-link) was weighted by the inverse of the International Commission on Non-Ionizing Radiation Protection (ICNIRP) reference levels and expressed

TABLE 18.2

Studies about Mobile Phone Base Stations and Well-Being

Study	Study Type	Group Studied	Exposure Duration	Exposure Intensity/ Surrogate	Effect
Santini et al. 2002, 2003	Cross-sectional	Adults (46 ± 20 years)	Chronic	<300 m (self-rated)	Reduced well-being
Navarro et al. 2003	Cross-sectional	Adults (>15 years)	Chronic	1.1 mW/m^2 average	Reduced well-being
Zwamborn et al. 2003	Crossover	EHS and controls	45 min	2.7 mW/m^2	Reduced well-being after UMTS
Hutter et al. 2006	Cross-sectional	Adults (18–91 years)	Chronic	>0.5 mW/m^2	Headache, vegetative symptoms, concentration difficulties increased
Regel et al. 2006	Crossover	EHS and controls	45 min	UMTS 2.7 and 265 mW/m^2	No effect on well-being
Abdel-Rassoul et al. 2007	Cross-sectional	Adult employees	Chronic	<1 mW/m^2 (inside and opposite building with base station)	Headache, sleeping problems, concentration difficulties, neurological symptoms increased
Heinrich et al. 2007	Field intervention	Adult employees	Short term	0.03 mW/m^2 average, 0.75 mW/m^2 max	Tendency for increased symptom score
Riddervold et al. 2008	Crossover	Adolescents and adults	45 min	2.1–12.8 mW/m^2	Increase in headaches (adolescents and adults combined)
Blettner et al. 2008	Cross-sectional	Adults (14–70 years)	Chronic	<500 m (geocoded)	Higher symptom score
Leitgeb et al. 2008	Field intervention	EHS (55 ± 11 years)	Short term	Unknown	No effect
Thomas et al. 2008	Cross-sectional	Adults (18–65 years)	Short term and chronic	~0.05–0.4 mW/m^2	Headache, concentration difficulties not significantly elevated
Augner et al. 2009	Field intervention	Adults (18–67 years)	Short term	0.15–2.1 mW/m^2	Increased calmness
Berg-Beckhoff et al. 2009	Cross-sectional	Adults (15–71 years)	Chronic	≥0.03 mW/m^2	No effect on well-being
Eltiti et al. 2007, 2009	Crossover	EHS and controls	50 min	GSM 10 mW/m^2 and UMTS 10 mW/m^2	Reduced well-being after GSM and especially UMTS in EHS individuals
Furubayashi et al. 2009	Crossover	Women	30 min	265 mW/m^2	No effect on comfort and cognitive function

Continued

TABLE 18.2 (*Continued*)

Studies about Mobile Phone Base Stations and Well-Being

Study	Study Type	Group Studied	Exposure Duration	Exposure Intensity/ Surrogate	Effect
Danker-Hopfe et al. 2010	Field intervention	Adults (18–81 years)	Short term	50–500 m	No effect on subjective and objective sleep quality
Heinrich et al. 2010	Cross-sectional	Children and adolescents (8–17 years)	Short term	~0.03–0.2 mW/m^2	Headache, irritation, concentration difficulties increased
Mohler et al. 2010	Cross-sectional and 1-year follow-up	Adults (30–60 years)	Chronic	>0.05 mW/m^2	Sleep disturbances not significantly increased
Röösli et al. 2010b	Cross-sectional and 1-year follow-up	Adults (30–60 years)	Chronic	>0.05 mW/m^2	No effect on well-being and sleep
Thomas et al. 2010	Cross-sectional	Children and adolescents (8–17 years)	Chronic	~0.03–0.5 mW/m^2	Conduct problems increased
Baliatsas et al. 2011	Cross-sectional	Adults (≥18 years)	Chronic	Average 347 m	No effect on physical symptoms
Heinrich et al. 2011	Cross-sectional	Children and adolescents(8–17 years)	Chronic	~0.03–0.5 mW/m^2	No effect on well-being
Wallace et al. 2010, 2012	Crossover	EHS and controls	50 min	10 mW/m^2	No effect on well-being, cognitive and physiological responses
Bortkiewicz et al. 2012	Cross-sectional	Adults (≥18 years)	Chronic	101–150 m	Headache increased
Frei et al. 2012	Cross-sectional and 1-year follow-up	Adults (30–60 years)	Chronic	>0.05 mW/m^2	No effect on well-being
Mohler et al. 2012	Field study nested within cross-sectional study	Adults (30–60 years)	Short term	>0.05 mW/m^2	No effect on sleep quality

as a percentage. Adults, children, and adolescents in the highest exposure quartile showed a tendency for reduced well-being; and in children and adolescents, significantly increased conduct problems were reported (based on the Strengths and Difficulties Questionnaire). Due to the problems with discriminating between up- and down-link signals, exclusion of nighttime exposure, and especially due to the averaging procedure with very low levels of about 0.03 mW/m^2 as the lower limit of the highest exposure category, substantial exposure misclassification could have biased results toward lower effect sizes of exposure.

Another series of investigations were published in the context of the QUALIFEX study (Mohler et al. 2010, 2012; Röösli et al. 2010b; Frei et al. 2012). This study had a design combining a cross-sectional approach with a follow-up investigation and a nested field study. Exposure to sources of RF-EMFs was determined using validated calculation methods. Participants were randomly selected adults between 30 and 60 years of age from the Basel area in Switzerland. In none of these investigations was a significant relationship between exposure to stationary sources of exposure and well-being or objective and subjective sleep quality obtained. However, similar to other investigations relying on an unstructured random sample, exposure was too low to expect an effect. The 90th percentile was as low as 0.05 mW/m^2, demonstrating the very low exposure of the population to signals from stationary transmitters of RF fields.

The first studies on this issue were cross-sectional studies with exposure determined based on the distance to the base station as estimated by the participants themselves (Santini et al. 2002, 2003). Although criticism against this approach and also the recruitment methods is justified, results about decreased well-being in the vicinity of base stations may in part be valid. Of note is the observed relationship between distance and symptoms that is not monotonously declining as would be expected if only concerns about adverse effects are responsible for the increased prevalence of symptoms in the vicinity of base stations. Rather, there is a trend that corresponds to the actual exposure patterns that is unknown by the public and typically shows highest exposures at a distance of 50–100 m from the base station. Navarro et al. (2003) applied an improved design with measurements in the homes of participants. Both analysis by distance and analysis by measured field strength revealed an association with symptom scores.

A large population-based cross-sectional study was conducted in two phases (Blettner et al. 2008; Berg-Beckhoff et al. 2009) in Germany. In the initial phase, 30,047 persons from a total of 51,444 (58% response rate) who took part in a nationwide survey also answered questions about mobile phone base stations. A significantly increased symptom score was found in those living within 500 m (obtained by geocoding of addresses) of a mobile phone base station after correction for concerns about adverse effects from exposure to EMFs. In the second phase, a sample of those residing in preselected areas was chosen for a more comprehensive assessment of well-being and for EMF measurements in the homes. In this subsample, no association between measured exposure and well-being was found. Similarly to other studies, this study was compromised by the very low levels of exposure, with only 34% above the sensitivity level of the dosimeter (~0.007 mW/m^2) and a 99th percentile of 0.3 mW/m^2.

The study to date with a design having the highest potential to detect an effect, if there is one, avoiding too low exposure levels was carried out by us (Hutter et al. 2006). We took the low average exposure of the population into account and selected participants randomly within predefined areas next to 10 preselected base stations. Base stations were selected in urban and rural areas according to specific criteria (no protests of the neighbors against the base station, no other nearby base station, only GSM 900 network). Based on the antenna characteristics provided by the network operators, the area was divided into segments with an expected large exposure gradient at each distance from the antenna. Measurements in the sleeping rooms were conducted after the data about sleep, well-being, and cognitive performance were obtained. Despite the preselection of the study areas, it transpired that the measured exposure to base station signals was only about half the expected power density. Consequently, the two lower quartiles had levels too low to be meaningfully discriminated and were thus combined into one category (categories were ≤0.1, 0.1–0.5, and >0.5 mW/m^2). Results indicated increased prevalence of headaches, vegetative symptoms, and difficulties with concentration at increasing exposure levels. Sleep problems were associated also with higher levels of exposure, but this association was no longer significant after correcting for concerns about adverse effects from base stations in the analysis. Although this study has the advantage of a meaningful gradient of exposures within the same area, a limitation is the cross-sectional approach. A longitudinal study with a similar design is still missing.

Taking account of the evidence available to date, adverse effects on well-being from long-term exposure to radiofrequency EMFs from mobile phone base stations exceeding levels around 1 mW/m^2 are likely.

Given the evidence of an impact on well-being, with an increase in the prevalence of so-called unspecific symptoms, the siting of mobile phone base stations in future should consider not only aspects of radio engineering and economics but also exposure of the neighbors in the context of a strategy of prudent avoidance. Large investigations such as the QUALIFEX study in Switzerland (Mohler et al. 2010, 2012; Röösli et al. 2010; Frei et al. 2012) and the study by Berg-Beckhoff et al. (2009) in Germany demonstrate that the proportion of the population exposed to levels that have been associated with long-term effects on well-being is very low, not exceeding 1%. It should therefore be feasible to site base stations in such a way that exposure levels around 1 mW/m^2 are not exceeded.

The large variation in study design, exposure levels, and methods of exposure assessment, outcome definition and measurement makes it impossible to derive at a meta-analytical effect estimate. Considering the studies of Hutter et al. (2006), Heinrich et al. (2010), and Bortkiewicz et al. (2012) as well as those of Santini et al. (2002, 2003) and Navarro et al. (2003), an about twofold increased prevalence of some unspecific symptoms at comparatively high chronic exposure levels (above 1 mW/m^2) can be expected. Because almost everybody experiences one or the other of these symptoms (e.g., headaches, dizziness, nausea, irritability, concentration difficulties) once in a while, the possible increase in prevalence can hardly be noticed. Due to the low fraction of the population exposed to levels expected to increase the prevalence of these symptoms, the population-attributable fraction is low too (see Equation 18.2). However, to protect public health it is recommended to take precaution during erection of base stations to choose places and antenna configurations as well as technical specifications that guarantee levels of exposure of the neighbors as low as possible and not exceeding a target value of about 1 mW/m^2 at places where people are living or are spending considerable amounts of time.

References

Abdel-Rassoul G, El-Fateh OA, Salem MA, Michael A, Farahat F, El-Batanouny M, et al. 2007. Neurobehavioral effects among inhabitants around mobile phone base stations. *Neurotoxicology*. 28(2): 434–440.

Ahlbom A, Day N, Feychting M, Roman E, Skinner J, Dockerty J, et al. 2000. A pooled analysis of magnetic fields and childhood leukaemia. *Br J Cancer*. 83: 692–698.

Augner C, Florian M, Pauser G, Oberfeld G, Hacker GW. 2009. GSM base stations: short-term effects on well-being. *Bioelectromagnetics*. 30(1): 73–80.

Baliatsas C, van Kamp I, Kelfkens G, Schipper M, Bolte J, Yzermans J, et al. 2011. Non-specific physical symptoms in relation to actual and perceived proximity to mobile phone base stations and powerlines. *BMC Public Health*. 11: 421.

Berg-Beckhoff G, Blettner M, Kowall B, Breckenkamp J, Schlehofer B, Schmiedel S, et al. 2009. Mobile phone base stations and adverse health effects: phase 2 of a cross-sectional study with measured radio frequency electromagnetic fields. *Occup Environ Med*. 66(2): 124–130.

Bernard N, Alberdi AJ, Tanguy ML, Brugere H, Helissey P, Hubert C, et al. 2008. Assessing the potential leukemogenic effects of 50 Hz magnetic fields and their harmonics using an animal leukemia model. *J Radiat Res*. 49(6): 565–577.

Blettner M, Schlehofer B, Breckenkamp J, Kowall B, Schmiedel S, Reis U, et al. 2008. Mobile phone base stations and adverse health effects: phase 1 of a population-based, cross-sectional study in Germany. *Occup Environ Med*. 66(2): 118–123.

Bortkiewicz A, Gadzicka E, Szyjkowska A, Politański P, Mamrot P, Szymczak W, et al. 2012. Subjective complaints of people living near mobile phone base stations in Poland. *Int J Occup Med Environ Health*. 25(1): 31–40.

Danker-Hopfe H, Dorn H, Bornkessel C, Sauter C. 2010. Do mobile phone base stations affect sleep of residents? Results from an experimental double-blind sham-controlled field study. *Am J Hum Biol*. 22(5): 613–618.

Deltour I, Auvinen A, Feychting M, Johansen C, Klaeboe L, Sankila R, et al. 2012. Mobile phone use and incidence of glioma in the Nordic countries 1979–2008: consistency check. *Epidemiology*. 23(2): 301–307.

Deltour I, Johansen C, Auvinen A, Feychting M, Klaeboe L, Schüz J. 2009. Time trends in brain tumor incidence rates in Denmark, Finland, Norway, and Sweden, 1974–2003. *J Natl Cancer Inst*. 101(24): 1721–1724.

de Vocht F, Burstyn I, Cherrie JW. 2011a. Time trends (1998–2007) in brain cancer incidence rates in relation to mobile phone use in England. *Bioelectromagnetics*. 32(5): 334–339.

de Vocht F, Burstyn I, Cherrie JW. 2011b. Authors' reply to Kundi's comments on deVocht et al. "Time trends (1998–2007) in brain cancer incidence rates in relation to mobile phone use in England." *Bioelectromagnetics*. 32(8): 675–676.

Dobes M, Shadbolt B, Khurana VG, Jain S, Smith SF, Smee R, et al. 2011. A multicenter study of primary brain tumor incidence in Australia (2000–2008). *Neuro Oncol*. 13(7): 783–790.

Eltiti S, Wallace D, Ridgewell A, Zougkou K, Russo R, Sepulveda F, et al. 2009. Short-term exposure to mobile phone base station signals does not affect cognitive functioning or physiological measures in individuals who report sensitivity to electromagnetic fields and controls. *Bioelectromagnetics*. 30(7): 556–563.

Eltiti S, Wallace D, Ridgewell A, Zougkou K, Russo R, Sepulveda F, et al. 2007. Does short-term exposure to mobile phone base station signals increase symptoms in individuals who report sensitivity to electromagnetic fields? A double-blind randomized provocation study. *Environ Health Perspect*. 115(11): 1603–1608.

Frei P, Mohler E, Braun-Fahrländer C, Fröhlich J, Neubauer G, Röösli M, et al. 2012. Cohort study on the effects of everyday life radio frequency electromagnetic field exposure on non-specific symptoms and tinnitus. *Environ Int*. 38(1): 29–36.

Furubayashi T, Ushiyama A, Terao Y, Mizuno Y, Shirasawa K, Pongpaibool P, et al. 2009. Effects of short-term W-CDMA mobile phone base station exposure on women with or without mobile phone related symptoms. *Bioelectromagnetics*. 30(2): 100–113.

Greenland S, Sheppard AR, Kaune WT, Poole C, Kelsh MA. 2000. A pooled analysis of magnetic fields, wire codes, and childhood leukemia. Childhood Leukemia–EMF Study Group. *Epidemiology*. 11: 624–634.

Hardell L, Carlberg M, Hansson Mild K. 2006a. Pooled analysis of two case-control studies on the use of cellular and cordless telephones and the risk of benign brain tumours diagnosed during 1997–2003. *Int J Oncol*. 28: 509–518.

Hardell L, Carlberg M, Hansson Mild K. 2006b. Pooled analysis of two case-control studies on use of cellular and cordless telephones and the risk for malignant brain tumours diagnosed in 1997–2003. *Int Arch Occup Environ Health*. 79(8): 630–639.

Heinrich S, Ossig A, Schlittmeier S, Hellbrück J. 2007. Elektromagnetische Felder einer UMTS-Mobilfunkbasisstation und mögliche Auswirkungen auf die Befindlichkeit—eine experimentelle Felduntersuchung. *Umwelt Med Forsch Prax*. 12: 171–180.

Heinrich S, Thomas S, Heumann C, von Kries R, Radon K. 2010. Association between exposure to radiofrequency electromagnetic fields assessed by dosimetry and acute symptoms in children and adolescents: a population based cross-sectional study. *Environ Health*. 9: 75.

Heinrich S, Thomas S, Heumann C, von Kries R, Radon K. 2011. The impact of exposure to radio frequency electromagnetic fields on chronic well-being in young people—a cross-sectional study based on personal dosimetry. *Environ Int*. 37(1): 26–30.

Huss A, Röösli M. 2006. Consultations in primary care for symptoms attributed to electromagnetic fields – a survey among general practitioners. *BMC Public Health*. 30(6): 267.

Hutter H-P, Moshammer H, Wallner P, Kundi M. 2006. Subjective symptoms, sleeping problems and cognitive performance in subjects living near mobile phone base stations. *Occup Environ Med*. 63: 307–313.

IARC (International Agency for Research on Cancer). 2002. *Monographs on the Evaluation of Carcinogenic Risks to Humans: Volume 80. Non-Ionizing Radiation, Part 1: Static and Extremely Low Frequency (ELF) Electric and Magnetic Fields*. Lyon, France: IARC Press.

IARC (International Agency for Research on Cancer). 2013. *Monographs on the Evaluation of Carcinogenic Risks to Humans: Volume 102. Non-Ionizing Radiation, Part 2: Radiofrequency Electromagnetic Fields*. Lyon, France: IARC Press.

INTERPHONE Study Group. 2010. Brain tumour risk in relation to mobile telephone use: results of the INTERPHONE international case-control study. *Int J Epidemiol*. 39(3): 675–694.

Johansen C, Boice J, Jr, McLaughlin J, Olsen JH. 2001. Cellular telephones and cancer–a nationwide cohort study in Denmark. *J Natl Cancer Inst*. 93: 203–207.

Kheifets L, Ahlbom A, Crespi CM, Draper G, Hagihara J, Lowenthal RM, et al. 2010. Pooled analysis of recent studies on magnetic fields and childhood leukaemia. *Br J Cancer*. 103(7): 1128–1135.

Kranzinger M, Jones N, Rittinger O, Pilz P, Piotrowski WP, Manzl M, et al. 2001. Malignant glioma as a secondary malignant neoplasm after radiation therapy for craniopharyngioma: report of a case and review of reported cases. *Onkologie*. 24: 66–72.

Kundi M. 2010. Essential problems in the interpretation of epidemiologic evidence for an association between mobile phone use and brain tumours. *C R Physique*. 11: 556–563.

Kundi M, Hutter HP. 2009. Mobile phone base stations-effects on wellbeing and health. *Pathophysiology*. 16(2–3): 123–135.

Langholz B. 2001. Factors that explain the power line configuration wiring code-childhood leukemia association: what would they look like? *Bioelectromagnetics*. (Suppl 5): S19–S31.

Leitgeb N, Schröttner J, Böhm M. 2005. Does "electromagnetic pollution" cause illness? An inquiry among Austrian general practitioners. *Wien Med Wochenschr*. 155(9–10): 237–241.

Leitgeb N, Schröttner J, Cech R, Kerbl R. 2008. EMF-protection sleep study near mobile phone base stations. *Somnologie*. 12: 234–243.

Little MP, Rajaraman P, Curtis RE, Devesa SS, Inskip PD, Check DP, et al. 2012. Mobile phone use and glioma risk: comparison of epidemiological study results with incidence trends in the United States. *BMJ*. 344: e1147.

Lönn S, Klaeboe L, Hall P, Mathiesen T, Auvinen A, Christensen HC, et al. 2004. Incidence trends of adult primary intracerebral tumors in four Nordic countries. *Int J Cancer*. 108(3): 450–455.

Milham S, Ossiander EM. 2001. Historical evidence that residential electrification caused the emergence of the childhood leukemia peak. *Med Hypotheses*. 56: 290–295.

Mohler E, Frei P, Braun-Fahrländer C, Fröhlich J, Neubauer G, Röösli M, et al. 2010. Effects of everyday radiofrequency electromagnetic-field exposure on sleep quality: a cross-sectional study. *Radiat Res.* 174(3): 347–356.

Mohler E, Frei P, Fröhlich J, Braun-Fahrländer C, Röösli M, QUALIFEX-Team. 2012. Exposure to radiofrequency electromagnetic fields and sleep quality: a prospective cohort study. *PLoS One.* 7(5): e37455.

Mohyuddin A, Vokurka E, Evans D, Ramsden R, Jackson A. 2003. Is clinical growth index a reliable predictor of tumour growth in vestibular schwannomas? *Clin Otolaryngol Allied Sci.* 28: 85–90.

Navarro EA, Segura J, Portoles M, Gomez-Perretta C. 2003. The microwave syndrome: a preliminary study in Spain. *Electromagn Biol Med.* 22: 161–169.

Popper K. 1945. *The Open Society and Its Enemies.* London: Routledge.

Regel SJ, Negovetic S, Röösli M, Berdinas V, Schuderer J, Huss A, et al. 2006. UMTS base station like exposure, well being and cognitive performance. *Environ Health Perspect.* 114: 1270–1275.

Riddervold IS, Pedersen GF, Andersen NT, Pedersen AD, Andersen JB, Zachariae R, et al. 2008. Cognitive function and symptoms in adults and adolescents in relation to RF radiation from UMTS base stations. *Bioelectromagnetics.* 29(4): 257–267.

Röösli M. 2008. Radiofrequency electromagnetic field exposure and non-specific symptoms of ill health: a systematic review. *Environ Res.* 107(2): 277–287.

Röösli M, Frei P, Mohler E, Hug K. 2010a. Systematic review on the health effects of exposure to radiofrequency electromagnetic fields from mobile phone base stations. *Bull World Health Organ.* 88(12): 887–896F.

Röösli M, Mohler E, Frei P. 2010b. Sense and sensibility in the context of radiofrequency electromagnetic field exposure. *C R Physique.* 11: 576–584.

Sadetzki S, Chetrit A, Freedman L, Stovall M, Modan B, Novikov I. 2005. Long-term follow-up for brain tumor development after childhood exposure to ionizing radiation for tinea capitis. *Radiat Res.* 163: 424–432.

Salvati M, D'Elia A, Melone GA, Brogna C, Frati A, Raco A, et al. 2008. Radio-induced gliomas: 20-year experience and critical review of the pathology. *J Neurooncol.* 89(2): 169–177.

Santini R, Santini P, Danze JM, Le Ruz P, Seigne M. 2002. Enquête sur la santé de riverains de stations relais de téléphonie mobile: I/Incidences de la distance et du sexe. *Pathol Biol (Paris).* 50: 369–373.

Santini R, Santini P, Danze JM, Le Ruz P, Seigne M. 2003. Enquête sur la santé de riverains de stations relais de téléphonie mobile: II/Incidences de l'âge des sujets de la durée de leur exposition et de leur position par rapport aux antennes et autres sources électromagnétiques. *Pathol Biol (Paris).* 51: 412–415.

Thomas S, Heinrich S, von Kries R, Radon K. 2010. Exposure to radio-frequency electromagnetic fields and behavioural problems in Bavarian children and adolescents. *Eur J Epidemiol.* 25(2): 135–141.

Thomas S, Kühnlein A, Heinrich S, Praml G, Nowak D, von Kries R, et al. 2008. Personal exposure to mobile phone frequencies and well-being in adults: a cross-sectional study based on dosimetry. *Bioelectromagnetics.* 29: 463–470.

Umansky F, Shoshan Y, Rosenthal G, Fraifeld S, Spektor S. 2005. Radiation-induced meningioma. *Neurosurg Focus.* 24: E7.

Wallace D, Eltiti S, Ridgewell A, Garner K, Russo R, Sepulveda F, et al. 2010. Do TETRA (Airwave) base station signals have a short-term impact on health and well-being? A randomized double-blind provocation study. *Environ Health Perspect.* 118(6): 735–741.

Wallace D, Eltiti S, Ridgewell A, Garner K, Russo R, Sepulveda F, et al. 2012. Cognitive and physiological responses in humans exposed to a TETRA base station signal in relation to perceived electromagnetic hypersensitivity. *Bioelectromagnetics.* 33(1): 23–39.

WHO (World Health Organization). 2006. *Research Agenda for Radio Frequency Fields*: WHO.

WHO (World Health Organization). 2007. *Environmental Health Criteria 238: Extremely Low Frequency Fields.* Geneva: WHO.

Wöhrer A, Waldhör T, Heinzl H, Hackl M, Feichtinger J, Gruber-Mösenbacher U, et al. 2009. The Austrian brain tumour registry: a cooperative way to establish a population-based brain tumour registry. *J Neurooncol*. 95(3): 401–411.

Zwamborn APM, Vossen SHJA, van Leersum BJAM, Ouwens MA, Mäkel WN. 2003. *Effects of Global Communication System Radio-Frequency Fields on Well Being and Cognitive Functions of Human Subjects with and without Subjective Complaints*. The Hague, the Netherlands: TNO. FEL-03-C148.

19

An Outlook into Future EMF Exposure Scenarios

Norbert Leitgeb

CONTENTS

Introduction

The electromagnetic environment is characterized by an ongoing dramatic change within the entire frequency range of nonionizing electric, magnetic, and electromagnetic fields (EMFs), such as by using yet unused frequencies and intensifying use of already used frequencies. This development is driven by emerging technologies and new appliances, in particular, various kinds of smart devices and technologies such as smart phones, smart cars, smart homes, smart clothing, smart medical applications, and even smart power generation and distribution. New EMF-emitting sources are applied inside and at our body, within our homes, at workplaces, in warehouses, in streets, inside and outside vehicles, and public transport.

Telecommunication

In telecommunication, future development is characterized by two main streams: main stream one and mainstream two.

Main stream one is characterized by evolution from mobile personal oral communication to inclusion of mobile data exchange and communication (such as mobile Internet access) to upcoming mobile entertainment such as by video-on-demand services. As a consequence, this development is characterized by an increasing demand on data rates from initial low data rates for speech communication (at about 10 kbit/s) to high data rates, such as for high definition video transmission, at up to 6 Mbit/s. This is accompanied by an increasing need of higher transmission capacities to serve the increasing number of clients requesting higher data rates. Technically, higher data rates require larger bandwidths, and higher transmission capacity requires additional frequencies. However, the range of frequencies suitable for telecommunication is limited by physical constraints that make frequencies a rare and hence precious good. The growing hunger for data rates and transmission capacity is reflected by the technical evolution of mobile communication systems from the first to the currently fourth generation (Figure 19.1):

- The first generation (1G) mobile communication system was based on *analog* frequency-modulated signals and was introduced in 1980. It needed two frequencies allocated to each user for uplink and downlink, operated at about 450 MHz with handset-radiated power of 600 mW and data rates of about 28–56 kbit/s.
- This system was replaced by the second generation (2G) digital Global System for Mobile (GSM) communication that was first implemented in 1991. By allocating to clients one of repeatedly transmitted time slots (time division multiple access [TDMA]), it was possible for eight participants to simultaneously communicate at the same carrier frequency. This allowed more efficient use of available frequencies. This generation is still in use and operates at 900 MHz and 1.8 GHz with 250 kHz bandwidth per channel. Depending on the quality of radio connection, dynamic power regulation allows down-regulation of emitted power from 250 to about 10 mW. Speech-coded data rates are 9.6 kbit/s per allocated time slot (user).
- To meet the demand for higher data rates by an improved GSM system, the (2.5G) General Packet Radio Service (GPRS) was introduced in 2000. It allowed increasing

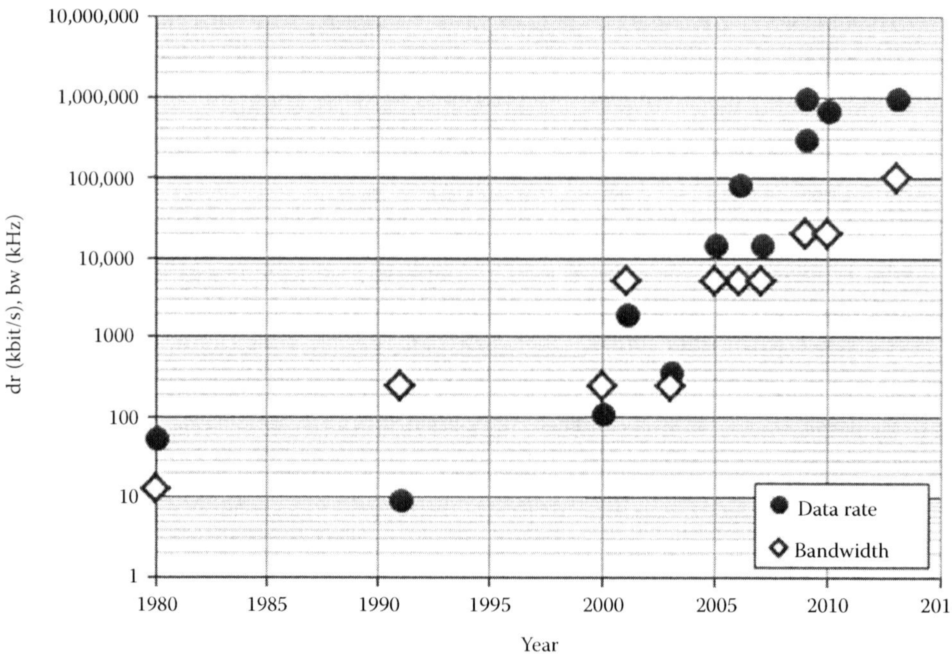

FIGURE 19.1
Development of bandwidth (bw, squares) and maximum data rate (dr, circles) of mobile communication systems.

data rates by improved coding and by allocation of more than one time slot to users with data rates up to 56–115 kbit/s, however, at the expense of proportional increase of handset-transmitted power.

- The system Enhanced Data rate for GSM Evolution (EDGE) was a further step toward higher data rates. It is based on the GSM system and led to a further improved second generation (2.75G) system and was realized in 2003. It allowed increasing data rates per time slot by improved encoding. Depending on the quality of the transmission link, data rates of 48–69 kbit/s per time slot could be achieved that could be further increased to up to about 384 kbit/s by allocating several time slots to a user.
- The third generation (3G) named Universal Mobile Telecommunication System (UMTS) was deployed in 2001 and operates at 1.9 and 2.1 GHz. To achieve even higher data rates, the coding scheme was changed to Code Division Multiple Access (CDMA). This allows simultaneous communication of several users by allocating to them codes rather than time slots. In addition, the bandwidth per channel was increased by 20-fold up to 5 MHz. With this approach, data rates up to 2 Mbit/s could be realized.
- Because the demand for data rates in telecommunication is asymmetric and higher for downloading data, High-Speed Downlink Packet Access (HSDPA) has been launched in 2006. It is coined 3.5G system (or 3G+ or turbo 3G) and enhances the 3G communication protocol. Depending on front-end devices, it allows data transfer downlink rates of 1.8, 3.6, 7.2, and 14.0 Mbit/s. Higher data rates were provided by High Speed Packet Access plus (HSPA+) with data rates up to 42 Mbit/s downlink and with High Speed Download Packet Access plus (HSDPA+) even with 84 Mbit/s. Improved uplink data rates up to 200 kbit/s were achieved by High Speed Uplink Packet Access (HSUPA) coding.
- The (almost) fourth generation (3.9G) Long Term Evolution (LTE) is an evolution of the UMTS/GPRS system. It was launched in 2009. In Europe, it operates at frequencies that differ among countries. In Germany, 800 MHz, 900 MHz, 1.8 GHz, and 2.4 GHz are allocated for this service. It uses ultrawide bandwidths scalable in the range 1.4–20 MHz and asymmetric data rates of up to 300 Mbit/s (downlink) and up to 75 Mbit/s (uplink) and allows high-speed Internet access and high definition (HD) video transmission.
- The fourth generation (4G) offered for Internet access by mobile phones was realized by adopting the Worldwide interoperability for Microwave Access (WiMAX) scheme. It operates at frequency bands within 2–66 GHz, for example, in the band 2.5–3.5 GHz and is specified for data rates up to 1 Gbit/s with a bandwidth of 1.25–20 MHz.
- LTE-advanced, another approach to fourth generation (4G) mobile telecommunication, was deployed in 2013 and, due to increased bandwidth (in the range 20–100 MHz) and improved coding, allows data rates up to 1 Gbit/s at frequencies that became available from former analog broadcasting such as in the bands 790–862 MHz (digital dividend 1) and 694–790 MHz (digital dividend 2).

Main stream two is characterized by the evolution of Internet use from the "Internet of computers" (connected via the world wide web) to the "Internet of people" (connected via social networks) to the upcoming "Internet of things" (connected via radiofrequency identification [RFID] technology). The Internet of things will allow addressing of and communicating with devices that may be equipped with sensors and/or actuators and hence allow remote data acquisition and/or device operation (Shi et al. 2010; Coetze et al. 2011).

Emerging technologies are going to make life "smart" by introducing smart technologies, such as smart power grids, smart homes, smart cars, smart TV, and smart cards, or smart applications, such as smart phones, smart radios, and smart clothing.

This development is associated with a considerable change of applying RF-EMF sources in terms of use frequency, duration per use and use pattern.

Tele-Entertainment

The conventional phone use is dramatically changing. In Germany, by 2009, 69% of the general population used their phone for calls, 69% also for short message services (SMS), and 32% also for multimedia messaging services (MMS). At that time, already 45% of visits of social networks were done via mobile phones, and from 2009 to 2010 the motivation to use Internet for entertainment increased by 2.6-fold (TNS 2010, 2011). In addition, an increasing number of applications (apps) are installed on smart phones that substantially contribute to data traffic. This change in use is accompanied by a change of technical requirements with regard to data rates and transmission capacity. Therefore, existing and future communication needs high-speed and high-capacity systems and, most importantly, sufficient bandwidth (Figure 19.1). It can be expected that "video-on-demand," or watching videos and movies any time anywhere, has the potential to become a major application comparable to the initial boom of talking any time anywhere to (almost) anybody, however, at the expense of a manifold increased capacity of data exchange.

This development results in an enormous hunger for new frequencies that, however, is facing spectral limitations. Frequencies suited for (tele-)communication are limited and hence are a precious good. Consequently, there is a strong motivation to use the limited commodity "frequency" as efficiently as possible. This also has an impact on environmental EMFs. One reason is the extension of the used frequencies far beyond 2 GHz such as for Wireless Fidelity (WiFi) and Wireless Gigabit (WiGig) short-range communication with frequencies up to some 60 GHz. Another reason is the introduction of new signal coding schemes to multiply the use of single frequencies. This was the basis for changing from analog to digital mobile communication and from analog to digital video (and audio) broadcasting, and for freeing frequencies for new and/or more services. Because the hunger for bandwidth for large-scale video communication can hardly be satisfied by allocating enough frequencies, it can be expected that this service be provided by a mix of wired (optical fiber) transmission and bridging the last gap to users via short-range telecommunication both outdoors and within homes.

The transition from analog to digital video broadcasting is accompanied by changed broadcasting infrastructure. The digital antennas transmit with reduced power output but are situated at additional sites closer to urban areas. Although the required power per transmitted program is considerably lower for digital versus analog broadcasting, this reduction is compensated by increasing the number of programs. Overall, the associated change of EMF exposure is nonuniform, with winners and losers, but on average exposure levels remain orders of magnitude below ICNIRP's reference levels (ICNIRP 1998; Bornkessel et al. 2007). Although in rural regions EMF immissions may be reduced, they may increase in urban areas. Measurements by Schubert et al. (2007) showed that the increase with terrestrial digital video broadcasting (DVB-T) can be sevenfold, whereas it may be 10-fold with digital audio broadcasting (DAB). Overall, the more intensive use of the allocated frequency spectrum will result in an elevation of the EMF background level.

Although broadcasting antennas cause widespread low-level immissions associated with low-level whole-body exposure, higher public exposure is caused by partial body or local exposure due to portable devices that are no longer limited to mobile phones, but now also include more versatile devices such as iPods, smart phones, notebooks, ipads, and tablets with different use options. Consequently, local exposure will no longer be restricted to the head and will involve other parts of the body, in particular, the lower abdomen, including gonads and thighs. Typically, exposure from WLAN, Bluetooth, or WiFi devices is lower than that of mobile phones, but its duration per use often lasts considerably longer. Except from devices contacting the thighs, the distance to head and thorax is larger and, consequently, their exposure to RF-EMF intensity is smaller.

However, there is one emerging new group of devices with antennas operated close to the head, namely, electronic goggles for 3D visualization. These are used to create a virtual reality for gaming or to augment reality such as for clinical use by assisting the surgeon in navigation and identifying the object of interest (Azimi et al. 2012), and they may also serve as guidance devices for pedestrians or drivers. Electronic video goggles or data goggles are equipped with integrated WiFi or Bluetooth connectivity. Due to duration and frequency of use, and the potential use by children for video gaming, goggles could merit particular attention.

Smart Technologies

RFID, although not receiving similar public attendance as mobile communication, has already considerably changed our life and will continue to do so. Storing detailed information in cheap chips, labels, and tags, and reading it remotely by wireless communication, enables identifying and controlling articles (electronic article surveillance [EAS]); tracking goods, objects, pets and livestock; identifying persons (personal identification number [PIN]) and monitoring and/or charging access to regions and services. RFID labels and tags are read either from some distance when passing examination gates or more closely when held directly to a reader of a near field communication (NFC) system.

Consequently, it has already become frequent in daily life to pass gates of EAS systems or PIN systems, and to be exposed, transiently almost, by whole body to gate fields or partial body to stray fields of NFC readers. Depending on the working principle, operating frequencies are those allocated for industrial, scientific, and medical (ISM) use and hence may belong to the Extremely Low-Frequency (ELF), intermediate (IF), or RF range. A review revealed that emitted field quantities of existing EAS devices varied considerably. Electromagnetic emissions can be rather high, and in particular high enough to adversely interfere with implanted pacemakers or cardioverter defibrillators (Leitgeb et al. 2012, 2013). Measured field maxima already exceed the former ICNIRP reference levels (ICNIRP 1998) up to 32-fold (Leitgeb et al. 2013). However, target identification reliability increases with increasing field amplitudes. Therefore, it can be expected that future EAS devices will generate higher fields as a consequence of the elevated reference levels of ICNIRP's revised exposure guidelines (ICNIRP 2010). In contrast to transient exposures of shoppers passing gates of shops, employees working close to EAS gates might be much longer exposed. Cashiers, for example, can be repeatedly and considerably higher exposed both to EAS gate fields and to the high electromagnetic pulses used for deactivating RFID labels. Also workers within the vicinity of large EAS gates of storehouses can be higher exposed when loading or checking transport vehicles such as trucks.

NFC applies mainly to the ISM frequency 13.56 MHz, and is used for exchanging of (comparably small amounts of) data stored on RFID chips, such as for payment at cashiers; ticketing at touch points, such as in public transport; or providing personal data, such as that contained in passports. Because an increasing number of phones are enabled for NFC, they can transmit data to and read them from NFC tags. However, users are locally exposed to low values of about 0.5% (Cecil et al. 2010) of ICNIRP's basic restriction (ICNIRP 1998).

Medical Applications

EMFs have gained increasing importance in health care (Leitgeb 2012). The major applications are based on direct interaction of EMFs either to deliver energy for treatment such as for transcranial magnetic stimulation by pulsed magnetic fields; by converting RF-EMF energy into heat for surgery, diathermy, and tumor ablation; or to acquire diagnostic information on distribution and binding status of nuclei by magnetic resonance imaging and spectroscopy. In addition, magnetic sources injected into tissues, such as magnetic nanoparticles, might justify attention in case of high EMF exposures. Emerging new applications are used for the purpose of monitoring, targeting, tracking, and navigating incorporated objects that may act themselves as EMF sources, such as electronic implants, capsular endoscopes (i.e., swallowed pill cameras with incorporated data-transmitting capability), and inserted micromanipulators. Therefore, such applications are associated with new, yet-unexplored exposure scenarios. Incorporated EMF sources in this manner expose tissues directly with no benefit from amplitude decrease due to separation distance. An overview on magnetic emissions from medical devices is presented in Figure 19.2.

NFC is becoming a widespread technology in health care. It may be used for several purposes, including for assisting or performing treatment. NFC allows communicating with electronic implants, for example, to read or change their functional settings, or read out stored bioevents. Consequently, the occurrence of associated exposures is increasing. As an example, the prevalence of pacemaker patients is already high and still increasing. With an average device lifetime of 9.8 years, the prevalence of patients with implanted cardiac pacemakers or cardioverter defibrillators can be estimated at about 0.8% of the population (Markewitz 2009). However, apart from acute cases, NFC-based pacemaker checks are performed at regular intervals, about once or twice a year. In contrast, frequent or even continuous NFC application is associated with enabling and controlling intracorporal electrical stimulation by implanted RFID transponders, used for spinal cord stimulation in chronic pain treatment; for deep brain stimulation to treat Parkinson's disease or epilepsy; for gastric stimulation of the pathologically obese; or for functional stimulation, such as breath stimulation in case of respiratory paralysis (Freudenthal et al. 2007).

Other NFC applications in health care include checking NFC-enabled patient RFID tags for patient identification to enable storage of health records or to verify patients for radiation treatment. Another widespread application is electronically reading and/or storing patient data on e-cards for administrative purposes and quickly and comprehensively identifying a patient's specific medical background and restrictions in delivering adequate treatment.

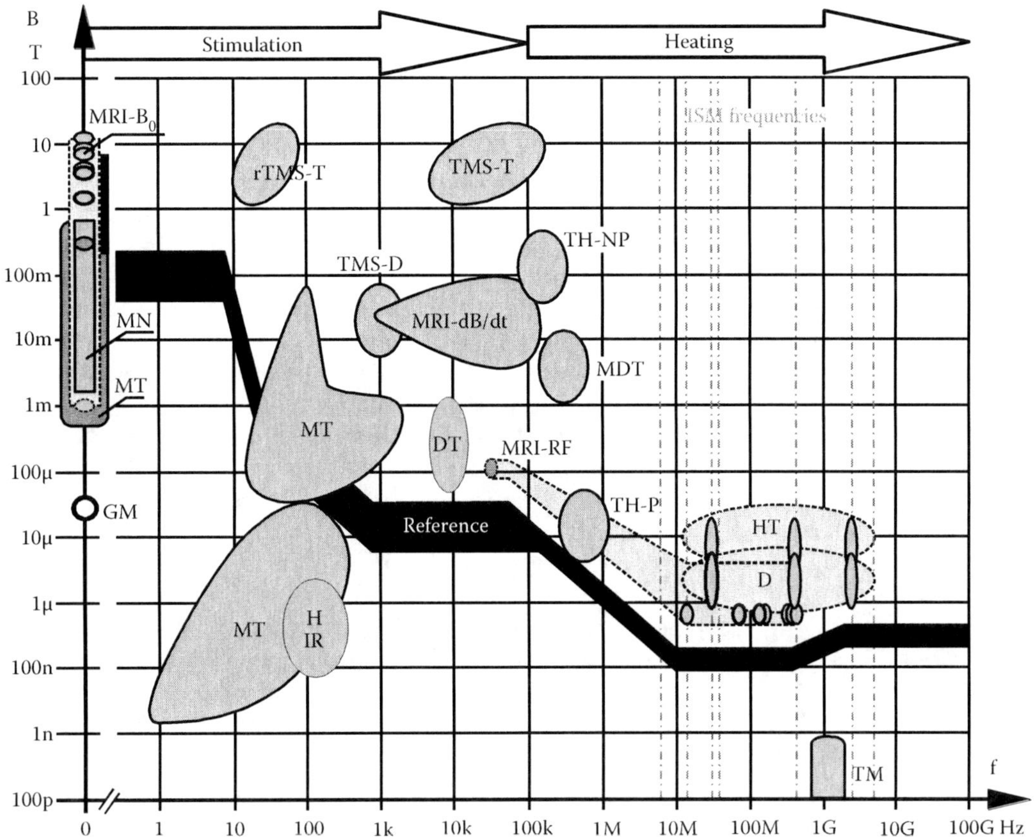

FIGURE 19.2
Magnetic emissions from medical devices. Reference values limiting exposure of the general population (lower boundary) and workers (upper boundary); GM, geomagnetic field; MT, magnet therapy;MN, magnetic navigation; TMS-T, transcranial therapeutic magnetic stimulation; rTMS-T, transcranial therapeutic magnetic stimulation; TMS-D, transcranial diagnostic magnetic stimulation; TH-NP, teleheating of nanoparticles; TH-NP, teleheating of ferromagnetic parts; MDT, magnetic drug targeting; DT, magnetic device tracking; D, diathermy; HT, hyperthermia; TM, telemedicine (Leitgeb 2012).

Magnetic resonance imaging (MRI) has become a major diagnostic tool, with still increasing pervasion and applicability. As an example, in Germany within the 10 years from 1999 to 2009, the frequency of diagnostic MRI has increased by 2.5-fold. On average, MRI is currently applied annually once per 10 inhabitants (BfS 2010). The operation principle is based on alignment of intracorporal nuclei (e.g., hydrogen) by very strong external static magnetic fields, excitation of the nuclei within a particular body slice by irradiation with resonant RF-EMF, and scanning for computed image reconstruction with the aid of rapidly switched magnetic gradients. Consequently, MRI is associated with increased prevalence of high combined exposures to static magnetic fields, transient magnetic fields (switched gradients), and RF EMFs with frequencies (f_{RF}) directly depending on the flux density (B_0) of the static magnetic field. For hydrogen imaging, the applied RF must meet the resonance equation

$$f_{RF} = 42.58 \times 10^6 . B_0$$

Currently, most widely used MRI devices apply 64 and 128 MHz corresponding to 1.5 T and 3 T static fields, respectively. Whole-body exposures may reach the averaged specific absorption rate (SAR) of 2 W/kg in normal operation mode and 4 W/kg in controlled operation mode stage 1. In controlled operation mode stage 2, the SAR may even exceed 4 W/kg. Regional exposures (of the exposed part of the body) are allowed to reach the regionally averaged SAR of 10 W/kg. The exposures of the head may reach a head-averaged SAR of 3.1 W/kg in normal and controlled operation mode stage 1; or in controlled operation mode stage 2, may go even beyond this limit (EN 60601-2-33). On average, exposures last for about 30 min. Consequently, the basic restriction of the general population in terms of the whole-body averaged SAR of 0.08 W/kg may be exceeded by 25- to 50-fold in normal and controlled operation stage 1, respectively. Medical staff may also experience high exposures before as well as during imaging, in particular with ultra high-field MRI (i.e., because of their presence in the vicinity of the devices for anesthetic monitoring).

MRI development leads to increasing magnetic fields up to ultrahigh magnetic flux densities. Introduction of 7 T devices into clinical routine is already ongoing; 9.4 T devices are upcoming and already used in clinical research; and 16.4 T devices, currently used in animal studies, are envisaged for human diagnostics in the future. The reasons for this development is that ultra high-field MRI enables generation of better images with improved spatial resolution and better signal-to-noise ratio, and quicker dynamic imaging as well as molecular imaging. Consequently, exposure to high static magnetic fields and high-intensity RF-EMFs up to several hundred megahertz will become more and more relevant.

Smart home technology is going to offer the possibility to wireless control appliances and installations such as heating, lighting, windows, or blinds. A central control unit allows device management by remote control via short-range wireless communication or mobile telecommunication, so as to enable activating the oven before coming home, or responding to visitors ringing at the front door even if not at home. In addition, RFID may enable checking for the presence of or need for goods during shopping by communicating with the refrigerator's interface. Even lost items could be found by this technology. Based on input signals of sensors, the central control unit may manage heating dependent on environmental conditions; the presence of persons and intended use of rooms; or control temperature, light, and air-conditioning according to a diurnal or environmentally dependent pattern. Together with smart meters, these technologies can manage energy consumption by adapting to fluctuating energy prices or avoiding expensive peak load charges.

Electromagnetic emissions from wireless home devices are low; however, they may reach levels similar to nearby mobile phone base stations (Kühn et al. 2007).

Lighting

In Europe, the directive demanding ecodesign for energy-consuming products (directive 2005/32/EC) together with the ban of popular bulbs (Commission Regulation 244/2009) has triggered the ongoing replacement of bulbs. Consequently, their low single (power-) frequency magnetic emissions are being replaced by those of energy-saving fluorescent lamps that, in addition to their power frequency component, emit intermediate frequency electric and magnetic fields at fundamental frequencies (dependent on the particular device type) between 25 and 77 kHz together with a series of harmonics extending into the RF range. At a distance of 30 cm, electric emissions of some fluorescent lamps reach 55% of

ICNIRP's reference level (Nadakuduti et al. 2010). Overall, the change to fluorescent lamps will considerably increase the prevalence of enhanced levels of IF background electric and magnetic fields within homes (Kerr et al. 2001).

Traffic

Vehicle electronics are going to increasingly make cars smarter. Although large-scale implementation of driver-assistive systems such as anti-lock breaking systems (ABSs), electronic stability programs (ESPs), tire pressure control systems, and break assist systems already exist, features for advanced driver assistance are still restricted to a small subgroup of vehicles. These actively gather external data information to help control the car, for example, by using nonionizing radiation for parking assistance; visual light for lane detection, control, and change; for leap frog assistance; traffic sign recognition; and driver drowsiness detection. They can also apply infrared laser-based radar (lidar) for collision avoidance and pedestrian protection, and electromagnetic radar for adaptive cruise control and collision prevention. Although advanced driver assistance systems are already implemented in luxury and middle-class cars, broad-scale pervasion can be expected in the near future. Consequently, antennas from radar anticollision systems in the front and the rear of cars emitting electromagnetic radiation of 24 GHz and 77–81 GHz will soon elevate the background RF-EMF level in the streets and public places, and in particular expose cyclists and pedestrians.

In addition, pervasion of **electric motors** in electric bicycles and, in particular, in hybrid and electric cars is growing, with a series of consequences. Exposure of the general public to magnetic and EMFs is no longer restricted to public transport such as trains, trams, buses, and subways. In the European community, **e-bikes** are increasingly used. It is expected that their market share will increase from 4% to 35% within the next decade. They are defined as "cycles with pedal assistance which are equipped with an auxiliary electric motor having a maximum continuous rated power of 0.25 kW, of which the output is progressively reduced and finally cut off as the vehicle reaches a speed of 25 km/hr (16 mph) or if the cyclist stops pedaling" (EU 2002/24/EC). The assisting motor power is frequently limited to 750 W. During operation, battery cables and the motor expose the lower leg to magnetic stray fields in the microtesla range.

Hybrid cars and **e-cars** are already in serial production and are gaining increased market shares, although not as quickly as optimistically forecasted. Exposures of the driver and passengers to magnetic emissions vary considerably with the drive cycle, both with regard to amplitude and frequency content. Frequencies extend from several hertz to several kilohertz. In addition to the speed-dependent magnetic field components originating from the rotating magnetized tires that are found in every car, in hybrid as well as e-cars the electric power consumption and/or battery recharging produces additional magnetic fields. Measurements showed that the lower part of the body, including the lower abdomen of driver and passengers may be locally exposed to peak magnetic fields up to 35% of the ICNIRP's reference levels. In hybrid buses close to electric cables and/or power electronics, body parts of passengers (including their head) may be exposed up to 20% of the ICNIRP's reference levels (Schmid et al. 2009; Concha et al. 2012).

The increased use of electrically operated vehicles will also have an impact on energy consumption and distribution and, consequently, on magnetic immissions locally at homes, at loading stations, and around power lines. Batteries may be recharged in two ways, either with plug-in connectors or via contactless inductive loading coils. Charging cables might be single-phase, carry 16 A and provide 3.7 kW power, or three-phase allowing 28 A and providing 11 kW power. At quick-loading stations, higher loading power might be provided.

Due to the lifetime of conventional cars with turnover periods of >10 years, and the limited motivation to change to e-cars or hybrid cars, the percentage of electric vehicles will take time to increase. In Germany, depending on the assumed scenario, a percentage of 1%–4% e-cars (including hybrid cars) of the 43×10^6 existing cars is expected in 2020 and 4%–23% in 2030. If simultaneously loaded, this would require provision of up to 36.6 GW of electric power. Numerical models show that existing power grids could manage an e-car fraction of up to 2.3% if loading power is restricted to 3.7 kW, but with no restrictions on loading time. Higher e-car shares and/or increased loading power could be allowed in case of controlled loading, for example, in allocated periods during nighttime (Linssen et al. 2012). Overall, the electric energy requirement of e-cars will cause increased current flow in distribution power grids and subsequently elevated ambient magnetic field levels, in particular, in urban areas.

Wireless energy transfer is an upcoming option. Most power will be needed for recharging e-car batteries. Wireless charging will be a particular alternative to prevent damage of loading cables and plugs, in particular, due to vandalism. Because efficiency of energy transfer depends on the frequency of magnetic fields, frequencies in the IF band between 110 and 205 kHz are applied. For recharging e-car batteries, induction coils with an area of about 1 m^2 will be used. They will be submerged into the road cover and may be capable of transmitting power up to some kilowatts. In proximity to these coils, stray fields may exceed reference levels, especially in the case of poor car-to-coil adjustment. Normal charging will require several hours. Quick charging at higher power will need about 30 min. During loading, persons remaining inside cars, or children playing outside close to cars, and pets may be higher exposed. Therefore, particular safety precaution could be needed to ensure protection.

Energy transfer to portable devices and household appliances is also no longer restricted to power cables. In addition to the now commonly available induction cooking, wireless energy transfer based on magnetic induction coils is increasingly used also for charging devices such as mobile phones, smart phones, tablets, cameras, and electric toothbrushes, with frequencies allocated for ISM use (e.g., 6.8 or 13.6 MHz).

Smart Power Grids

The ongoing change to alternative renewable energy sources poses two problems in managing electric energy supply, namely, the longer distance between power generation and consumption, and the difficult match of availability and need of power due to the volatile nature of power sources such as wind and solar irradiation.

The first problem, the increased distance between generation and consumption of electric energy, requires efficient long-distance power transfer capacity. As a consequence, in addition to conventional AC high-voltage power lines, DC high-voltage power lines (HVDC-PLs) are going to be built. These HVDC-PLs will become a major new source of environmental static electric and magnetic fields. Static magnetic field flux densities may reach several 10 μT and hence stay in the order of the geomagnetic field that already varies considerably due to presence of field-distorting ferromagnetic parts in our living environment, such as iron radiators, door frames, and armored constructions. However, although magnetic flux densities will stay well below the existing reference level of 400 mT (ICNIRP 2009, 2010), the situation is different with electric fields. Due to ionic clouds as a consequence of corona effects, immissions of static electric fields from HVDC-PLs are higher than those of similar AC power lines and they extend to larger regions. In addition, hybrid power lines with both HVDC and HVAC systems will cause simultaneous exposure to AC and DC electric and magnetic fields. At present, guidelines for limiting static electric fields and combined DC and AC fields are lacking. Although perception and annoyance cannot be excluded in the vicinity of HVDC-PLs, living rooms are protected from electric fields by shielding of houses, which is even more efficient for DC than for ELF electric fields.

The ongoing change in requirements for energy supply also needs a general improvement in the efficiency and capacity of energy distribution. The existing power grids are not able to distribute all the generated energy at any given time. Apart from supplementing power grids by new AC and DC power lines, another option of power supply management is to allow increased current load in power grids with amplitudes even beyond present technical restrictions. This will be possible by making energy transfer smarter via sensor-based load management of power lines. In fact, the actual maximum permitted current load, and hence the maximum magnetic fields, are limited by heating of power line conductors as specified in their technical data sheets under normal conditions. However, because cooling increases with wind and lower ambient temperatures, environmental monitoring allows load up-regulation above nominal specifications. As a consequence, in the vicinity of power grids average magnetic fields will increase accordingly and thus may exceed precaution limits set by authorities at the time of power line admission. Currently, the reported median ELF magnetic field exposure of the general population typically remains <0.2 μT, with only few percent of the population being exposed to higher levels (SCENIHR 2009).

The second problem associated with volatile energy sources is the mismatch between power generation and power consumption. One way to improve matching is leveling out local and regional consumption peaks through smart meter–assisted active power consumption management in real time. This is going to be done at two levels.

The first level is smart energy **distribution** by energy consumption management of utility networks in addition to power line load management as described above. Consumption management can be realized by implementing supervising power control units and two-way communication systems for data transfer from and to remote smart meters situated in the distribution boxes at the premises of consumers. In addition to pure reading and billing energy consumption, the data exchange allows utilities also to manage power grids by remotely connecting and disconnecting client's electric circuits. As an additional advantage for utilities, smart meters can be used for safety and maintenance purposes, and for failure localization.

The second level is smart energy **consumption.** This will be realized in homes, and managed there by home area networks (HANs), and will allow implementing cost-saving strategies such as peak levelling to avoid expensive peak charges, or in-house load adjustment to benefit from floating prices by activating, deactivating, or changing the operation mode of high-power appliances and installations.

For the bidirectional data exchange with smart meters required for smart energy management, there are two methods available. One method is to base it on wireless technology; the other is using power line communication (PLC) in terms of utilizing the existing wired network of the power supply infrastructure including the in-home installation. Large-scale deployment of smart meters is on the way with 212 million devices expected to be operating worldwide by 2014 (Depuru et al. 2011).

Wireless smart meters operate at telecommunication frequencies with random frequency hopping and transmitted power similar to mobile phones. Measurements at the device reported instantaneous peak intensity values during transmission periods of up to 140 μW/cm². EMF intensities rapidly decrease with distance. Peak values were found close to background levels at about a 90-cm distance from smart meters. However, because of the duty cycle of only about 1%, the average intensity reduces to 1.4 μW/cm (Tell et al. 2012; VDH 2012).

In the *wired* case, depending on the required data rate, PLC may be performed either with narrow bandwidth (NBPLC) at 9–95 kHz, with a bit rate of up to 128 kbit/s, or with broad bandwidth (BBPLC) at 1.8–28 MHz that allows a bit rate of up to 220 Mbit/s (Lin et al. 2002; Jamian et al. 2011; Aalamifar et al. 2012). Depending on the size of utility-managed regions, the required data rates are moderate (up to Mbit/s) but with frequent data exchange. For in-house energy management, low data rates are sufficient, and data transfer would only happen occasionally (i.e., several times per day).

Apart from communication with smart meters, power lines also offer the possibility to be used for other home services, for example, for supply with radio programs and advertisements. If PLC is used to provide digital entertainment and Internet access, higher data rates and consequently broadband communication would be required. This could be achieved by BBPLC at frequencies up to some 100 MHz.

PLC may also be used inside vehicles both for transmission of sensor data and passenger entertainment. In spite of unfavorable conditions such as noise, load changes, and low-impedance connections, this can be achieved via battery cables.

Wearable Electronics

Electronics are coming closer to the body and soon may even enter it. Smart fabric technology provides new options for lighting, heating, and cooling of clothes for sensing, measuring, monitoring, and communicating. New applications are already available, such as smart clothes with washable electronic circuits, incorporated sensors, actuators, antennas, or devices such as iPod jackets with built-in headphones and wearable computers. Other clothes contain integrated sensors for interactive gaming. Wearable bio-monitoring systems for elderly and health care may help assess physiologic parameters such as heart rate, pulse, blood pressure, temperature of skin or body core, oxygen saturation, electrocardial and myocardial signals, body motion including fall detection, or even biochemical information on immune response or of wound healing (Rais et al. 2009; Mitcheson 2010; Pantelopoulos et al. 2010; Cho et al. 2011). Data may be communicated to the personal area network (PAN) and subsequently transmitted from the body to a local controller or to a medical center via the fabric-integrated textile antenna or a wearable personal digital assistant device. The required electric power may be provided by batteries and/or by harvested energy. As an example, biokinetic or biothermal energy can produce some microwatts per square centimeter of fabric (Jung et al. 2003). Although electric currents within electronic fabrics are small and the available power low, due to the proximity to the body, exposure of superficial body regions may not be negligible.

Fabric-based body temperature management can be realized either chemically by phase-change materials that absorb excess heat and deliver it when needed, or electrically by battery-fed clothing with integrated electric conductive metallic or carbonic fibers. This can be used for sportswear undervests, jackets, shoes, and gloves designed to manage body heat. Fabric-based temperature management may also be expected in vehicle seats. Long-term functioning may be provided by energy harvesting as described above. Clothing for special applications with excessive heat, such as protective clothing for fire-workers or for hot environments, can be equipped with additional features such as integrated electric fans for enforced evaporative cooling. In addition, actively lighted clothing is increasingly used for fashion, allowing for changing intensity and color depending on person's mood or present music. Electronically lighted clothes may also be used for safety with integrated electroluminescent designs or with optionally incorporated light-emitting diodes (LED).

Summary

With regard to EMF exposure, the electromagnetic environment is becoming more and more complex in almost every aspect, in terms of EMF frequency content, bandwidth, and time course. This applies to the entire frequency range, from extremely low frequencies, to intermediate frequencies and radiofrequencies, to ultrahigh frequencies. EMF frequencies which are suitable for communication have become a precious good. Consequently, previous gaps of unused frequencies are quickly being filled by new applications, and existing frequencies are used more efficiently. Although background levels remain well below existing reference levels, they are increasing with regard to both amplitude and bandwidth in the RF range as well as in the IF and ELF ranges. EMF sources are becoming more diverse and pervasive, and most importantly, they are entering all parts of daily life. People are going to be exposed anywhere, at their homes, in streets, vehicles, and workplaces. They may carry EMF sources at the body either as particular devices; PANs; or even as part of their clothing, including goggles.

The future development will be characterized by omnipresent exposure to still weak but increased EMF levels and will be accompanied by ongoing rapid changes in technology and device use patterns, both in terms of exposed body region and of duration per use and frequency of use. These exposures will likely be by more frequent short-term exposures to higher field levels, such as whole-body exposures in EAS gate fields or partial body exposures of diverse body parts (including the head) to emissions from smart RF devices and electric appliances. Exposure scenarios will also include occasional exposure to ultrahigh static and gradient magnetic and RF EMFs, such as applied in medical imaging and therapy.

With regard to health risk assessment and epidemiologic studies, the following aspects need consideration.

- First, how to account for high short-term exposures such as to emissions from EAS gates or from medical MRI. As an example, if hypothesized nonthermal effects with a potential dose response in terms of summation over time existed, a half-hour exposure to MRI (e.g., 2 W/kg × 0.5 hr = 1 W hr/kg) would be equivalent to a >15-month exposure to environmental RF-EMF mean values.
- Second, how to assess exposures in terms of broadband versus narrowband EMF (such as from 20 MHz 4G handsets compared to 0.25 MHz 2G devices) or single-frequency, continuous versus short-term or repetitive versus transient exposures.
- Third, how to assess or compair partial body exposures at different locations (e.g., temporal head, lower abdomen, thighs, and gonads) if associated with the use of the same device such as smart phones.

- Fourth, how to assess relevant simultaneous exposures to EMF emissions in various frequency ranges, for example, from base stations, smart meters, energy-saving lamps, and electric appliances, with their emissions of multiple harmonics (including the fundamental power-frequency).

The situation is complicated by the fact that EMFs in the entire range from very weak ELF magnetic fields to RF-EMFs are classified as possibly carcinogenic class 2B (IARC 2002, 2011). This makes it difficult to restrict investigating EMF carcinogenicity to one kind of exposure only of a particular device or frequency without accounting for the remaining contributions except this would be justified by a clear-cut rationale.

Overall, with regard to epidemiological studies classification of EMF exposure will require more comprehensive exposure assessment including EMF-related lifestyle parameters, occupation, transport, homes, leisure, and medical history.

References

Aalamifar F, Hassanein HS, Takahara G. Viability of powerline communication for the smart grid. *Proceedings of the Symposium Communications (QBSC)*, Kingston, Ontario: Canada, 2012, pp. 19–23.

Azimi E, Doswell J, Kazanzides P. Augmented reality goggles with an integrated tracking system for navigation in neurosurgery. *Proceedings of the IEEE Virtual Reality Conference*, 2012, Orange County, California: USA, pp. 123–124.

BfS. *Parlamentsbericht: "Häufigkeit und Dosis röntgendiagnostischer Untersuchungen in Deutschland."* 2010. Available from: http://www.umweltbundesamt-daten-zur-umwelt.de/umweltdaten/public/theme.do?nodeIdent=2662 (accessed 17 January 2013).

Bornkessel C, Schubert M, Wuschek M, Schmidt P. Determination of the general public exposure around GSM and UMTS base stations. *Radiat Prot Dosimetry*. 2007; 124(1): 40–47.

Cecil S, Schmid G, Lamedschwandner K, Morak J, Schreier G, Oberleitner A, et al. Numerical Assessment of Specific Absorption Rate in the Human Body Caused by NFC Devices. *Second International IEEE Workshop on Near Field Communication (NFC)*, Helsinki: Finland, 2010, pp. 65–70.

Cho G, Jeong K, Paik MJ, Kwun Y, Sung M. Performance evaluation of textile-based electrodes and motion sensors for smart clothing. *IEEE Sensors J*. 2011; 11(12): 3183–3193.

Coetze L, Eksteen J. The internet of things- promise for the future. An introduction. *Proceedings of the IST-Africa*, Gaborone: Botswana, 2011, pp. 1–9.

Concha P, Lourd J, Lafoz M, Arribas J. Evaluation of the magnetic field generated by the inverter of an electric vehicle. *IEEE Trans Magn*. 2012; 49(2):837–844.

Depuru, SSSR, Wang L, Devabhaktuni V, Gudi N. Smart meters for power grid—challenges, issues, advantages and status. *Power Systems Conference and Exposition (PSCE), IEEE/PES*, Phoenix, Arizona: USA, 2011, pp. 1–7.

EN 60601-2-33. Medical electrical equipment. Part 2-33: particular requirements for the basic safety and essential performance of magnetic resonance equipment for medical diagnosis. *CENELEC Standard*, Brussels, 2010.

EU 2002/24/EC. Type-approval of two or three-wheel motor vehicles and repealing. Council Directive 92/61/EEC. *EU Directive*, OJEC L 124/1.

EU 2005/32/EC. Establishing a framework for the setting of ecodesign requirements for energy-using products. *EU Directive OJEU* L 255/11, 2005.

EU 244/2009 (EC). Implementing directive 2005/32/EC of the European Parliament and of the Council with regard to ecodesign requirements for non-directional household lamps. *EU Commission Regulation OJEU* L 76/3, 2009.

Freudenthal E, Herrera D, Kautz F, Natividad C, Ogrey A, Sipla J, et al. Suitability of NFC for medical device communication and power delivery. *Proceedings of the IEEE Engineering in Medicine and Biology Workshop*, Dallas, 2007, pp. 51–54.

IARC. *IARC Classifies Radiofrequency Electromagnetic Fields as Possibly Carcinogenic to Humans. Press release 31/05/2011*. Available from: http://www.iarc.fr/en/media-centre/pr/2011/pdfs/pr208_E.pdf (accessed 17 January 2013).

IARC. *Monograph on the Evaluation of Carcinogenic Risks to Humans. Volume 80: Non-Ionizing Radiation, Part 1: Static and Extremely Low-Frequency (ELF) Electric and Magnetic Fields*. Lyon: IARC, 2002.

ICNIRP. Guidelines for limiting exposure to time-varying electric, magnetic and electromagnetic fields (up to 300GHz). *Health Phys.* 1998; 96(4): 504–514.

ICNIRP. Guidelines on limits of exposure to static magnetic fields. *Health Phys.* 2009; 96(4): 504–514.

ICNIRP. Guidelines for limiting exposure to time-varying electric and magnetic fields (1Hz to 100kHz). *Health Phys.* 2010; 99: 818–836.

Jamian JJ, Mustafa MW, Mokhlis H, Bahamdin MA. Conceptual data management and communication for smart distribution system. *Proceedings of the IEEE Conference on Clean Energy and Technology (CET)*, Kuala Lumpur: Malaisia, 2011, pp. 364–368.

Jung S, Lauterbach C, Strasser M, Weber W. Enabling technologies for disappearing electronics in smart textiles. *Proceedings of the IEEE International Conference on Solid State*, San Francisco, California: USA, 2003, pp. 22.1.

Kerr LN, Boivin WS, Boyd SM, Coletta JN. Measurement of radiated electromagnetic field levels before and after a changeover to energy-efficient lighting. *Biomed Instrum Technol.* 2001; 35(2): 104–109.

Kühn S, Lott U, Kramer A, Kuster N. Assessment methods for demonstrating compliance with safety limits of wireless devices used in home and office environments. *IEEE Trans Electromagn C.* 2007; 49(3): 519–525.

Leitgeb N. Medical devices and systems exposure and dosimetry. In: Lin JC, ed. *Electromagnetic Fields in Biological Systems*. CRC Press, Boca Raton, Florida: USA, 2012, pp. 331–368.

Leitgeb N, Niedermayr F, Fuchs C. Impact of a radio frequency electronic article surveillance (EAS) system on active implants. *J Electromagn Anal Appl.* 2012; 4: 353–357.

Leitgeb N, Niedermay F, Loos, G. Electromagnetic compatibility of electronic article surveillance (EAS) systems with active electronic implants. *J Electromagn Anal Appl.* 2013; 5: 67–73.

Lin Y-J, Latchman HA, Lee M, Katar S. A power line communication network infrastructure for the smart home. *IEEE Wireless Comm.* 2002; 9(6): 104–111.

Linssen J, Bickert S, Hennings W, Schulz A, Marker S, Waldowski P, et al. Netzintegration von Fahrzeugen mit elektrifizierten Antriebssystemen in bestehende und zukünftige Energieversorgungsstrukturen (NET-ELAN). *Energ Umwelt/Energ Environ. 150.* Forschungszentrum Jülich, 2012. Available from: http://hdl.handle.net/2128/4695 (accessed 17 January 2013).

Markewitz A. Annual report on the German cardiac pacemaker register. 2009. Available from: http://www.pacemaker-register.de (accessed 17 January 2013).

Mitcheson PD. Energy harvesting for human wearable and implantable bio-sensors. *Conf Proc IEEE Eng Med Biol Soc.* 2010; 2010: 3432–3436.

Nadakuduti J, Douglas M, Capstick M, Kühn S, Benkler S, Kuster N. *Assessment of EM Exposure of Energy Saving Bulbs & Possible Mitigation Strategies*. IT'IS Foundation, Zurich: Switzerland, 2010.

Pantelopoulos A, Bourbakis NG. A survey on wearable sensor-based systems for health monitoring and prognosis. *IEEE Trans Syst Man Cybern.* 2010; 40(1): 1–12.

Rais NHM, Soh PJ, Malek F, Ahmad S, Hashim NBM, Hall PS. Are view of wearable antenna. *Antennas & Propagation Conference LAPC*, Loughborough: UK, 2009.

SCENIHR. *Health Effects of Exposure to EMF*. SCENIHR, 2009. Available from: http://ec.europa.eu/health/ph_risk/committees/04_scenihr/docs/scenihr (accessed 17 January 2013).

Schmid G, Überbacher R, Cecil S, Petric B, Göth P. Bestimmung der Exposition durch Magnetfelder alternativer Antriebskonzepte. *Ressortforschungsbericht* BfS-RESFOR-21/09, 2009; urn:nbn:de:0221-2009082182.

Schubert M, Bornkessel C, Wuschek M, Schmidt P. Exposure of the general public to digital broadcast transmitters compared to analogue ones. *Radiat Prot Dosimetry*. 2007; 124(1): 53–57.
Shi Z, Liao K, Yin S, Ou Q. Design and implementation of the mobile Internet of things based on TD-SCDMA network. *IEEE International Conference on Information Theory and Information Security*, Beijing: China, 2010, pp. 954–957.
Tell RA, Kavet R, Mezei G. Characterization of radiofrequency field emissions from smart meters. *J Expo Sci Environ Epidemiol*. 2013; 23(5): 549–553.
TNS. *E- Communication Household Survey*. Special Eurobarometer 335/Wave 72.5. Brussels, Belgium: TNS Opinion & Social, 2010.
TNS. *Mobile Life: Die Zukunft der Mobilkommunikation in Deutschland*. 2011. Available from: http://www.tns-infratest.comn (accessed 17 January 2013).
VDH. *Radio Frequency Radiation and Health: Smart Meters*. Vermont Department of Health, 2012. Available from: http://healthvermont.gov/pubs/ph_assessments/radio_frequency_radiation_and_health_smart_meters.pdf (accessed 17 January 2013).

20

Future of EMF Epidemiology

Mireille B. Toledano and Rachel B. Smith

CONTENTS

Introduction

Radiofrequency-electromagnetic fields (RF-EMFs) and extremely low frequency (ELF)-magnetic fields have been classified as possibly carcinogenic to humans (Group 2B) on the basis of limited evidence (IARC 2002; Baan et al. 2011). Methodological limitations in epidemiological research and the absence of established biological mechanisms for either RF-EMFs or ELF-MFs to support causal relationships make it difficult to reach firm conclusions. For intermediate frequency (IF)-electromagnetic fields, data on health effects are sparse and too limited for risk assessment (SCENIHR 2009). Without established mechanisms to direct efforts to health endpoints of most relevance for investigation, what is driving continuing research in EMF epidemiology? The one certainty in this research field is that there is significant public concern regarding EMF and potential human health effects. In a survey published in 2010, 46% of the public expressed concern about the potential health risks of EMFs, with more than two-thirds of people believing that their health is affected *to some extent* by high-voltage power lines, mobile phone masts, and mobile phone handsets, and with 35%, 33%, and 26%, respectively, believing that these EMF sources affect their health *to a large extent* (Eurobarometer 2010). Is it legitimate for this public concern to drive environmental epidemiological research? The National Institute of Environmental Health Sciences defines environmental public health as "the science of conducting and translating research into action to address environmental exposures and health risks *of concern to the public*" (NIEHS 2012). Research can provide evidence—based on data rather than speculation—with which to both evaluate the public health question and address legitimate public and policy concerns (Elliott and Wartenberg 2004).

In this chapter, we discuss the particular challenges facing EMF epidemiology, and we put forward a new approach for EMF epidemiological research that we believe will address these challenges. We then set out the research questions relating to adult and child health that should be addressed in future EMF epidemiology, and we discuss study designs that are being, or should be, used in future to address these questions.

What Challenges Does Future EMF Epidemiology Face?

EMF epidemiology today needs to address specific challenges to move forward:

- Lack of established biological mechanisms
- Very low, ubiquitous, whole-body exposures from a wide range of EMF sources
- Rapid changes in technology and exposure-related user behaviors
- Multiple simultaneous and/or correlated EMF exposures

Traditionally, EMF epidemiology has been broken down into RF-, IF-, and ELF-EMF exposures. It is clearly important to understand which specific EMF frequencies and sources have adverse health effects in order that appropriate health protection policy can be developed and applied if necessary. Specific priority research questions are proposed and discussed by frequency range later in this chapter. However, researchers need to remember that EMF is a continuous single spectrum, and the division of EMF health research into almost separate, disconnected, research fields of RF-, IF-, ELF-EMF may now hinder progress in this field. In real-life experience, humans are not exposed to each EMF frequency and source in isolation, but rather to an increasingly multiplex EMF environment over a lifetime, as discussed in Chapter 19.

Ideally, future epidemiological research requires a more comprehensive integrated approach taking into account all sources of RF-, IF- and ELF-EMF to reflect people's daily environment. This will enable investigation of potential health effects from multiple low-level exposures in combination, for example, to identify additive or multiplicative effects of combined exposures on health that would be missed by analysis of isolated sources/frequencies, and any possible modification of exposure–disease associations by other EMF exposures. To achieve this comprehensive integrated approach, the challenges above need to be addressed.

Lack of Established Biological Mechanisms

Despite considerable research to date, there is a lack of evidence for biologically plausible mechanisms for RF-, IF-, or ELF-EMF, which makes it difficult to target epidemiological research and focus on specific health outcomes. Thus, epidemiologists are often left "fishing," that is, searching without strong *a priori* hypotheses, thereby increasing the likelihood of chance findings and making interpretation of results more difficult.

Very Low, Ubiquitous, Whole-Body Exposures

Human beings live in a complex and fluid EMF environment, comprising RF-, IF-, and ELF-EMFs from multiple sources; near fields and far fields; ubiquitous low-level exposures and acute high-level exposures; partial- and whole-body exposures; and simultaneous exposures to EMFs at different frequencies. A key challenge to EMF epidemiology is that humans now experience very low, ubiquitous, whole-body exposures from a wide range of sources (e.g., RF-EMF exposures from mobile phone base stations, IF-EMF exposures from energy-saving fluorescent light bulbs, and ELF-EMF exposures from electric appliances and high-voltage overhead power lines). Assessing the health effects of these exposures is therefore difficult because there are no unexposed people to compare them with, and we have to compare people according to their level of exposure, that is, low exposure versus high exposure. The ability to accurately discriminate between individuals with sufficiently different levels of exposure is critical in environmental epidemiology, and this relies on there (1) being sufficient exposure variability in the study population and (2) sufficiently detailed and accurate exposure assessment. Failure to assess exposure accurately may result in exposure measurement error and/or misclassification leading to loss of statistical power and/or biased results.

Rapid Changes in Technology and Exposure-Related User Behavior

Rapid technology development over the past two decades has resulted in a constantly evolving EMF exposure environment (see Chapter 19). For example, magnetic resonance imaging (MRI) scanner technology has developed to use increasing magnetic fields, up to ultrahigh magnetic flux densities that allow improved image quality. The EMF frequency used is proportional to magnetic field, so as magnetic fields increase, the EMF exposure from these devices changes and may range from 15 MHz up to 200–400 MHz, depending on device used (AGNIR 2012), thereby spanning both IF- and RF-EMFs. As described in Chapter 19, mobile communications technology has evolved continuously since the first-generation analog systems were launched in the 1980s. For example, second-generation (2G) digital GSM systems (900 MHz, 1.8 GHz) were introduced in the early 1990s, followed by third-generation (3G) UMTS systems (1.9 and 2.1 GHz) in 2001, and then the recently introduced fourth-generation (4G) systems (operating at and allocated various frequencies, e.g., 800 MHz, 1800 MHz, and 2.6 GHz in the United Kingdom [Ofcom 2013]) that increase data speeds and transmission capacity further. With each new "technology generation," there have been changes or additions to the RF-EMF frequencies being used, and capabilities of the technology. Exposure assessment needs to account for changes, over time, in levels of EMF emissions from devices resulting from technological developments, for example, average power output of calls made on second-generation GSM network is approximately 100–500 times higher than calls made on the third-generation UMTS network (Gati et al. 2009; Persson et al. 2012). The recent introduction of terahertz whole-body security scanners at airports introduces exposures in the extremely high-frequency (EHF) range (30–300 GHz) about which little is known with regard to exposure levels and/or potential health effects (AGNIR 2012).

There have been concomitant developments in the way technologies are being used. In the medical setting, MRI scans have become increasingly common and are now a major diagnostic tool, and their use for MRI-guided interventional medical procedures is growing. Widespread introduction of smart phones is expected to have resulted in a considerable change of exposure-related user behavior, for example, from use of mobile phones for voice calls only to use of phones for mobile data exchange/tele-entertainment. These changes affect the ways in which people are exposed to RF-EMFs that, in turn, exposure assessment needs to address.

In such a rapidly changing environment, exposure assessment in EMF epidemiology is constantly trying to catch up. The sheer number of exposure sources we encounter in everyday life is increasing rapidly, as new applications for these technologies are developed and adopted and become commonplace in our environment. In addition, exposure levels and/or specific body areas exposed may change as users take up the new technologies and modify their behaviors to make full use of the technological capabilities. This results in a variety of exposure modes for epidemiology to assess, that is, continuous versus short-term exposures, repetitive versus transient exposures, whole- versus partial-body exposures, and high short-term exposures. Epidemiologists appear to have made little headway to improve clarity in EMF epidemiology, and this may be partly because the natural sequence of epidemiological investigation from relatively inexpensive and quick studies (e.g., ecological or cross-sectional) to more detailed, time-consuming, and expensive studies (e.g., prospective cohort) does not work well for EMF exposures that change so rapidly. The EMF exposure environment now being investigated using detailed prospective cohort study designs is no longer the same as that originally investigated using cheaper, quicker study designs. Thus, establishing a consistent body of scientific evidence over time is particularly difficult.

Multiple Simultaneous and/or Correlated EMF Exposures

In today's complex environment, humans may experience simultaneous and/or correlated EMF exposures of different types and frequencies. For example, an MRI scanner uses a static field, RF-EMF in the 10–100 MHz range and a magnetic field gradient that is switched on and off at a frequency of around 1 kHz (IF-EMF) during imaging (IOP 2008). As described in Chapter 19, emissions from energy-saving fluorescent light bulbs comprise both power-frequency (ELF) magnetic emissions and IF electric and magnetic fields (e.g., 25–77 kHz), with harmonics extending into the RF-EMF range. Indeed, any electrical device plugged into and operating from a mains electricity socket will generate electric and magnetic fields, for example, 50 Hz in the United Kingdom (HPA 2013), so common devices such as DECT phones, baby monitors, microwave ovens, and wireless networked PCs will produce both low-level ELF-EMF exposure in addition to RF-EMF exposures. In addition, behavioral determinants of exposure may be correlated, for example, heavy users of mobile phone technologies may be early adopters and/or heavy users of other/new technologies. These may give rise to an exposure assessment profile for an individual that is highly intercorrelated that, in turn, may introduce problems of exposure collinearity in analysis.

New Approaches in Future EMF Epidemiology

Exposure-related challenges can be addressed by adopting a cumulative and integrated exposure assessment approach within prospective study designs. Cumulative and integrated estimates of exposure to each of RF-, IF-, and ELF-EMFs, from environmental and occupational sources, will address both complexity and change in the EMF exposure environment and individuals' exposure profiles. This approach comprises two main components: (1) the collection of exposure and determinant information from a range of different sources, including data about exposure-relevant behaviors (e.g., time spent cooking with an induction hob, or use of hands-free when talking on a mobile phone) and (2) dosimetric modeling.

Within the context of a prospective cohort study, information on exposures to environmental and occupational EMF sources and exposure-relevant behaviors would be collected by questionnaire (e.g., use of mobile and cordless phones, wireless internet, induction hobs, MRI scans, induction welding), and questionnaires should adapt over follow-up to encompass rapidly changing and emerging technologies (e.g., 4G mobile communications, radiofrequency identification device [RFID] applications, terahertz whole-body scanners). Self-reported mobile phone use from questionnaires must be supplemented and validated by objective data, through regular prospective linkage to individual mobile phone traffic data from mobile network operators (see Chapter 4). Moreover, through use of smart phone apps that automatically collect information on mobile phone use, modes of use such as hands-free and laterality, output power, and frequency bands. In addition, models should be used to estimate cohort members' RF-EMF exposure from mobile phone base stations, radio and TV stations, and ELF-EMF exposure from nearby power lines (see Chapter 8). Exposure assessment of prospective cohort members should be complemented by integrated RF-, IF-, and ELF-EMF exposure evaluation and dosimetry at the wider population level, to estimate population exposure distributions. Specifically, personal RF-EMF exposure monitors, activity diaries, and smart phone apps should be used to collect integrated personal RF-EMF exposure and exposure-relevant behavior data in population samples. However, validated personal dosimeters do not span the whole EMF spectrum, and a comprehensive survey of potential IF-EMF sources with source-based measurements is required to understand the extent and level of IF-EMF exposure from devices such as induction hobs, energy-saving lightbulbs, and RFID devices. These data should then be used in dosimetric calculations to derive transfer functions for both near-field and far-field RF-, IF-, and ELF-EMF exposures, to quantify the energy absorbed in tissues, organs, or the whole body under particular exposure conditions, such as making a mobile phone call, cooking on an induction hob, or undergoing an MRI scan. These transfer functions will allow the calculation of integrated whole-body, partial-body, and organ-specific exposures from multiple RF-, IF-, and ELF-EMF sources. Cumulative and integrated exposure estimates for each of RF-, IF-, and ELF-EMFs for prospective cohort members can then be calculated for use in epidemiological analyses.

Aside from advancing exposure assessment, it is essential that there is an interface between epidemiology and complementary scientific methods, to assess the potential biological effects of exposure and identify possible mechanisms. This can be done by carrying out experimental studies involving systems biology approaches to analyze blood samples from exposed human volunteers, and *in vitro* studies on cells.

Systems biology approaches to study the health effects of environmental exposures have emerged as a recent tool offering insight for epidemiological research. Systems biology is a relatively new science that seeks to understand the complex interactions in biological systems. It uses systematic high-throughput "omics" technologies and bioinformatics to study all components and interactions within a cell or organism in a holistic manner. So far, studies conducted using high-throughput techniques have been too few and too heterogeneous to identify genes or proteins that reliably or consistently respond to EMF exposures, and there is a dearth of studies using human volunteers (Leszczynski et al. 2012). Some of these validated "omics" technologies, such as transcriptomics and metabonomics, can be used in human experiments to evaluate thoroughly any biological effects of RF-EMF exposures. Human provocation experiments would expose volunteers to GSM-RF, UMTS-RF, or sham-RF under double-blind conditions, with pre- and postexposure blood sampling, to discover any biological changes resulting from RF-EMF exposures. Analysis of blood samples could be undertaken to explore the human transcriptome and prespecified pathways (e.g., inflammation, oncogenes, DNA repair), as well as to conduct untargeted (global) metabolic profiling. This may provide potential to find biomarkers of exposure (see Chapter 4) and/or effect that could enhance future epidemiological investigation with more focused analysis.

In vitro studies of RF-EMF exposure provide no robust evidence of genotoxic or carcinogenic effects with exposure below guideline values, and various reported findings on cell membranes and proteins require independent replication (AGNIR 2012). *In vitro* studies of ELF magnetic fields provide limited evidence of effects on calcium ion, reactive oxygen species, and genotoxicity (EFHRAN 2010). Cellular and molecular events can be examined in real time under various exposures, for example, RF- and IF-EMFs, using recently developed specific spectroscopic, for example, bioluminescence and fluorescence resonance energy transfer (BRET and FRET), and electrochemical techniques, to understand the mechanisms behind any potential effects.

In general, EMF epidemiology requires improved and more comprehensive exposure data, particularly for the general population, and systems biology techniques to evaluate potential biological effects of RF-, IF-, and ELF-EMFs. Future EMF epidemiology must be enhanced by better exposure assessment and by integrating dosimetry and systems biology, to advance the field and make it possible to identify/exclude small risk increases in epidemiological studies.

What Specific Research Questions Should Future EMF Epidemiology Address?

Various recommendations for epidemiological research have been made by expert groups for RF- (MTHR-PMC 2007; SCENIHR 2009; WHO 2010), IF- (SCENIHR 2009), and ELF-EMFs (WHO 2007; SCENIHR 2009). Toxicology is an important complimentary science that can help provide clues as to which health endpoints may be fruitful candidates for epidemiological investigation. Taking into account the most up to date toxicological evidence available, the content of the second part of the book (Chapters 7–16) and previous recommendations by expert groups, we outline below what we consider to be the highest priority epidemiological research questions, with suggestions as to how these might be addressed. Questions on RF-EMFs are addressed first, followed by IF-EMFs and finally ELF-EMFs; each section focuses on adults first and, where appropriate, also children. Broad research recommendations for RF-EMF exposures in children reflect the paucity of evidence available, specifically in children, to date. Recommendations to limit children's mobile phone use (e.g., IEGMP 2000; NRPB 2004) are based on potential susceptibility and the precautionary principle, rather than scientific evidence. Considering rapid uptake of mobile phones by children in recent years (Böhler and Schüz 2004; Dimonte and Ricchiuto 2006; Mezei et al. 2007; Soderqvist et al. 2007, 2008; Ofcom 2012), and their potential for a longer lifetime of exposure compared with the previous generation, reducing scientific uncertainty is a high priority.

Radiofrequency-Electromagnetic Fields

- Adult Question 1. Is RF-EMF exposure from long-term and/or heavy use of mobile phones (and other wireless technologies) associated with increased risk of cancers, and neurological outcomes such as Alzheimer's and Parkinson's diseases, in adults?

Research questions to address cancer are not based on toxicological evidence (see Chapters 12–15). There is no compelling evidence from animal or cell studies that exposure to RF-EMFs below guideline values is either genotoxic or carcinogenic. Indeed, several large-scale animal studies investigating cancer have all been robustly negative (AGNIR 2012). This suggests the long-held epidemiological focus on cancer may be off target; however, as public concern remains, it is important to reduce uncertainty regarding long-term or heavy mobile phone use and risk of cancer. The ongoing COSMOS study (Schüz et al. 2011) addresses recommendations for a long-term prospective cohort study of adult mobile phone users. It is examining RF-EMF exposures and possible risk of cancers and neurological outcomes such as Alzheimer's and Parkinson's diseases. It will overcome limitations of INTERPHONE (the largest study to date on brain tumors and mobile phones), and is specifically designed to address all currently important unresolved questions concerning risk to the public from RF-EMFs.

In relation to neurodegenerative diseases, toxicology has so far provided few clues for the epidemiologist. Effects on brain and nervous system cellular physiology are largely null. Moreover, there is no clear evidence of non-thermal effects on brain electrical activity, and most recent toxicological studies on RF-EMF effects on the blood–brain barrier are robustly negative (AGNIR 2012). Several animal studies report increased glial fibrillary acidic protein (GFAP) expression, a marker of neuroinflammation that is related to neurodegenerative damage, after acute and/or repeated RF-EMF exposures (AGNIR 2012; Maskey et al. 2012), although it is not clear whether the increase is clinically significant. This may possibly represent a response to mild heating, and the relationship between markers of neuroinflammation such as GFAP and Alzheimer's disease may be age dependent (Hoozemans et al. 2011), making interpretation of animal models difficult. At present, there is too much uncertainty to draw any conclusions; however, further toxicological research into GFAP and other markers of neuroinflammation is encouraged and may provide clues to help direct future epidemiology, perhaps even to the point of using GFAP as a biomarker.

- Adult Question 2. Is RF-EMF exposure associated with adverse effects on reproductive function and fertility?

This remains a knowledge gap in EMF epidemiology. Evidence on RF-EMF exposures and non-thermal effects on fertility is limited. Epidemiological studies have mostly been small, predominantly on males from infertility clinics or military settings, and have been cross-sectional by nature; thus, the time sequence of events prevents causal interpretation (AGNIR 2003, 2012). Further cross-sectional studies are not recommended; instead, this question should be addressed by investigating RF-EMF exposures and time to pregnancy (TTP) in both males and females in a prospective cohort. The ideal study design would be a prospective TTP cohort study recruiting couples who are about to start or have recently started trying to conceive, to provide prospective exposure assessment for both parents, to avoid a long period of recall of mobile phone use and to allow objective traffic data to be obtained from operators for the relevant exposure window. However, because such a design would be costly, an alternative would be to incorporate retrospective and ongoing assessment of TTP into an existing cohort with prospective RF-EMF exposure assessment including objective traffic data, such as the COSMOS study. This would provide prospective exposure assessment for one parent, and exposure for the partner for the relevant time window could be obtained indirectly via the existing participant. Retrospective ascertainment of TTP requires only a few questions and has been previously validated (Joffe et al. 2005). Analysis should be limited to attempts at pregnancy occurring since baseline to avoid the use of retrospective exposure data.

- Children Question 1. Is RF-EMF exposure from maternal mobile phone use during pregnancy associated with behavioral and neurological disorders in childhood/adolescence?
- Children Question 2. Is RF-EMF exposure from children's mobile phone use associated with behavioral and neurological disorders in childhood/adolescence?

Studies in three different birth cohorts in Europe have examined maternal mobile phone use during pregnancy and neurobehavioral disorders in childhood. Exposure to mobile phones prenatally and, to a lesser degree, postnatally was associated with behavioral difficulties in 7-year-old children (Divan et al. 2008, 2012), but no associations were observed with behavioral difficulties at age 5 (Guxens et al. 2013), neurodevelopmental delays at 6 and 18 months (Divan et al. 2011), or neurodevelopment at 14 months (Vrijheid et al. 2010). With exception of the study by Vrijheid et al. (2010), all exposure assessment has been retrospectively self-reported and may be subject to recall bias. The studies are heterogeneous, each examining various outcome measures at different ages, and they are not driven by clear *a priori* hypotheses. There is no clear mechanism for effect, or any indication of how long the effect of prenatal exposure is expected to persist into childhood. Without an indication of the most relevant outcomes and ages in childhood in relation to a potential effect of maternal mobile phone use during pregnancy, future epidemiology needs to follow a range of outcomes prospectively from birth through childhood and adolescence to assess any potential risk. This could be achieved in a large prospective birth cohort, or by pooling recently established birth cohorts if they have prospective RF-EMF exposure assessment, to overcome the limitations of retrospective exposure assessment. However, a limitation of such a design would be the inability to control for potential confounding due to unmeasured maternal familial (genetic and/or environmental) factors that may be particularly important when studying outcomes such as attention-deficit/hyperactivity disorder (ADHD), a familial and highly heritable disorder (Faraone et al. 2005). Sibling-pair analyses could be conducted to control for unmeasured familial (genetic and environmental) factors shared by siblings, either as a stand-alone matched case–control analysis within sibling pairs, or nested within a large prospective birth cohort study. Such approaches have been used to detect and control for unmeasured familial confounding, for example, in studies of maternal smoking during pregnancy and educational achievement (Lambe et al. 2006) and suicidal acts (Cnattingius et al. 2011). The most important requirement for sibling-pair analyses is a sufficient number of siblings discordant for exposure status (Susser et al. 2010), that is, maternal mobile phone use during pregnancy, however, given the ubiquity of mobile phone use this may also be the greatest challenge in such a study.

Overall, the evidence from toxicology for non-thermal effects of low-level RF-EMF upon cognitive function and behavior is weak (AGNIR 2012). However, findings from animal behavioral studies suggesting improvements in learning and memory, and protective effects against age-related behavioral deficits (Kumlin et al. 2007; Arendash et al. 2010) in association with RF-EMF exposures, although still requiring replication (AGNIR 2012), suggest potential avenues for future epidemiology. Based on this current knowledge, it is worth pursuing an epidemiological study with a behavioral and neurological focus in relation to children's mobile phone use.

A small prospective cohort study of self-reported mobile phone use and cognitive ability in teenagers (median age, 13 years) suggested that those with higher numbers of voice calls had faster and less accurate responses to high-level cognitive tasks (Abramson et al. 2009). However, differentiating a possible "brain-training" effect from frequent mobile phone use and a potential RF-EMF exposure effect is not possible based on self-reported use alone; and changes in response time associated with increased number of voice calls and short message service (SMS) at 1-year follow-up may represent regression to the mean rather than an effect of mobile phone exposure per se (Thomas et al. 2010).

Children's mobile phone use and behavioral and neurocognitive outcomes could be investigated efficiently using a school-based prospective cohort study design. It requires participation of children at ages when personal mobile phone use is substantial but at which cognitive functions and their neural correlates are still maturing, such that performance is predicted to improve during follow-up. Exposure assessment must incorporate objective mobile phone traffic data from network operators to be able to discern whether any observed associations are likely due to exposure to RF-EMFs from mobile phones, a combination of all RF-EMF sources (e.g., DECT phones, wireless internet), or from a possible "brain-training" effect. Rapid uptake of smartphones among children (>50% by age 12 years in the United Kingdom [Ofcom 2012]) makes it feasible to use smartphone apps to capture objective mobile phone use data continuously over study follow-up to enhance exposure assessment. This design is being used in ongoing studies in Switzerland and the United Kingdom.

Investigation of the processes of learning, memory, and cognitive functioning in toxicology and pediatric epidemiology may inform and further develop understanding of possible associations between RF-EMFs and Alzheimer's disease, dementia, or other age-related cognitive functioning.

- Children Question 3. Is RF-EMF exposure associated with cancers in children/adolescents?

RF-EMF exposure from mobile phone base stations is an area of high public concern; and for very young children, it is likely to be one of their main sources of RF-EMF exposure together with WiFi in the home. However, there is no evidence of increased risk of childhood cancer associated with RF-EMF exposure from mobile phone base stations to date (Elliott et al. 2010). Given that exposure to RF-EMF from a mobile phone base station is negligible compared with exposure from mobile phone use (Neubauer et al. 2007), studies of base station exposures in isolation are unlikely to further understanding about RF-EMFs and carcinogenicity/childhood cancer risk. However, RF-EMF exposures from base stations need to be taken into account, but it is probably more appropriate to do so when considering cumulative and integrated multiple source exposures in our daily environment (as outlined earlier in this chapter). When considering childhood cancer and RF-EMF exposure specifically from mobile phone use, the MOBI-KIDS case–control study (see Chapter 13) is the largest to date, overcoming size limitations of previous research. However, it is our opinion that whatever the results from MOBI-KIDS, questions will remain unanswered due to inherent biases (e.g., recall bias) that will limit interpretation of results as seen in the INTERPHONE study (see Chapter 13). As whole-body and part-body RF-EMF exposures become more relevant as mobile phone technology and exposure-related behavior evolve, MOBI-KIDS' narrow focus on brain tumor risk may already be outdated and limit its capability to further understanding of carcinogenicity and RF-EMF exposures.

To overcome these limitations, we recommend a prospective cohort study design to resolve questions regarding RF-EMF exposure and risk of cancer in children and adolescents. In our opinion, this should not be delayed until results of the MOBI-KIDS study are available. However, because childhood cancer is rare (e.g., in the United Kingdom, incidence in childhood [ages 0–14 years] of all cancers, excluding nonmelanoma skin cancer, is 13 per 100,000; and incidence of brain and nervous system cancers is 2.9 per 100,000 [Ferlay et al. 2010]), a stand-alone prospective cohort study would require a sample size of several hundred thousand children with follow-up from birth to adulthood to have sufficient statistical power, and this would be prohibitive both practically and financially. However, an alternative is to pool internationally recently established or new large-scale birth cohorts with prospective collection of RF-EMF exposure data, and ongoing RF-EMF exposures and cancer follow-up, throughout childhood and adolescence. Furthermore, we recommend that childhood and adolescent cancer incidence data be used to conduct time trend studies, because these data provide population coverage and large numbers for rare tumors. Even without exposure information, time trend studies would provide an important form of health monitoring (see below on Complementary Research).

- Children Question 4. Is long-term RF-EMF exposure starting in early childhood associated with higher risk of adverse health outcomes in adulthood?

This question could be investigated fairly quickly through an existing adult cohort, for example, the ongoing COSMOS study (Schüz et al. 2011), but such an investigation would be reliant on retrospective exposure data that brings with it limitations. Alternatively, this question could be addressed prospectively through recently established large-scale birth cohorts with prospective collection of RF-EMF exposure data and ongoing follow-up continuing into adulthood; however, such a design would obviously take longer to provide answers.

Intermediate Frequency-Electromagnetic Fields

- Adult Question 3. Is IF-EMF exposure associated with adverse health outcomes?

In vivo, in vitro, and epidemiological data on potential health effects from IF-EMFs are extremely sparse (SCENIHR 2009), with little evidence to help prioritize future research directions and target specific health outcomes. However, increasing exposure to IF-EMFs, particularly in occupational settings (e.g., from induction welding, dielectric heating devices, industrial induction ovens, interventional MRI, RFID tag deactivation devices at shop checkouts), and, increasingly, among the general population (e.g., induction hobs, electronic article surveillance [antitheft] gates in shops, medical MRI scans), makes epidemiological (and experimental) studies in this field a high priority (SCENIHR 2009). Initially, occupational epidemiology is likely to be the most promising approach given higher occupational exposures.

Extremely Low Frequency-Electromagnetic Fields

- Adult Question 4. Is ELF magnetic field exposure associated with increased risk of neurodegenerative diseases, in particular, amyotrophic lateral sclerosis (ALS) and Alzheimer's disease?

There is consistent evidence of an association between electrical occupations and increased risk of ALS, but whether ELF magnetic fields are the relevant exposure or a proxy for another exposure is unresolved (see Chapters 3 and 11). There is evidence of an association between occupational ELF magnetic field exposure and Alzheimer's disease (see Chapter 11), but it is currently insufficient to support any causal interpretation, and results may be influenced by publication bias (Vergara et al. 2013). In relation to residential exposure to ELF magnetic fields from power lines, there is some evidence for an association with Alzheimer's disease mortality, but not for ALS mortality, nor incidence for either disease (see Chapter 11).

To understand what is driving associations between "electric occupations" and ALS, future job–exposure matrices (JEMs) must not only improve quantification of ELF magnetic field exposures but also assess electric field and electric-shock exposures, to tease out any effects of magnetic fields from other occupational exposures or confounders (WHO 2007). New JEMs for occupational electric shocks may advance the field; however, high concordance (≥67%) for ELF magnetic field exposures and risk of electric shock (Vergara et al. 2012; Huss et al. 2013) may necessitate a focus on occupations discordant for ELF magnetic field exposure level and risk of electric shock (i.e., high shock risk/low ELF magnetic field level and vice versa), to disentangle any effects of the two exposures. Exposure assessment/JEMs must go further to address severity of electric shocks, because potential health consequences may differ between mild versus severe nonfatal shocks. In terms of future epidemiology, there is little justification for establishing new occupational studies to examine ALS. A recommendation for pooled analysis of previous cohorts examining ALS combined with refined JEMs (Kheifets et al. 2009) should be prioritized, with the addition of JEMs for electric field and electric-shock exposures. A nested case–control study of ALS within a prospective cohort collecting information on occupational history and lifetime electric shocks may be able to provide insight, but it should not be considered the priority effort.

Research on Alzheimer's disease is important because this disease outcome is likely to increase with the ageing population in Europe. Poor outcome ascertainment, on top of likely exposure misclassification and a negative result, probably explains publication bias observed for Alzheimer's disease (Vergara et al. 2013). Identification of Alzheimer's disease from mortality data is inadequate for epidemiological purposes, because mortality is a poor proxy for incidence and dementias are underreported on death certificates (Kuller and Ives 2009). Outcome ascertainment must therefore improve by using Alzheimer's disease registries that exist in some parts of the United States and Europe, to identify cases. This will require care with case definitions, because most registries include both Alzheimer's disease and other dementias. Epidemiologists will need to scrutinize and sift through cases to reduce disease misclassification, and analyses distinguishing specific Alzheimer's diagnoses from all dementia diagnoses may be advisable.

Future epidemiological investigation of neurodegenerative outcomes is underpinned by some *in vivo* evidence for nervous system effects (on memory, anxiety-related behavior, and human EEG) from ELF magnetic fields above 0.1–1.0 mT (SCENIHR 2009). *In vivo* and *in vitro* studies suggest short-term ELF magnetic field exposure causes mild oxidative stress and possibly activates anti-inflammatory processes (Mattsson and Simko 2012).

If long-term exposure and possible long-lasting mild oxidative stress can also be indicated in toxicology, epidemiology could use biomarkers of inflammation among occupational cohorts with chronic long-term exposure to ELF magnetic fields to investigate this potential causal pathway to Alzheimer's disease.

- Children Question 5. How can we improve the evidence-base for the observed weak associations between ELF magnetic field exposures and risk of childhood leukemia to reduce scientific uncertainty?

ELF magnetic fields are designated as possibly carcinogenic in humans based on a consistent epidemiologic association with childhood leukemia, but no biologic mechanism has yet been identified (see Chapter 9). The ongoing ARIMMORA project aims to examine underlying biophysical mechanisms and may give new leads for future toxicological and epidemiological research or perhaps provide a definitive mechanistic understanding with which to explain observed epidemiological associations. Limitations of existing study designs, including possible selection bias, confounding, and exposure misclassification, prevent causal interpretation of the observed associations (see Chapters 5, 6, and 9). Although potential sources of bias were avoided in a large national registry based case–control study that reported a dose–response association between childhood leukemia and distance to high-voltage power lines (Draper et al. 2005), this finding has not yet been fully explained by magnetic field exposures (Kroll et al. 2010). Extensions to this study, for example, to include power lines operating at a wider range of voltages and also high-voltage underground power lines, are ongoing (John Swanson, personal communication, 2013). However, as with previous studies, potential for exposure misclassification remains.

New epidemiological approaches are now required. Novel cohort study designs that minimize bias and target populations with larger proportions in highly exposed groups or those with higher risk of childhood leukemia (to overcome prohibitive cohort size requirements given low disease incidence), have been outlined in Chapter 9. For example, the TransExpo study aims to explore incidence of leukemia in a cohort of children living in buildings with built-in transformers (Kheifets et al. 2010).

Complementary Research

Detailed epidemiologic research using a cumulative and integrated exposure assessment approach within prospective study designs and to address the specific questions set out above should be complemented by surveillance of routine health statistics. This will facilitate rapid epidemiological response to the emergence of new technologies, specifically to exclude large potential risks from these new technologies. For monitoring purposes, ecological (time trends) analyses (see Chapter 2) could be used to assess correlations between changing cancer incidence and changing technology and use, for both adults and children. Routine data are useful for well diagnosed and registered diseases (e.g., brain tumors) but less so for other outcomes such as symptoms and Alzheimer's disease. However, the establishment and development of new disease registries for neurodegenerative diseases, for example, ALS, dementias, and Parkinson's disease, may improve surveillance capabilities for these outcomes. To complement, inform, and enhance future EMF epidemiology, investigation of biological effects of EMF is required at human and cellular levels.

This will further our understanding of potential biological effects and mechanisms and has the potential to identify biomarkers of exposure and/or effect that could strengthen future epidemiological investigation. This will enhance our ability to identify risks and interpret epidemiological results, ultimately improving risk assessment.

Conclusions

Rapid growth of devices using EMFs, and the unceasing emergence of new EMF technologies and applications, may in part foster public concern regarding possible adverse effects of EMF exposure. In order that health and policy authorities can address public concerns, robust evidence on effects and risks of EMF exposure is required from scientists, to form the knowledge base for policy actions. New scientific evidence from epidemiological studies with cumulative and integrated EMF exposure assessment, and complementary experimental work on biological effects and mechanisms, will fuel improved risk assessment that, in turn, will underpin actions by regulatory bodies and the development of public health policies regarding EMF exposures. In particular, prospective cohort studies with cumulative and integrated exposure assessment will have the ability to calculate whole-body and organ-specific exposure estimates to determine individual dose. This will reduce exposure misclassification and improve reliability of health risk estimates from epidemiological studies. To communicate these health risk estimates appropriately and to apply them effectively in risk management, detailed knowledge of the extent, level and distribution of RF-, IF-, and ELF-EMF exposures in the general population is also required. In this way, epidemiology can play a vital role in reducing the discordance between current high levels of public concern regarding EMF and the current scientific knowledge base that although uncertain, does not suggest large elevated risks of adverse health effects associated with EMF.

Accurate estimates of exposure to RF-, IF-, and ELF-EMFs, and knowledge of the relative contributions of different EMF sources, will enable evidence-based adaptation approaches to be implemented, as part of policy development by regulatory bodies to reduce population exposure to EMF, as necessary. The provision of evidence-based advice on optional measures to reduce personal EMF exposure will allow informed choices to be made at the individual level.

References

Abramson, M. J., Benke, G. P., Dimitriadis, C., Inyang, I. O., Sim, M. R., Wolfe, R. S., et al. 2009. Mobile telephone use is associated with changes in cognitive function in young adolescents. *Bioelectromagnetics* 30(8): 678–86.

AGNIR. 2003. *Health effects from radiofrequency electromagnetic fields*. Report of an independent Advisory Group on Non-ionising Radiation. *Doc NRPB* 14(2): 1–177.

AGNIR. 2012. *Health Effects from Radiofrequency Electromagnetic Fields*. Report of the independent Advisory Group on Non-ionising Radiation. Doc HPA, Chilton: HPA, RCE-20, pp. 1–333.

Arendash, G. W., Sanchez-Ramos, J., Mori, T., Mamcarz, M., Lin, X., Runfeldt, M., et al. 2010. Electromagnetic field treatment protects against and reverses cognitive impairment in Alzheimer's disease mice. *J Alzheimers Dis* 19(1): 191–210.

Baan, R., Grosse, Y., Lauby-Secretan, B., El Ghissassi, F., Bouvard, V., Benbrahim-Tallaa, L., et al. 2011. Carcinogenicity of radiofrequency electromagnetic fields. *Lancet Oncol* 12(7): 624–6.

Böhler, E. and Schüz, J. 2004. Cellular telephone use among primary school children in Germany. *Eur J Epidemiol* 19(11): 1043–50.

Cnattingius, S., Svensson, T., Granath, F. and Iliadou, A. 2011. Maternal smoking during pregnancy and risks of suicidal acts in young offspring. *Eur J Epidemiol* 26(6): 485–92.

Dimonte, M. and Ricchiuto, G. 2006. Mobile phone and young people. A survey pilot study to explore the controversial aspects of a new social phenomenon. *Minerva Pediatr* 58(4): 357–63.

Divan, H. A., Kheifets, L., Obel, C. and Olsen, J. 2008. Prenatal and postnatal exposure to cell phone use and behavioral problems in children. *Epidemiology* 19(4): 523–9.

Divan, H. A., Kheifets, L., Obel, C. and Olsen, J. 2012. Cell phone use and behavioural problems in young children. *J Epidemiol Community Health* 66(6): 524–9.

Divan, H. A., Kheifets, L. and Olsen, J. 2011. Prenatal cell phone use and developmental milestone delays among infants. *Scand J Work Environ Health* 37(4): 341–8.

Draper, G., Vincent, T., Kroll, M. E. and Swanson, J. 2005. Childhood cancer in relation to distance from high voltage power lines in England and Wales: a case-control study. *BMJ* 330(7503): 1290.

EFHRAN. 2010. *D3—Report on the Analysis of Risks Associated to Exposure to EMF: In Vitro and In Vivo (animals) Studies*. Available from: http://efhran.polimi.it/docs/IMS-EFHRAN_09072010.pdf (accessed 29 October 2012).

Elliott, P., Toledano, M. B., Bennett, J., Beale, L., de Hoogh, K., Best, N., et al. 2010. Mobile phone base stations and early childhood cancers: case-control study. *BMJ* 340: c3077.

Elliott, P. and Wartenberg, D. 2004. Spatial epidemiology: current approaches and future challenges. *Environ Health Perspect* 112(9): 998–1006.

Eurobarometer. 2010. *73.3 Special Eurobarometer 347 Electromagnetic fields: TNS Opinion & Social*. Available from: http://ec.europa.eu/public_opinion/archives/ebs/ebs_347_en.pdf (accessed 14 February 2013).

Faraone, S. V., Perlis, R. H., Doyle, A. E., Smoller, J. W., Goralnick, J. J., Holmgren, M. A., et al. 2005. Molecular genetics of attention-deficit/hyperactivity disorder. *Biol Psychiatry* 57(11): 1313–23.

Ferlay, J., Shin, H. R., Bray, F., Forman, D., Mathers, C. and Parkin, D. M. 2010. *GLOBOCAN 2008 v2.0: Cancer Incidence and Mortality Worldwide: IARC CancerBase No. 10*. Lyon, France: International Agency for Research on Cancer.

Gati, A., Hadjem, A., Wong, M.-F. and Wiart, J. 2009. Exposure induced by WCDMA mobiles phones in operating networks. *IEEE Trans Wireless Comm* 8: 5723–7.

Guxens, M., van Eijsden, M., Vermeulen, R., Loomans, E., Vrijkotte, T. G., Komhout, H., et al. 2013. Maternal cell phone and cordless phone use during pregnancy and behaviour problems in 5-year-old children. *J Epidemiol Community Health* 67(5): 432–8.

Hoozemans, J. J., Rozemuller, A. J., van Haastert, E. S., Eikelenboom, P. and van Gool, W. A. 2011. Neuroinflammation in Alzheimer's disease wanes with age. *J Neuroinflammation* 8: 171.

HPA. 2013. *HPA Website: Electric and Magnetic Fields > Sources and Exposure 2013*. Available from: http://www.hpa.org.uk/webw/HPAweb&Page&HPAwebAutoListName/Page/1317133146663 (accessed 2 May 2013).

Huss, A., Vermeulen, R., Bowman, J. D., Kheifets, L. and Kromhout, H. 2013. Electric shocks at work in Europe: development of a job exposure matrix. *Occup Environ Med* 70(4): 261–7.

IARC. 2002. IARC Working group on the evaluation of carcinogenic risks to humans: non-ionizing radiation, part 1, static and extremely low-frequency (ELF) electric and magnetic fields. *IARC Monogr Eval Carcinog Risks Hum* 80: 1–395.

IEGMP. 2000. *Mobile Phones and Health. Report of an Independent Expert Group on Mobile Phones. Chairman, Sir William Stewart*. Chilton: NRPB.

IOP. 2008. *MRI and the Physical Agents (EMF) Directive: An Institute of Physics Report*. London: Institute of Physics (IOP).

Joffe, M., Key, J., Best, N., Keiding, N., Scheike, T. and Jensen, T. K. 2005. Studying time to pregnancy by use of a retrospective design. *Am J Epidemiol* 162(2): 115–24.

Kheifets, L., Bowman, J. D., Checkoway, H., Feychting, M., Harrington, J. M., Kavet, R., et al. 2009. Future needs of occupational epidemiology of extremely low frequency electric and magnetic fields: review and recommendations. *Occup Environ Med* 66(2): 72–80.

Kheifets, L., Breslow, N., Mezei, G. and TransExpo Study Group. 2010. *TransExpo: International Study of Childhood Leukemia and Residences Near Electrical Transformer Rooms*. Palo Alto CA: EPRI, Protocol 2010.

Kroll, M. E., Swanson, J., Vincent, T. J. and Draper, G. J. 2010. Childhood cancer and magnetic fields from high-voltage power lines in England and Wales: a case-control study. *Br J Cancer* 103(7): 1122–7.

Kuller, L. H. and Ives, D. G. 2009. Vital records and dementia. *Neuroepidemiology* 32(1): 70–1.

Kumlin, T., Iivonen, H., Miettinen, P., Juvonen, A., van Groen, T., Puranen, L., et al. 2007. Mobile phone radiation and the developing brain: behavioral and morphological effects in juvenile rats. *Radiat Res* 168(4): 471–9.

Lambe, M., Hultman, C., Torrang, A., Maccabe, J. and Cnattingius, S. 2006. Maternal smoking during pregnancy and school performance at age 15. *Epidemiology* 17(5): 524–30.

Leszczynski, D., de Pomerai, D., Koczan, D., Stoll, D., Franke, H. and Albar, J. P. 2012. Five years later: the current status of the use of proteomics and transcriptomics in EMF research. *Proteomics* 12(15–16): 2493–509.

Maskey, D., Kim, H. J., Kim, H. G. and Kim, M. J. 2012. Calcium-binding proteins and GFAP immunoreactivity alterations in murine hippocampus after 1 month of exposure to 835 MHz radiofrequency at SAR values of 1.6 and 4.0 W/kg. *Neurosci Lett* 506(2): 292–6.

Mattsson, M. O. and Simko, M. 2012. Is there a relation between extremely low frequency magnetic field exposure, inflammation and neurodegenerative diseases? A review of in vivo and in vitro experimental evidence. *Toxicology* 301(1–3): 1–12.

Mezei, G., Benyi, M. and Muller, A. 2007. Mobile phone ownership and use among school children in three Hungarian cities. *Bioelectromagnetics* 28(4): 309–15.

MTHR-PMC. 2007. *Mobile Telecommunications and Health Research Programme: Report 2007*. Chilton: HPA.

Neubauer, G., Feychting, M., Hamnerius, Y., Kheifets, L., Kuster, N., Ruiz, I., et al. 2007. Feasibility of future epidemiological studies on possible health effects of mobile phone base stations. *Bioelectromagnetics* 28(3): 224–30.

NIEHS. 2012. *The National Institute of Environmental Health Sciences Web Site* 2012. Available from: http://www.niehs.nih.gov/research/supported/dert/sphb/programs/peph/ (accessed 28 November 2012).

NRPB. 2004. Mobile phones and health 2004. *Doc NRPB* 15(5): 1–114.

Ofcom. 2012. *Children and Parents: Media Use and Attitudes Report*. Available from: http://stakeholders.ofcom.org.uk/binaries/research/media-literacy/oct2012/main.pdf (accessed 10 January 2013).

Ofcom. 2013. *Ofcom Website: Ofcom Announces Winners of the 4G Mobile Auction* 2013. Available from: http://media.ofcom.org.uk/2013/02/20/ofcom-announces-winners-of-the-4g-mobile-auction/ (accessed 2 May 2013).

Persson, T., Tornevik, C., Larsson, L. E. and Loven, J. 2012. Output power distributions of terminals in a 3G mobile communication network. *Bioelectromagnetics.* 33(4): 320–25.

SCENIHR. 2009. *Health Effects of Exposure to EMF*. Brussels: European Commission, Health & Consumer Protection DG.

Schüz, J., Elliott, P., Auvinen, A., Kromhout, H., Poulsen, A. H., Johansen, C., et al. 2011. An international prospective cohort study of mobile phone users and health (Cosmos): design considerations and enrolment. *Cancer Epidemiol* 35(1): 37–43.

Soderqvist, F., Carlberg, M. and Hardell, L. 2008. Use of wireless telephones and self-reported health symptoms: a population-based study among Swedish adolescents aged 15–19 years. *Environ Health* 7: 18.

Soderqvist, F., Hardell, L., Carlberg, M. and Hansson Mild, K. 2007. Ownership and use of wireless telephones: a population-based study of Swedish children aged 7–14 years. *BMC Public Health* 7: 105.

Susser, E., Eide, M. G. and Begg, M. 2010. Invited commentary: the use of sibship studies to detect familial confounding. *Am J Epidemiol* 172(5): 537–9.

Thomas, S., Benke, G., Dimitriadis, C., Inyang, I., Sim, M. R., Wolfe, R., et al. 2010. Use of mobile phones and changes in cognitive function in adolescents. *Occup Environ Med* 67(12): 861–6.

Vergara, X., Kheifets, L., Greenland, S., Oksuzyan, S., Cho, Y. S. and Mezei, G. 2013. Occupational exposure to extremely low-frequency magnetic fields and neurodegenerative disease: a meta-analysis. *J Occup Environ Med* 55(2): 135–46.

Vergara, X. P., Kheifets, L., Silva, M., Bracken, T. D. and Yost, M. 2012. New electric-shock job exposure matrix. *Am J Ind Med* 55(3): 232–40.

Vrijheid, M., Martinez, D., Forns, J., Guxens, M., Julvez, J., Ferrer, M., et al. 2010. Prenatal exposure to cell phone use and neurodevelopment at 14 months. *Epidemiology* 21(2): 259–62.

WHO. 2007. *WHO Research Agenda for Extremely Low Frequency Fields*. Geneva: WHO.

WHO. 2010. *WHO Research Agenda for Radiofrequency Fields*. Geneva: WHO.

Index

Index page numbers in italics refer to figures and tables

C

G

H

I

N

S